Behavior of Enzyme Systems

BEHAVIOR OF ENZYME SYSTEMS

SECOND EDITION

JOHN M. REINER

DEPARTMENTS OF BIOCHEMISTRY AND PATHOLOGY
ALBANY MEDICAL COLLEGE, UNION UNIVERSITY
ALBANY, NEW YORK

VNR

Van Nostrand Reinhold Company
New York Cincinnati Toronto London Melbourne

Van Nostrand Reinhold Company *Regional Offices:*
Cincinnati New York Chicago Millbrae Dallas

Van Nostrand Reinhold Company *Foreign Offices:*
London Toronto Melbourne

Library of Congress Catalog Card Number: 69-19593

Manufactured in the United States of America

Published by VAN NOSTRAND REINHOLD COMPANY
450 West 33rd Street, New York, N.Y. 10001

Published simultaneously in Canada by
D. Van Nostrand Company (Canada), Ltd.

15 14 13 12 11 10 9 8 7 6 5 4 3 2

To Candida

She walks in beauty . . .

Preface To The Second Edition

Since the first edition of this book was written, a number of things have occurred:

1. A goodly number of people wrote to me. Some accepted my invitation to correct errors or discuss issues; some just wanted to be pleasant. I am grateful to all of them; and I hope the readers of this version will adopt the precedent.

2. I have met or corresponded with various persons who appeared to confirm that I had not boasted idly when I promised that my readers could make experts out of themselves if they wished. The finest compliment I ever received came after a period when, due to serious illness in my family, I neglected my correspondence—and other things—for quite a time. One of the things I neglected was a kind invitation from a couple of my readers to visit them; among other things, they were interested in consulting me about a problem they had in hand. When the crisis was over, and I began writing scores of letters of apology, I found that these people had proceeded very well without my sage advice, and had an excellent paper in print. I should have been and was ashamed to have appeared rude, even if unintentionally; but I was vastly pleased rather then piqued at the evidence of my complete dispensability. The function of a good teacher is to make himself unnecessary to his students; and I am rather proud to have, in at least one instance, such impressive evidence that I succeeded.

3. When the first edition went out of print, my correspondence began to consist of urgent requests for a source of the book, not only for personal use, but for courses in the subject. I am sorry that I was able to be of so little help

to so many; but their inquiries helped to convince me that there was some reason for a second edition.

4. The mathematical orientation and training of biological scientists have improved vastly. It still seems reasonable to assume that many of those who will, I hope, read this edition may need some help over the rough spots; but I have felt emboldened to prune some of the very elementary and repetitious explanatory material that seemed to be necessary when the first edition was written. I do not overestimate the mathematical powers of my fellow biologists (I don't think I do!); and I have tried as hard as possible never to sacrifice clarity. But it seems to me that the task of writing a book about mathematical theory for such an audience has become materially easier.

The content of the book has not undergone a really radical change. I have included a discussion of so-called allosteric effects; the work on this topic seems to me one of the few really new and significant additions to enzyme kinetics. I have also included some discussion of the analysis of turnover by means of isotopic tracers, making large use of my own work of 1953; I had supposed that this work should be old hat to everyone, but my patient editor, Dr. Felix Bronner, assures me that the sermon still needs to be preached. I have elected to ignore an innovation suggested by Alberty, in which the number of intermediate compounds arising in an enzyme-catalyzed reaction can be any number whatever rather than the simple one or two usually postulated. With all due deference to one of the leading experts in this field, I must confess myself unconvinced that this generalization amounts to more than an elaborate algebraic exercise.

A major change in organization is that I have pulled together the discussion of theoretical methodology and mathematical techniques, previously somewhat scattered, into a first chapter which I hope will prove more useful as well as more coherent. I have reorganized parts of the early discussion of transient and steady states, and extensively reorganized the treatment of activators and inhibitors. Another innovation is the addition of exercises where it seemed feasible.

I have tried to catch and correct errors that appeared in the old edition; and I pray that I have not introduced new ones. But man is fallible; and I trust that my readers will continue to do me the favor of telling me where I have slipped.

Since this is still primarily a textbook rather than a handbook or encyclopedia, I must once again plead guilty of not even trying to compile an all-inclusive bibliography, and ask those whom I have failed to cite to believe that I have not gone out of my way to ignore them. What I can do, at this point, is to observe in a purely personal vein that the theorists who impress me most in this field are John Z. Hearon and Manuel Morales, and that the best combinations of theory and experiment seem to me to have issued from Robert Alberty and Paul Boyer and their associates. What my own work owes more or less directly to these or others is, in all honesty, very difficult to assess—when one has lived with a field, in study and classroom, for more than twenty years, everything one has read or heard has been metabolized and mixed with one's own pool of ideas. In any case, what I assuredly owe those I have mentioned is an acknowledgment of the high standards they have set.

An indebtedness that can never be cancelled is that to my preceptors in applied mathematics, Carl Eckart and Nicolas Rashevsky—not merely for what they taught me how to do but even more for serving as models for that combination of integrity, clarity, and passion that is the essence of the life of the mind. Two other great teachers touched me in a similar way, though the subject-matter of their lectures was less directly involved in my now life work: two philosopher-logicians—Rudolf Carnap, prince of philosophers, and Bertrand Russell, advocate of mankind. One could not sit at the feet of such men without feeling pride in belonging to a species that could produce them.

The instigation for the original edition of this book came from my delightful but all too brief association with Kenneth K. Tsuboi and Morris London. I miss them both.

Let me finally thank two friends: William Bender, physicist and indomitable swimmer against the current, for all the definable and undefinable things that have kept us friends, despite our wanderings, for thirty-five years; and John Muntz, Chairman of the Biochemistry Department of Albany Medical College, for giving me the first opportunity to test the material of this book in the classroom, even more for being what he is—intelligent, a gentleman and a gentle man.

JOHN M. REINER

Albany, New York
January, 1969

A Foreword For Timid Souls

(First Edition)

Enzymes and enzyme systems have increasingly taken a deserved place in the forefront of biological and medical research. Those who do not study enzymes as such are studying metabolism—that is to say, the behavior of complex systems of enzymes. The embroyologist and cytologist are trying to analyze development and differentiation in terms of enzymes, their changes with time, their redistribution, their characteristic products. The pharmacologist is looking to enzymes as the specific targets of drugs. The physiologist is seeking in enzymatic mechanisms the key to such different processes as renal function, muscle contraction, and nerve activity. The research clinician is studying enzyme systems in search of fundamental understanding of disease processes, not only in such settings as cancer and muscular dystrophy but also in mental disease and mental deficiency.

To analyze and interpret metabolic data in a rough qualitative way, it is often possible to view them in a framework of existing knowledge about metabolic paths and cycles. To analyze and interpret such data quantitatively and with confidence, a mathematical formulation of enzyme behavior is necessary.

This book is written for those research workers who feel the need of tools for quantitative interpretation of their work. It is addressed very particularly to those who not only feel the need, but also are in some doubt as to whether their mathematical training and facility are good enough for the mastery and use of a mathematical analysis of enzyme activity. Looking over the field for help, such workers will find that standard texts of biochemistry dismiss this important technique with a lick and a promise.* They will find a review or two; and these would be very convenient compendia of formulas, provided one already knew what this field is all about, or providing one were willing to use the formulas without understanding them too well. Finally, resorting to the "literature", they will find a scattering of papers, of which the most impressive—and indeed some of the best—ones appear to have been written by experts for experts alone. For themselves and others like them, they will find nothing.

The foremost purpose of this book, accordingly, is: To make it possible for anyone to begin the book knowing substantially nothing, and to finish it an expert for all practical purposes. The manner of presentation is based on the explicit assumption that most of those who should know the contents of the book will have had a relatively meagre preparation in mathematics, and that even this little has probably become rusty from lack of use. The only prerequisite is the desire to know what there is to be known, and the willingness to work at it just a little.

The operative principle behind this is a simple one: There is no such thing as a "mathematical mind"—or the lack of one (I speak now of learning and use, not of creation). Anyone with sense enough to design an experiment properly has all the mind he needs for mathematics. If the presentation is clear and simple, he will master the material at any level whatever. Of course, it is possible to treat this book like any other book which deals with an application of mathematics: namely, "lift" the formulas, and ignore the discussion and derivations. But we would warn the prospective reader that to do so will be to cheat himself out of a large part of his money's worth.

If mathematical analysis of enzyme behavior is to be conducted intelligently, it must be completely understood. We have therefore stated explicitly and often examined rather critically the assumptions and approximations underlying each piece of theory. The range of applicability of each formula is considered. The pitfalls of a superficial use of the formulas are frequently indicated.

We have tried to make this book very much of a practical handbook for the analysis and diagnosis of metabolic behavior. The analytical and above all the graphical methods for seeing what a theory means are presented at great length. Necessary bits of mathematical background are in virtually every case derived or reviewed—nothing has been taken for granted. Here again the aim held perpetually in view has been: To give the reader every conceivable technique for applying theoretical equations instead of simply admiring them—to furnish the tools for finding out what his experiments mean and for designing better experiments.

* Note added in press: A recent and honorable exception to this is Dixon, M., and Webb, E. C., "Enzymes," Academic Press, 1958.

While this book was not written primarily for the experts, they will find it as useful as will the novice. Besides covering the principal results of published work in the field, a considerable portion of the material covers hitherto unpublished work which for the past 18 years has represented something of a hobby for the author. The book does not pretend to be exhaustive, however; and we hope that no one will feel seriously hurt if he fails to find one of his papers in the bibliography.

We will welcome correspondence from readers who have suggestions, criticisms, or complaints. We have tried our utmost within the limitations of time and space to be completely clear. But if someone will show us where we have failed, we will be grateful, and the next edition will be that much better.

John M. Reiner

Emory University,
Atlanta, Georgia

Contents

I

Mathematical Theories: Some Principles and a Few Tricks

Enzyme kinetics is a branch of mathematical biology; it is one of the oldest mathematical theories in biology, yielding priority only to Mendelian genetics. Before we proceed to expound the theory, it would seem helpful to make sure that everyone is clear about the nature of mathematical theories, the way in which they are invented, and what they are good for. For those whose training in mathematics has been both limited and conventional, it may also be useful to set down some of the standard devices of the applied mathematician, since these usually get short shrift in undergraduate texts. For those readers who feel sure they know all they need to know about both of these topics, my advice is to skip to the next chapter; alternatively, you may enjoy a pleasing sense of superiority while skimming the following paragraphs.

MODELS, THEORIES, AND MATHEMATICS

Everyone is familiar with the idea of a natural law: a general statement about some regularity observed in nature. In this class fall statements like "If a piece of iron is exposed to a magnetic field, it will subsequently behave like a magnet" or "The pressure of a gas is proportional to the absolute temperature if the volume is kept constant". There is much to be said about this simple concept—but not here. For those who are interested in problems like that of the validity of induction, for instance, I heartily recommend some outside reading (1, 2).

Everyone has also heard the word "theory", perhaps in some such context as "Einstein's theory of relativity". But here is where the fog usually descends. In the biological sciences, "theory" has frequently had (and often deserved) a pejorative sense—that of "idle speculation". In the physical sciences, the word has a quite different and much more precise connotation.

A useful bridge between "law" and "theory" is the currently popular term "model". Ignoring much of the slop that has been written (and also ignoring some admirable discussions), let me propose the simple statement: A model is a mental image (or an artefact based on such a mental image) of the imagined "machinery" of a phenomenon. For example, a boxful of particles buzzing around at random, each of them subject to the laws of classical mechanics, is an elementary model of a gas. A string of beads is a crude model of the machinery of heredity, as the subject was discussed in American high schools before DNA became fashionable. Some of you may have seen a commercial model designed to explain the genesis of a Gaussian distribution in statistics: an inclined board studded with pegs, with balls released at the top of the board and rolling down among the pegs to pile up at the bottom, like a sort of tilted pinball machine.

The essence of a model is that someone has imagined a way in which a phenomenon works. Suppose someone hands you a metal box with some dials, knobs, and levers on it; you watch the needles move across the dials as you turn the knobs and push the levers back and forth. Now you try to imagine what kind of stuff inside the box might produce the events you have been watching; you dream up trains of gears, cams, electrical circuits. Eventually you build your model—another box, this time one whose guts you have put together yourself. If your box can reproduce the antics of the original box, your model is a success.

What, then, is a theory? A theory is an intellectual construction in which the essential features of a model have been expressed very carefully, either in sentences or in equations, and in which the sentences or equations have been manipulated by the rules of logical inference until sentences expressing an observable phenomenon have been obtained.

What do I mean by the essential features of a model? Consider the string-of-beads model of heredity. It does not matter in the least what size we make the beads, what material we use for them, whether they are white or black or candy-striped. The essence of the model is that there are distinguishable **units**, that they are arranged in a definite **order** or **sequence**, and that they are **constrained**, except for certain rare accidents, to remain together and in the given order.

The "theory" or "mathematics" of the isolated string of beads is thus very simple. A string contains a certain number of units, say n. They may be numbered according to the order in which they occur, thus: 1, 2, 3, . . . , n.

And that is all, unless we want to do a little more imagining. For instance, we may specify how many strings are present in a living cell. We may admit that the unit in a certain position on one string may be "different" from the unit in the same position on another string (without necessarily specifying in great detail what the difference is). We may imagine that two strings may occasionally exchange sub-sequences of units, and propose a simple idea as to how often this accident occurs and how it relates to the length of the sub-sequence. We may suppose that, when the two sexes of an organism mate, a string of units is selected at random from one partner and a similar string from the other, and that these two strings specify the offspring.

Since everyone is familiar with the Mendelian theory, this example need not be prolonged. The point is that the essence of the model consists of certain formal or abstract properties of the mental picture with which we begin, that these properties can be stated very precisely, frequently in mathematical symbolism, as **propositions** (sentences of the English language, equations, inequalities, etc.) from which **consequences** can be inferred. Eventually, a further translation process operates on the symbols to give statements about observable properties of nature. For example, in the heredity model, we translate as follows (a completely fictitious example, please note!): The third unit of a string specifies eye pigment, unit "3" means no pigment, the corresponding but different unit "3′" means brown pigment; the eleventh unit in the string specifies height range, such that "11" means less than 5 ft 6 in. high and "11′" means greater than 5 ft 6 in. high. We now proceed to predict what kinds of progeny, and what percent of each kind, should be expected if a brown-eyed individual over 5 ft 6 in. is mated with an individual with colorless eyes and less than 5 ft 6 in. tall. These translations (the assignment of observable properties to various kinds of units) are part of the theory; they are hypotheses, just as the string of beads itself is a hypothesis. The whole theory, of course, will be judged by the success of its predictions.

Theories vary tremendously in the depth to which they dig into the imagined machinery of a phenomenon. Some barely scratch the surface. For example, consider the problem of a culture of cells, say bacteria, as they are increasing in numbers. We may consider the cells as units, about which we specify just two things; at any moment of time t, there is some number of units, say $N(t)$; and a unit, besides just being, may sometimes give rise to two units. Let us say that there is a constant probability that any unit will double in unit time; call this probability k. If doubling is a random event (like the toss of a coin) with probability per unit time k, then the expected number of doublings per unit time, when N units are available, is kN (a result of the theory of Bernouilli trials, which is derived in textbooks of probability theory—the best of which, in my opinion, is Feller's (3)). The number of doublings per unit time is obviously also the number of extra units appearing per unit time—that is, dN/dt. The basic equation of the theory is therefore

$$dN/dt = kN$$

whose solution is

$$N = N_0 e^{kt}$$

N_0 being the number of units present at the initial time $t = 0$ (when the observations were started), and e the customary base of the natural logarithms.

Now this is not a very profound theory; it says nothing about the machinery of doubling except that it occurs as a random event with a constant probability. It is also open to the criticism (which is frequently voiced about mathematical theories in biology) that it tells us only what we knew already—in this instance, the exponential relation between number of units and time. Isn't $dN/dt = kN$ just a needlessly fancy way of saying $N = N_0 e^{kt}$?

Theories are like cities; they are continually growing and spreading. The theory just summarized is minimal; it barely qualifies. The question is what it can grow into.

There is an important difference between the initial differential equation $dN/dt = kN$ and the equation $N = N_0 e^{kt}$. In the final equation, **considered as a natural law**, the factor k is just a constant, whose numerical value may be obtained from the experimental data. In the differential equation, however, k was defined as a doubling probability per unit time. If we consider some more common information, this leads to a number of growth possibilities for the theory.

We know from microscopic observation that the cells increase in size between doublings, and that they fall in a fairly standard size range. Very roughly, we may say that a cell has volume v_0 just after division, and that the next doubling occurs when it has grown to size $2v_0$ approximately. The probability k that a unit will double during the next unit of time is the probability that it has just grown to the doubling size $2v_0$. The doubling probability k is therefore in some way contingent on growth; and growth is contingent on nutrition, temperature, and other obvious factors. So we may proceed to ask about the enzymatic reactions which constitute the metabolism of the cell, and try to imagine how they might affect the probability that a cell arrives at its doubling size. We may also ask whether the cells are good Malthusians—whether the number of them at any time affects their supply of nutrition. And we may try to build this factor into the theory.

It is also a fact that various experimental tricks may be used to make a cell culture divide synchronously, so that the cell number is practically stationary for a certain interval and then doubles in a few moments. It is likewise a fact that such synchrony is unstable, and that synchronized cultures revert to the usual random pattern of doubling within a few generations. We may therefore ask what probabilistic factors affect the processes of cell growth and division to produce this inevitable fully randomized result.

So—we're off to the races! The merit of the primitive little theory is not that it is remarkable in itself, but that it suggests (with some help from factual information) what to try next—that it is so phrased that it tends to generate new and better theories. The difference between the theory and the well-known answer it produces is the difference between an idea—something that has parts, that has a logical structure, that you can tinker with on paper without dirtying a single test tube—and an opaque brute fact, that can stand for nothing but itself. As the saying goes—*Vive la différence*!

Why Mathematical Theories?

To this question there is an intellectually trivial but practically important answer: It is much easier to manipulate mathematical symbols than complex mental images—a mathematical theory is a highly convenient tool for getting results. There are, however, three more significant features:

1. Mathematical theories in the simplest sense (theories that deal with variables whose values are numbers) are important because they commit themselves more deeply than verbal theories, hence are easier to **disprove** experimentally if they are wrong (and the possibility of disproof is one of the most important aspects of a theory). To be concrete, consider the following relations:

$$N = N_0 e^{kt} \qquad N = N_0 + a(1 - e^{-kt})$$

$$N = N_0 e^{e^{kt}-1} \qquad N = N_0 e^{kt}/(1 - b + be^{kt})\,;\; b = N_0/(N_0 + a)$$

$$N = N_0 + kt \qquad N = N_0 + at/(k + t)$$

$$N = N_0 + kt^2$$

$$N = N_0 + \ln(1 + kt)$$ ("ln" stands for "natural log")

These are just a few of the infinitely many expressions which could be described as starting with the value N_0 when $t = 0$ and climbing steadily as t increases. The last three approach an upper limit $N_0 + a$, the others increase without bound. It seems obvious that not all of these expressions could be inferred from the same imaginary machinery of doubling. With adequate data, one can readily decide which equation predicts the actual behavior of a culture correctly, if any. But verbal theories give verbal conclusions and predictions, which **assert less** than mathematical conclusions. (Such statements as "N increases in proportion to the square of time", which are simply clumsy ways of describing an equation, I am not counting as "verbal" in this discussion.) If two verbal theories are going to conclude simply that N increases and eventually stops, what can you do to discriminate between them?

2. Mathematical theories give reasons for performing experiments that no one would dream of performing in their absence; and they dictate the precision and sometimes the very form of such experiments. There is an abundance of examples; but the Michelson-Morley experiment, of which every freshman has at least heard, is perhaps one of the clearest. The speed of light was actually known with fair accuracy when Athens flourished. Better figures still were obtained by early workers in optics, around the time of Newton—better, indeed, than would be required for most calculations that a physicist would want to perform. Yet Michelson invested his immense skill and ingenuity and considerable toil in repeating the measurement—accurate to eight significant figures. This experiment would have been too silly for the most naïve freshman to suggest—except that the "ether drift" required by the Lorentz electrodynamic theory demanded a minute effect—a difference between two measurements of velocity which would show up only in the last digit if at all. The silly experiment thus became, against the background of a mathematical theory, a critical and revolutionary experiment. (As so often happens, the outcome of the experiment was negative; the experiment undermined the theory which had inspired it and given it significance, and helped to prepare for the birth of relativity theory.)

3. Mathematics is much broader than the familiar use of number-valued variables would lead one to suppose. Mathematics is more the logic of **pattern** than of number—number is simply one of the commonest and most convenient ways of analyzing patterns. Mathematics also embraces concepts like "inside" and "outside", "before" and "after". (In the string model of heredity, for instance, the assignments of ordinal numbers to the beads was a mere matter of convenience—we could just as well have labelled them "a, b, c, ..." or "Joe, Charlie, Harry, ...", and the fundamental order relation would have been completely expressed by statements like "*d* is before *e*" or "Abe is between Charlie and Sam".) Since every science is concerned with relations among the things it studies, with pattern, regularity, order, structure, it follows that mathematics, in the wider sense just described, is the indispensable and unique intellectual tool of science. (For a beautiful example, see the analysis of the bandwagon effect in election predictions in Simon's Chapter 5 (4); see also Kemeny and Snell (5), especially Chapters II and VIII. An intriguing if narrower approach is that of Woodger (6), in which he makes use only of the non-numerical methods of symbolic logic.)

Theory Construction

Building a theory starts with imagining a model; and there is no really systematic prescription for successful imagination. Part of the recipe is the artistic equipment one is born with; part is experience—the theorist learns to look for and recognize patterns in nature, and to correlate them with mathematical patterns he has applied in the past. Fundamentally, however, he is on his own at this stage.

However, there are some commonsense considerations that are worth keeping in mind. They won't make any of us into a Maxwell or an Einstein; but they will help as we struggle to be reasonably competent garden-variety theorists. The most important of them is to look for key features of the pattern we are trying to explain, and ask what kinds of mathematical relations resemble these features.

For instance, take the cell-doubling theory we discussed earlier. We know that cell cultures do not increase indefinitely, but come to a stop with a finite number of units. We could simply pick one of the last three equations of the batch above, since we know that they have this property. But this is mere curve-fitting; we have no interpretation for the equations, no idea what kind of mechanism they represent.

Let us look at the basic $dN/dt = kN$. How could we modify it so that dN/dt, the rate of increase, would become zero when N arrives at some finite value? If doubling probability depends on cell growth, then doubling could be limited if the raw materials for growth were limited. How shall we express this? The simplest thing (not necessarily the best—but always start with the simplest!) is to say that the doubling probability is proportional to the available nutritional raw material at any time. How does this quantity depend on N? Again starting simply and crudely, let us say this: Suppose that the initial supply of raw material is enough to make A cells. After N cells have been made, the raw material left, in units of numbers of cells, is $A - N$. Then let us write the doubling probability, at the time when N cells have been made, not as k, but as $k'(A - N)$. The differential equation is now

$$dN/dt = k'(A - N)N$$

The solution (consult your calculus text for details) is

$$N = AN_0 e^{k'At}/(A - N_0 + N_0 e^{k'At})$$

which you should recognize as one of the equations we looked at previously, with the constants written in a different way. It is well known under the not very descriptive name of the logistic equation. It does not give the best possible description of the growth of a bacterial culture, but it does predict the cessation of growth when $N = A$.

The point of the example is that it shows, in a very simple way, how we carry out the exhortation to look for suitable patterns. We saw a pattern of growth-stops-at-a-finite-number, or more precisely, growth-slows-to-a-stop-as-numbers-increase-to-finite-level. We matched this, in the differential equation, with a pattern that says something about the machinery of growth in terms of nutrition: nutrition-slows-to-zero-as-numbers-increase-to-finite-level.

Sometimes, of course, you don't have enough information in advance to guide you. In such cases you must simply go ahead and imagine whatever appeals to you as plausible or elegant, and then see whether it works. Don't feel bad if it frequently doesn't—every theorist keeps a large wastebasket handy. (How many not very instructive experiments have you carried out?)

The process of going from model to theory, which was very simple and straightforward in the preceding example, is sometimes troublesome. Part of the problem is linguistic—it is the same problem that makes so-called "word problems" in mathematics texts so difficult for many students who are by no means dull-witted.

Assume that you have stated the essence of your model rather precisely in words. The crucial step is going from words to mathematical symbols—more precisely, going from verbal statements to mathematical statements (equations, inequalities). Part of the process is again art and experience; but there are some simple rules of thumb that tend to make things easier.

1. Break every compound sentence down into simple (declarative) sentences.

2. Try to rephrase every sentence in a form where the principal verb is "is". The simplest translation of "is" into mathematical symbols is "$=$". Thus, "the box has a volume of C units" becomes "the volume of the box is C units".

3. Write a glossary of symbols for key words **and phrases**. There is no mathematical equivalent for "of the box". But if you write in the text of your theory a definition: "Let 'V' stand for 'volume of the box'", the sentence in the preceding paragraph becomes the equation "$V = C$".

4. Some principal verbs simply will not come out "is". The translation of "is greater than" is, of course, the symbol "$>$". The translation of "is proportional to x" and of "varies as x" is "$= kx$", k being a constant. The translation of "varies inversely as x" is "$= k/x$". The verb "is not less than" can be reorganized into" is greater than or equal to", which translates into "$\geqslant$".

5. Some expressions should immediately suggest derivatives. In a context where time is explicitly involved, "rate" or "rate of change" becomes "d/dt", the derivative with respect to time. Expressions like "change of ... for a change in x" or "increase of ... due to increase in x" translate into "d/dx", the derivative with respect to x. Losses are treated as gains with a change of sign: If "gain for an increase in x" is "d/dx", then "loss due to a gain in x" becomes "$-d/dx$". Similarly, expressions like "the greatest possible" or "the most efficient" imply the existence of functions of one

or more variables, such that the functions have a maximum with respect to one or several or all of the variables—i.e., such that the corresponding derivatives or partial derivatives equal zero.

6. Not all statements are translatable into mathematical propositions. For example, if A and B are the names of chemical compounds, "A becomes B" or "A reacts to give B" translate into the **chemical** or **stoichiometric** equation "A → B". (As we shall see in Chapter 2, such equations have correlated with them certain mathematical **rate equations**; but the chemical and the mathematical equations are not logically identical.)

7. Some expressions are translatable into non-numerical mathematical expressions. Thus, "a collection of units with a definite sequential arrangement, the first of these being labelled '*a*,' the second '*b*', etc." translates into "the ordered set (a, b, c, ...)" or "the sequence: *a*, *b*, *c*, ...".

8. It is sometimes very useful to abandon the strict goal of translation, and to employ symbols that have no direct interpretation, either at the level of observation or at the level of mechanisms. For example, suppose we perform a sequence of tests, such as microscopic examination of cells for mitotic figures. We may define a set of variables x_1, x_2, x_3, ... with the following property: If the *i*th test comes out positive (we see mitotic figures), $x_i = 1$, otherwise $x_i = 0$ (for example, if the 10th cell examined shows mitotic figures $x_{10} = 1$). The individual variables mean no more than a "Yes" or "No", a check or cross on a tally sheet. On the other hand, the sum $x_1 + x_2 + x_3 + \cdots$ **is** significant— it is the number of cells with mitotic figures. What is the good of this beating about the bush? As it happens, if one knows the probability law for each of a set of random variables, it is easy to find the law for their sum. Hence this device is frequently used in applications of probability theory. I have used a similar device for the analysis of order in macromolecules. The pretty thing about such tally variables is the combination of a yes-or-no function with an arithmetic function—the individually uninterpreted variables can be combined by the ordinary arithmetic laws to give interpretable and significant functions.

Interpretation and Application of Theories

I have already indicated, in connection with the heredity-string model, the general nature of the interpretation process: It is necessary to stipulate that certain of the units, alone or in definite combinations, correspond to certain observable traits of the organisms. In receptor theory in pharmacology, the units are either enzymes or are treated as formally similar to enzymes; and the clauses necessary for interpretation say in effect that the activity of certain nerve pathways or of certain effectors is proportional to (or at least is an increasing function of) the activity or the output of these units.

The complexity and importance of interpretation clauses will vary with the nature of the theory and particularly with the distance between theory variables and observation variables. In some theories the bulk of the interpretation is already built in, as it were. For instance, in Maxwell's kinetic theory of the gas laws, the final equation of the theory comes out in terms of a variable p, the pressure exerted on the walls of the container by the atoms or molecules. The result is obtained by noting that a pressure is defined as a force per unit area, and that force is rate of change of momentum, so that the transfer of momentum from molecules to wall by collisions appears in the final equation as a pressure. The only interpretation needed is to say that, if you substitute a pressure gauge for part of the wall, the response of the pressure gauge will be the same as the pressure on the wall. A major part of the interpretation has been included in the theory by going from the kinetic variable, mean rate of momentum transfer, to the mechanical variable, pressure.

In the same way, if a theory of heredity were to include an analysis of development, the eventual variables might include a function P_e, the amount of eye pigment. The interpretation might simply read: If $P_e = 0$, the eyes are colorless, and appear blue; if $P_e > 0$, the eyes are brown. The distance between the key variables of the theory and the actual observations would have been cut down by extending the theory to follow the consequences of the activity of the string units through various chemical reactions and cell movements, winding up with a quantity of accumulated pigment P_e.

I have stressed the problem of interpretation, because the process is frequently semiconscious, and tends to be overlooked.[1] Biochemists who were active in the heyday of research on intermediary carbohydrate metabolism will recall the problems of analyzing, say, the ATP content of tissues or tissue extracts. The theoretical variable was ATP content; the experimental variable was a dial reading on a colorimeter, following a standard series of physical and chemical manipulations. A simple identification of the two was frequently misleading; and some puritanical souls never referred to "ATP", but only to "10-minute-hydrolyzable phosphate". This did not solve the problem of interpretation, but merely evaded it; and even the evasion ignored some possible flaws. I recall the otherwise easygoing and charming Gerhard Schmidt saying emphatically that he would not trust a graduate

[1] The distinction between the formal core of a theory (mathematical and chemical equations, for instance) and the text of the theory, in which one speaks **about** the formal part, is familiar to logicians under such terms as "object language" and "metalanguage" (7). In a book that is at this writing still in manuscript, tentatively entitled "Metabiology," Prof. Luico Chiaraviglio of the Georgia Institute of Technology has performed a remarkable feat in analyzing this problem in all its ramifications, drawing most of his theoretical and experimental illustrations from contemporary work in molecular biology.

student to perform a phosphate analysis until he could demonstrate that he fully understood both the electrical and optical properties of the colorimeter and the chemistry of the phosphomolybdate reaction.

The **application** of a theory has two major aspects:

1. A theory has numerical applications, which are enormously convenient but otherwise trivial. For instance, if I know that a bacterial culture grows according to the exponential or the logistic law, and I have performed preliminary experiments to obtain the numerical values of the constants, I can inoculate a culture at 10 A.M., and decide that by 2 P.M. I will have a bacterial density of 3×10^8 cells per ml ($\pm$ some probable error), which will give me the number and density of cells that I want for starting a certain experiment. The astonomer's prediction of the polar angle and azimuth at which a certain star will be found at 11.30 P.M. next Friday or the engineer's calculations of materials and dimensions for a proposed structure are in the same category.

2. To the extent that a theory based on a detailed model works, it permits one (until a better theory comes along) to say that the machinery postulated in the model is in truth the reason why the observed facts are as they are. In common parlance, the theory **explains** the observations. It is beyond anyone's competence, at the moment, to tell why the ability to explain is so satisfying. It is also beyond the scope of this book to discuss the logical problems of the concept of explanation in any detail; I refer the interested reader once more to references 1 and 2. It is clear, at all events, that this application of a theory is the important one.

A major problem, therefore, is to decide how well a theory works, or which of two or more proposed theories works best. (We may profitably ignore, I think, the horrendous alternative that certain philosophers are so fond of brandishing like King Charles' head: the occurrence of two theories whose predictions are in every respect indistinguishable. This is indeed a logical possibility; but experience suggests that it is practically negligible. The standard remark about the equivalence of the Copernican and Ptolemaic theories is simply nonsense!)

One approach to this problem is the brute-force application of statistical methods, such as testing the fit of experimental data to theoretical curves with the help of χ^2 and similar criteria. I shall ignore such methods, first because an adequate treatment would consume too much space (standard texts should be consulted by the persistent) and secondly because I find them boring.

I prefer a method that is admittedly less systematic than statistical ditch-digging, but is, I think, much more promising as a day-to-day procedure. Its basis is best understood in terms of an example. If a vaudeville clairvoyant writes down on a little sheet of paper the prediction that the individual in seat 14D will rub his cheek or chin with his right hand in the next 10 minutes, no one but a simpleton would be vastly impressed by the fulfillment of this prediction. But if he predicted that the same individual would stand on his head and recite "Father William", a positive outcome, with collusion excluded, would be rather interesting; and I presume I need not offer a solemn lecture on why this is so. In the same way, if one can find an outstanding idiosyncrasy in a theory, finding the same idiosyncrasy in the data would tend to strengthen one's belief in the validity of the theory rather more than the confirmation of a more ordinary feature.

Of course, it is not a foregone conclusion that every theory one is interested in will have, so to speak, a strongly marked individuality that one can use as a handle. Fortunately, such happy occurrences seem to be more nearly the rule than the exception. However, it is usually necessary to manipulate the theory, to cast it in various forms until one finds one form that shows up the special features of the theory most sharply, and that will be most sensitive to departures of the data from the theory.

A large part of the space in this book is devoted to just such manipulations. Deriving the equations of the theory in some form or another is scarcely ever, in the case of enzyme kinetics, really difficult. Looking for the place where the individuality of the theory shows up clearly and strikingly may be more troublesome.

In the search for sensitive and revealing forms of a theory—what I have referred to as **diagnostic** forms—graphical methods play a conspicuous part; and the reason is not far to seek. The whole business could indeed be conducted analytically: A linear expression (or in general a polynomial) can be fitted to data by the method of least squares, and the estimated parameters can be used to evaluate the quality of the fit. But it is a fact about the human central nervous system that almost anyone can glance momentarily at a graph of experimental data and decide that they fall pretty well on a straight line, while someone else is spending 15 minutes with a desk calculator, or while a digital computer, which after half an hour of programming can perform the calculations in a fraction of a second, is racking up a charge of fifty dollars. Obviously, if it is really important to know whether the slope of the line is 1.2 or 1.3, the calculations will have to be performed sooner or later. But if the interesting question is, as it usually is, whether the points seem to fall on a straight line, the combination of the human eye and the human brain is a remarkable computer, as every computer expert will tell you.

It is also a fact about the human computer that its brilliance with straight lines is matched by its mediocrity in discriminating among nonlinear forms. Hence transforming the equations of a theory into linear forms is a major technique used in this book. Sometimes an equation simply will not linearize (or at least not in any useful way); but often one can find a form in which the curve is essentially linear over much of its extent, and deviates from linearity in a characteristic way at certain places.

There are, finally, a few instances where numerical computations have some significance beyond practical convenience; but in those cases rough calculations will usually suffice. Suppose, for example, that you have a theory about an enzyme with two distinct mechanisms of action (the product, however, being the same for both). To each mechanism there corresponds, let us say, a characteristic constant. If you find yourself working with an enzyme that appears to be in this category, you might sensibly ask the question: How much of the overall activity of the enzyme is determined by the one mechanism and how much by the other? Is the enzyme largely this kind, or largely that kind, or pretty evenly both kinds? The limiting cases may, as suggested in the foregoing paragraphs, be readily distinguishable by some typical pattern. If not, however, you will want to estimate the values of the two constants and compare them. But obviously it does not really matter, for the kind of conclusion you are after, whether k_1 has a value of 1.0 or 1.25, etc.—what you want to know is whether k_1 is five or ten or a hundred times as great as k_2, or the converse, or whether they are about the same.

It follows from all that I have said so far that the reader will find very little, in the chapters that follow, about the estimation of enzymatic constants from experimental data, and that he will find a great deal of algebraic manipulation and graphical analysis designed to find out what is the mechanism behind the observations. Some discussion of the necessary techniques will be found in the next section.

HELPFUL HINTS FOR HOMESPUN MATHEMATICIANS

Every mathematician, and particularly every applied mathematician, constantly and almost without thinking uses a host of techniques, not very profound in themselves but so convenient that he would be crippled without them. The techniques are implicit in the standard textbook versions of mathematics; but one is as unlikely to find them explicitly described and discussed as to find an exposition of the techniques of pipetting in a textbook of biochemistry. One learns the methods by observation and experience, and by getting occasional pointers from older and more experienced colleagues. Most of this section is devoted to summarizing and illustrating such techniques for the readers of this book. Nothing that will be discussed would be of the least interest to a professional mathematician; but this does not make the methods contemptible any more than shoes are contemptible because you walk on them instead of reading them.

Generalities

Mathematics is not a subtle discipline; its results are nothing like Houdini getting out of a locked safe.[2] It is essentially simple-minded; and this is precisely why it is so powerful. There are indeed numerous people below the intellectual level at which they could even understand any mathematics, let alone use it. But it is questionable that such people will normally be found working in a laboratory rather than sweeping it out.

What makes mathematics simple-minded is the fact that its operations are all analyzable into simple unitary operations, and that every operation that can be performed is justifiable by a rule—an axiom that stipulates the commutativity of algebraic operations, a theorem about subtraction that validates "transposition" across an equality sign, a syntactical rule about substitution, and so on. Experienced mathematicians perform many such simple operations mentally and never write them down; sometimes they do this with less simple operations, and complicate the result with a bit of one-upmanship by prefacing the skipping of ten steps with "It is obvious that . . . ." In a similar way—and justifiably—textbooks in higher-level courses skip steps that appear in full at lower levels, on the assumption that the students have by this time learned to make the right jumps almost automatically. (One might add that textbooks at all levels skip steps that a logician would regard as essential—I do not recall a mathematical proof whose rigor would seem unimpeachable to a logician.)

Taking these qualifications into account, we may say that a mathematician performs an operation only if it is justifiable in the manner specified. The highly important converse is this: If a step is justifiable, you may perform it. I have seen innumerable victims of *horror mathematicus* hesitate over a perfectly straightforward manipulation, with "Dare I do that?" written all over their faces. It is important for the beginner in applied mathematics to realize precisely how much freedom he has—the freedom to do anything for which he can find a rule, regardless of whether or not this is how the teacher or the textbook did it. It is important because—and I cannot stress this too strongly—before a result has been obtained, there is no logical way to decide whether the result is or is not valid, and no logical way to decide how it must be proved if it is valid (unless the result flatly and obviously contradicts a known valid result, in which case it is of course false). Once the proof has been found, of course, a computer can easily be programmed to check it.

[2] Here, as throughout this book, I exempt creation, as distinguished from application, from my comments. Creation is subtle and complex to us because we don't understand it very well. When we eventually come to understand it, it will probably seem simple. For readers who feel that they might profit from a more extensive view of how mathematical thinking comes about than I can give here, I recommend some books by Polya (8, 9, 10), who stands alone in having attempted to depict the process concretely and systematically.

In the strictest sense, then, in moving through mathematical *terra incognita*, you cannot tell where you are going until you get there; and you are never going to get anywhere until you try things. This is the area where, in addition to the indefinable element of creativity, experience counts heavily. The mathematician who has seen many patterns, including patterns of proof, will frequently see you starting off in a given direction and tell you that this will get you nowhere, or that this is the hard way to get the result. There is no magic about it—it is the same thing that causes an experienced investigator to observe things in the laboratory which may be missed completely by the most brilliant first-year graduate student. If you are ever going to be able to use mathematics, you must be prepared to flounder for a while, to follow hunches and see where they lead (and to make little mental notes about when and why they lead to results or to blind alleys). There is no magical recipe for the best way to do something new—and the most piddling exercise in proof is new when you haven't done it before. And remember, when you feel like a fool for the hundredth time—and assuredly you will!—that Gauss and Hilbert and Einstein had wastebaskets, and filled them with their failures, even as you and I; what distinguished them was not infallibility, but the quality of their successes.

This is all there is to say about the subject. Treat mathematics as a game: Learn to recognize the pieces, learn the rules, and proceed to play. Eventually you will enjoy yourself.

And now to some of our tricks.

Infinite Series and Approximations

Most well-behaved functions may be represented by infinite series of terms—sometimes by series of positive powers of the variable, sometimes by series of negative powers, sometimes by more complex kinds of terms. There is no space here to discuss how such series are obtained, how one tells whether a given series really converges to a limit, and the like; you should consult your calculus text. What is important for us here is a common application of such series, and especially of power series, for obtaining approximations to functions.

Successive terms of such series grow smaller in absolute magnitude; and sometimes they grow smaller so fast that the first few terms are a rather good approximation to the value of the function. A general rule is that the error in stopping with the nth term is about equal to the $(n+1)$st term.

For example, take the geometric series

$$1 + x + x^2 + \cdots$$

which converges and represents the function $1/(1-x)$ when $|x| < 1$. If $x = \frac{1}{2}$, the first three terms are, of course, $1, \frac{1}{2}, \frac{1}{4}$. Their sum is $7/4$; the actual value of the function is 2, so the approximation is crude, but not really bad. Take $x = 1/10$. The first three terms are $1, 1/10, 1/100$. Their sum is $111/100$, the actual value of the function is $10/9 = 1.11111\ldots$ The first three terms differ from the actual value of the function by about 0.1%. Even the sum of the first two terms differs from the value of the function by only about 1%.

Several instances of this kind of approximation will appear repeatedly in this book.

1. The binomial expansion, especially for negative and fractional powers:

$$(1 + x)^u = 1 + ux + \tfrac{1}{2}u(u - 1)x^2 + \cdots$$

If you actually have something like $(a + y)^u$, write it as $a^u(1 + y/a)^u$, and proceed as before, treating y/a like x. The rule is to **factor out the term that will be large** in the case you are considering, so that the other factor will be $1\pm$ a small term. If x (or y/a) is small, we approximate by keeping only the terms $1 + ux$. (Caution: If, as sometimes happens, the whole expression you are trying to simplify is $1 + ux - (1 + x)^u$, the first two terms of the binomial expansion will be cancelled by the terms preceding the minus sign. In that case, you must obviously keep the next term of the expansion, $\tfrac{1}{2}u(u - 1)x^2$; it may be small, but compared with zero it is still the largest term you have. A safe rule: Don't discard terms of a series until you see what is left when the dust settles.)

A specially important application is to finding the limit of an expression when the variable grows infinitely large. For instance, suppose you have

$$a + x - \sqrt{x^2 + bx + c}$$

Suppose you want to know what this becomes if x becomes infinitely large. You may argue that under this condition a is negligible compared with x, and that the lower power, bx, as well as the constant c in the radicand becomes negligible—i.e., that we wind up with $x - \sqrt{x^2} = x - x = 0$. But this is not true.

The careful way to do it is this: Factor x^2 out of the radicand, which becomes $x^2[1 + (bx + c)/x^2]$. The term $(bx + c)/x^2$ approaches zero as x approaches infinity; hence the binomial approximation is immediately applicable, treating $\sqrt{1 + (bx + c)/x^2} = [1 + (bx + c)/x^2]^{1/2}$ like $(1 + x)^u$ above. The square root becomes (remembering that $\sqrt{x^2} = x$ since $x > 0$):

$$x[1 + \tfrac{1}{2}(bx + c)/x^2 + \text{higher powers of } (bx + c)/x^2]$$

Multiplying out, neglecting the higher powers, we get $x + \tfrac{1}{2}b + \tfrac{1}{2}c/x$, which is subtracted from $a + x$ in the original expression. We get $a - \tfrac{1}{2}b - \tfrac{1}{2}c/x$, and now, noting that c/x will vanish with infinite x, it can be dropped. The limit is not 0, but $a - \tfrac{1}{2}b$. You can readily satisfy yourself that keeping higher powers of $(bx + c)/x^2$ will not alter the conclusion.

2. The logarithmic expansion. When $|x| < 1$:

$$\ln(1 + x) = x - \tfrac{1}{2}x^2 + \tfrac{1}{3}x^3 - \cdots$$

For small values of x, the first term is a good approximation. As in the case of the binomial, we convert an expression like $\ln(a + y)$ to $\ln[a(1 + y/a)] = \ln a + \ln(1 + y/a)$, and treat y/a like x.

In addition to finding limits of expressions for null or infinite values of the variable, these approximations are useful when mixtures of terms occur, and one wants to get an idea, at least for small and large values of the variable, how the various terms contribute to the overall result. The binomial expansion example was a mixture of linear term plus radical; and in Chapter II we will meet the case of linear term plus logarithm.

Note that the "variable" may be anything whatever that you want to investigate. The temperature of a chemical reaction mixture may be a constant in the sense that you keep it at one value throughout the reaction by the use of a thermostatically controlled waterbath. But it is also a variable in the sense that you can run the same experiment at different temperatures. (Such constants, as distinguished from numerical constants or universal physical constants like the charge on the electron, are frequently termed "**parameters**".) You may reasonably be curious as to how changing the temperature would affect the system you are studying, according to the theoretical formula you are using; and in the light of that curiosity the temperature is a "variable" whose effect on the equations is to be analyzed.

Finally, note that the foregoing cases constitute a small sample of the possibilities, selected for their frequency in this book. Consulting the list of series in any standard handbook of chemistry and physics will furnish you with other cases that may from time to time be useful in your work.

The Positive Square Root

It is elementary but not superfluous to state that $\sqrt{u}$ is defined to mean the positive value of the square root. (It is not superfluous because there are college algebra texts which altogether fail to be clear on this point.) It follows that $\sqrt{x^2} = |x|$, that is, x if $x \geqslant 0$, $-x$ if $x < 0$.

An important application of this will be used in several places in this book. To wit:

$$\begin{aligned} \sqrt{(x-a)^2} &= x - a \quad \text{if} \quad x \geqslant a \\ &= a - x \quad \text{if} \quad x < a \end{aligned}$$

Thus, while it is immaterial whether you write $(x - a)^2$ or $(a - x)^2$, the order of x and a becomes quite material when you extract the square root.

The form that occurs in later chapters follows immediately:

$$\tfrac{1}{2}[x+a+\sqrt{(x-a)^2}] = x \quad \text{if} \quad x \geqslant a$$
$$= a \quad \text{if} \quad x < a$$

Strategy in Solving Systems of Equations

Problems in enzyme kinetics involve the solution of systems of algebraic equations, which represent the steady-state (time-independent) version of a system of reactions. There is no inherent mathematical difficulty involved; and linear systems can be solved straightforwardly either by the elementary methods of elimination or by the more efficient method of determinants (for which, if you need a refresher, consult a text of college algebra). But sometimes the systems are not linear; and often the direct methods lead to complex and messy expressions which give ample scope for errors of transcription before you get them simplified. A little strategy in selecting the order of the steps of solution will frequently be helpful. A look at the chemical equations is particularly recommended; being a sort of picture of the system, they are more vivid than the algebraic equations, and salient features are often apprehended with great speed and ease.

To take some concrete examples, consider reactions which terminate reaction chains or branches of such chains. (These I have termed "dead-end reactions".) The equilibrium equation for the terminal product is easy to solve, precisely because the terminal product isn't going anywhere—it enters just one reaction. Moreover, the "steady-state" equation for the penultimate (next to final) compound will contain the two (forward and back reaction) terms of the dead-end reaction; and this combination of terms is identically zero. It can therefore, by simple inspection, be dropped from the equation for the penultimate compound.

Some compounds are involved in irreversible reactions. The equations for such compounds are often good places to begin the assault on a system of equations—the irreversibility, the absence of a back reaction, means that one unknown, the product of the irreversible reaction, will be absent from the equation, which will therefore be that much simpler.

A common combination of chemical equations is

$$A + B \underset{b}{\overset{a}{\rightleftharpoons}} C \qquad C \overset{c}{\rightarrow} X$$

The "steady-state" equation for C, as will be shown in the next chapter, is

$$k_a AB - k_b C - k_c C = 0$$

(I shall use A, B, C, etc., for concentrations as well as names of compounds, since the two uses never conflict.)

The solution is

$$C = AB/K$$

where K is an abbreviation for $(k_b + k_c)/k_a$. After you have seen this situation a few times, you will write down the solution by inspection.

A similar dead-end case would be

$$A + B \underset{b}{\overset{a}{\rightleftharpoons}} C$$

The equation is

$$k_a AB - k_b C = 0$$

and the solution is

$$C = AB/K$$

where K is an abbreviation for k_b/k_a. Thus the solution of the irreversible system above is formally the same as that of the dead-end case, the difference being in the definition of K. This is no accident: In the dead-end case, the compound C isn't going anywhere; in the irreversible case, it is going somewhere, but nothing is coming back.

Experience will suggest further shortcuts from time to time. The point of these remarks is to alert the reader to the possibility of strategies that will minimize his labor.

Graphical Representation of Relations

In high school we were taught that you get the graph of a mathematical relation by assuming numerical values for one variable, and then using the equation to calculate the corresponding values of the other variable. If we had $y = 2x^2$, we would set $x = 1$, and calculate $y = 2$; then set $x = 2$, and calculate $y = 8$, and so on till we got enough points to allow us to draw the curve—or until *rigor mortis* set in.

But what if you were given $y = ax^2$, with a numerical value for a unspecified? You might try various values for a, and arrive at a curve for each one; but the time and energy expended are, even more than in the numerical case, out of proportion to what you learn.

However, there are methods for studying the **general form** of the curve corresponding to an equation, so that the curve may be **sketched** rather than plotted. The basis of these methods is the fact that every curve is characterized by certain features—you might call them **landmarks**—in terms of which you can predict the general shape of the curve.

In this section I propose to discuss the most useful and simple ways of looking at functions to see what they are like, and in particular to enable one to sketch their graphs with ease. In so doing, I have selected illustrative examples and exercises on the basis of (a) their probable relevance to cases likely to arise later in this volume and (b) the fact that conventional courses in analytic geometry, with their heavy stress on the conic sections, traditionally neglect most algebraic functions, and that even the conics appear most frequently in a classical standard form, while other and often more useful forms are omitted or treated as secondary. Neither the selection of cases nor the treatment makes any pretense at being exhaustive.

1. Domain and Range. The set of admissible values of a variable, as it appears in a given function, is the **domain** of the function. This set may be considered, roughly speaking, to be at your disposal: You may, for your own reasons, be willing to let the variable assume all real values, or only positive real values, or only rational values, and so forth; or you may wish to confine yourself to an uninterrupted subset of such sets (for instance, all reals from -1 to $+1$), or to several such (non-adjacent and non-overlapping) subsets.

The domain being determined, the set of admissible values of the function, its **range**, is now fixed, and frequently restricted. Thus for $y = ax$, the domain being all reals, the range is the same; but for $y = ax^2$, the domain being all reals, the range is confined to the non-negative reals. For $y = a^2 - x^2$, admitting all real values for x implies that y ranges from a^2 to negative infinity; for $y = \sqrt{a^2 - x^2}$ the corresponding range includes pure imaginaries.

To examine this relation between domain and range is an important initial step in analysis of any function. In this volume, for instance, basic parameters such as rate constants and variables such as concentrations are intrinsically positive. In most of the cases considered as examples in this section, this restriction has been observed, and the functions are analyzed for a domain of positive reals. The extensions of the domains to negative values may then be regarded as exercises for the reader.

Examination of domain and range may lead you to place additional restrictions on the domain. Thus if the function is $y = \sqrt{a^2 - x^2}$, and if y stands for a quantity of a kind that could not conceivably be imaginary, the domain must now be restricted to real numbers from $-a$ to a inclusive. Similarly, if a reasonable choice of domain leads to a range that includes negative values, and the function you are interested in is intrinsically positive, you will want to restrict the domain so that the range is confined to positive values.

The admission or exclusion of **infinity** requires special comment, since "infinity" is not the name of a number in the same sense that "five" is. If a variable you are interested in is intrinsically such that you could imagine increasing it as much as you like, or beyond any preassigned bound, then you must be willing to talk about letting that variable approach infinity. For example, in Chapter 2 we will be interested in a function that relates a concentration variable to a reaction rate; and we commonly say that the rate approaches its highest value as the concentration approaches infinity. This is perfectly meaningful, and must not be misunderstood because of practical limitations. For example, it is true but trivial that every compound has a finite solubility, beyond which you cannot increase the concentration; it is trivial because the solubility, in the context of a kinetic problem, is a sort of arbitrary or accidental feature—if the solubility were much greater, you could perfectly well imagine studying higher concentrations. Again, one frequently reads that the saturation value of the concentration of substrate for an enzyme (value for which the rate is maximal) is something like $0.1M$—which does not seem to be very great, let alone approaching infinity. But all that this means is that the investigator is unable reliably to detect any additional increase in the rate if he raises the concentration beyond $0.1M$; the curve has not really stopped rising, but experimental error is as great as the increments that might be expected.

Thus you may see in a textbook the definition "$y = 1/(x-c)$ if $x \neq c$". This is good as far as it goes: It means that **substituting** c for x in the expression on the right of the "$=$" sign gives a meaningless result. But it is perfectly correct, in this case, to say that y **approaches infinity**, or increases without bound, as x **approaches** c. (See further discussion under *Poles*, below.) The limitation on the use of such phraseology and on the acceptance of such a function is set by the nature of y—whether y *is* in fact the kind of thing that you can imagine as increasing without bound, like a concentration.

Closely related to the rejection of parts of a domain because you cannot tolerate the corresponding range is the rejection of an entire function that may arise in your work. The most common example in this volume is the following: A concentration is given as "the" solution of a quadratic equation; and this solution is of course two functions, one with a plus sign before a radical, one with a minus sign. If you know that the concentration should be zero when one of the parameters in the solution is zero, it may turn out that only one of the roots of the quadratic will behave as desired; the other is therefore rejected *in toto*. Or one root may prove to have its range confined to negative values, and the consideration that concentrations are by nature positive will again lead you to discard that root.

2. Properties of Simple Powers. Consider $y = x^2$ (since $y = ax^2$ differs only by the constant factor a multiplying each ordinate). Since the square of a proper fraction is less than that fraction, $y = x^2$ lies below the straight line $y = x$ for $x < 1$. Since the square of a number greater than unity is greater than that number, $y = x^2$ lies above the line $y = x$ for $x > 1$. The sketch is shown in Fig. 1.1.

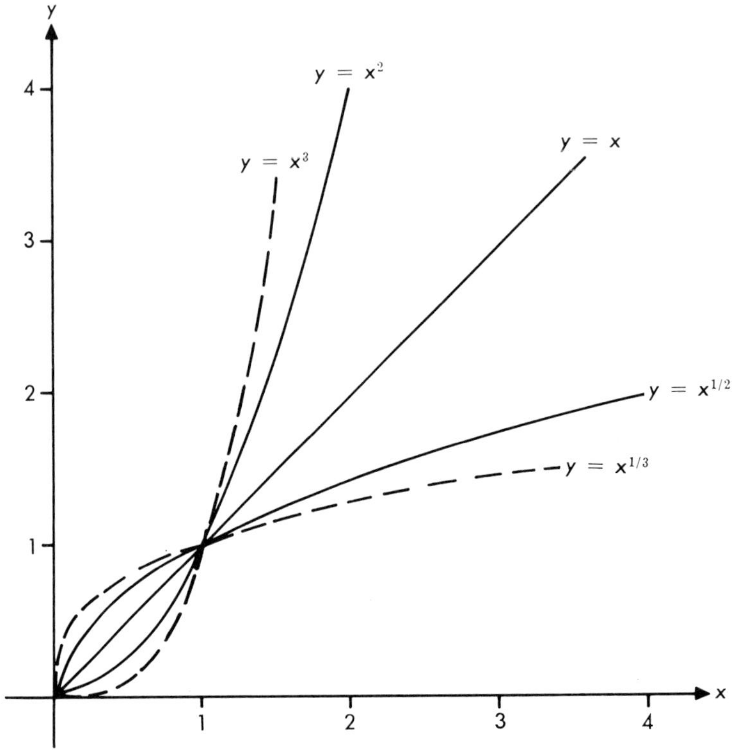

Fig. 1.1. Comparison of Straight Line, Simple Parabola, and Its Inverse.

Higher powers of x behave the same way, only more so. That is, they hug the x axis more closely for $x < 1$, and rise more sharply for $x > 1$. (Here as elsewhere I confine the discussion to positive values of x, since most chemical variables, such as concentrations, are intrinsically positive.)

The behavior of powers less than 1 (proper fractional powers) follows at once from that of powers greater than 1. Thus $y = x^{1/2}$ (I will use the fractional power notation rather than the radical sign most of the time) is equivalent to $y^2 = x$ (squaring both sides), which is equivalent to $x = y^2$. It follows that the curve of $y = x^{1/2}$ is the same as that of $y = x^2$ except for the orientation of the coordinate axes. Draw the curve of $y = x^2$ on tracing paper; turn the paper through a right angle, counterclockwise; flip the paper over as if turning the page of a book—and you will see the curve of $y = x^{1/2}$. Stated crudely, the square-root curve is the square curve lying on its side. The square and the square root are **inverse functions**.[3]

3. Slopes and Extrema. The upward concavity of $y = x^2$ is also demonstrable by calculating the slope, which is given by dy/dx.[4] For $y = x^2$, $dy/dx = 2x$. This increases with x; hence the slope of the curve increases with x—the curve bends upwards. For $y = x^{1/2}$, $dy/dx = \frac{1}{2}x^{-1/2} = 1/2x^{1/2}$. This decreases with increasing x, hence the curve bends downwards for $x > 0$.

If the slope dy/dx is increasing, **its** rate of increase, the second derivative, d^2y/dx^2, must be positive. For $y = x^2$, $d^2y/dx^2 = 2$, which is positive. If the slope is decreasing, the second derivative will be negative. For $y = x^{1/2}$, $d^2y/dx^2 = -\frac{1}{4}x^{-3/2}$, which is negative.

Maxima and minima are landmarks of major usefulness. Recall that **extrema** (collective term for maxima and minima) are points where

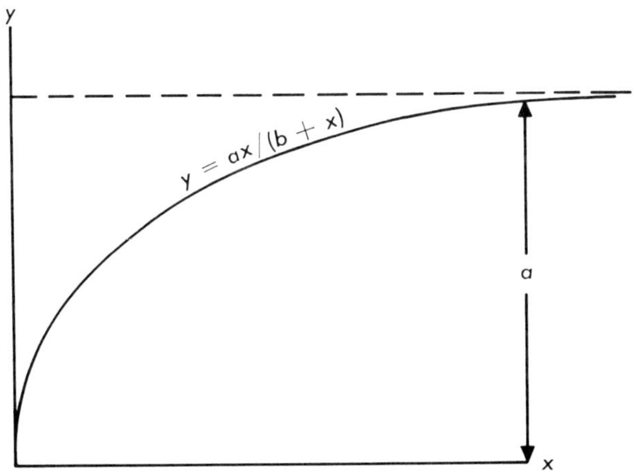

Fig. 1.2. Rectangular Hyperbola. Solid line: Curve $y = ax/(b + x)$. Dashed line: Asymptote $y = a$.

[3] The relation between curves and axes that holds for the square and square root holds for all pairs of inverse functions. To be precise: If $y = f(x)$, and if you could solve this equation for x, you would get some expression for x in terms of y—say $x = g(y)$. In this case, $g(y)$ is by definition the inverse of the function $f(x)$. The practical importance of the relation between the graphs (we use it later on) is this: In some cases it would be cumbersome or difficult, if not impossible, to obtain the actual solution, to arrive at a concrete expression for the function g. But if you can prepare the graph of $f(x)$, you can turn it around as I have described, and see what g would be like. Another procedure, in place of the rotate-and-flip method for getting $x = g(y)$, is to draw $y = g(x)$, which is really the same thing, on the original system of axes. To do this, we construct the mirror image of $y = f(x)$ through the line $y = x$. I omit the elementary proof that this is valid, leaving it to the ambitious reader (Hint: If corresponding points on the two curves are mirror images through $y = x$, then by definition they must be equal distances from a point on $y = x$). The construction is illustrated by the case $y = ax^{1/2}$ in Fig. 1-1.

[4] If you don't remember this, check your calculus text. You will find tables of formulas for the more common derivatives in any standard handbook of chemistry.

$dy/dx = 0$. Given $y = f(x)$, get dy/dx, set it equal to zero, solve for x, find the corresponding y by substitution in $y = f(x)$; the resulting pair (x_e, y_e) gives the coordinates of the extremum. Since a curve is concave down at a maximum, concave up at a minimum, the sign of the second derivative will tell you which kind of extremum you have. Get d^2y/dx^2, substitute x_e for x in it, and ascertain whether the result is positive or negative.

Inflection points are points where the curvature of the graph changes. Since the sign of the second derivative gives the direction of curvature, the second derivative must pass through the value 0 in changing sign (from positive to negative or the reverse). Hence inflection points are located by setting $d^2y/dx^2 = 0$, and solving for x, obtaining the corresponding y value from $y = f(x)$.

This rule must be qualified, as may be seen by considering $y = x^4$. The reader will readily satisfy himself that this curve is much like $y = x^2$—everywhere positive, symmetric about the y axis (see below), with a minimum at $x = 0$, and everywhere concave upward. But $d^2y/dx^2 = 12x^2$, which is zero at $x = 0$. The correct rule is as follows: If all successive derivatives of a function up to a certain one vanish at a point, the point is an inflection only if the first non-vanishing derivative is an odd one (for example, third derivative, fifth derivative); if the first non-vanishing derivative is even, the point is not an inflection point (11). The first non-vanishing derivative of $y = x^4$ at $x = 0$ is $d^4y/dx^4 = 24$; hence $x = 0$ is not an inflection point.

As an intuitively obvious example, consider $y = ax/(b + x^2)$, where as usual $a > 0$, $b > 0$ (Fig. 1.3). This is evidently zero at $x = 0$, and it will also become zero when x becomes infinitely great (see subsection 5 below), since

Fig. 1.3. Curve with Maximum, Showing Sign of Slope of Tangent in Various regions.

the higher power, x^2, in the denominator will outweigh the lower one, x, in the numerator when x is large. (For more precise argument, see subsections 4 and 5 for application of the binomial approximation.) Since the function is continuous, and is everywhere positive (for $x > 0$), it must increase and then decrease, and have at least one maximum in between. (How do we know that the function is continuous? If you're curious, return to a calculus text to refresh yourself on the definition of continuity, and apply it.)

We readily find $dy/dx = a(b - x^2)/(b + x^2)^2$. This can vanish if (a) its numerator is zero while its denominator is not zero or (b) if its denominator approaches infinity while its numerator does not (or if its denominator approaches infinity faster than its numerator). Condition (a) is fulfilled if $x = \pm b^{1/2}$; and we will consider only the positive solution, since we are confining our analysis to positive values of x. We confirm our intuitive decision that the extremum is a maximum by way of $d^2y/dx^2 = 2ax(x^2 - 3b)/(b + x^2)^3$, which becomes negative for $x = b^{1/2}$.

Condition (b) is rather trivially satisfied as x approaches infinity. Dividing numerator and denominator of dy/dx by x^4, we get $dy/dx = a(b/x^4 - 1/x^2)/(b/x^2 + 1)^2$. In the limit, all terms of the numerator approach 0; the fraction approaches $0/1 = 0$.

The second derivative vanishes for $x = (3b)^{1/2}$; there is an inflection point to the right of the maximum. This can also be seen intuitively; the only way to fit together (with continuity) a concave-downward curve with a maximum and a horizontal tail for large x is to have the curve change its direction of concavity in between.

4. *The Origin and Its Neighborhood.* We may ask two questions about the graphical behavior of a function at $x = 0$: What is its value **at** $x = 0$, and how does the function behave **near** $x = 0$?

For example, $y = ax$ vanishes at $x = 0$, so does $y = ax^2$. But $y = ax + b$ equals b at $x = 0$, and so on. Looking at more complex cases, $y = ax/(b + x)$ vanishes at $x = 0$, and so does $y = ax^2/(b + x)$, and so does $y = ax^2/(b + x^2)$. Also, $y = 1 - e^{-ax}$ vanishes at $x = 0$.

Now let us consider the question of behavior *near* the origin. Applying the binomial approximation to $y = ax/(b + x)$, we get $ax(1 - x/b)/b = ax/b - ax^2/b^2$. Thus as x becomes very small, the function approaches $y = ax/b$—that is, it behaves like a straight line in the neighborhood of the origin. To convince yourself that this is right, try a numerical example. Pick $a = 1$, $b = 1$, and take x small compared with b, say 0.05. Now y is seen to equal about 0.048. If we now take $x = 0.025$ (half the previous value), we get y about 0.024. Thus y changes in proportion to x, as should be the case for a straight line. If we double the original x to 0.1, we get y about 0.091, somewhat less than twice as much as 0.048; the linearity is still with us as x increases, but to a lesser extent.

Does $ax^2/(b+x)$ behave the same way? Again take a and b equal to 1. For $x=0.05$, y is about 0.0024; for $x=0.025$, y is about 0.0006—not half as great, but one-fourth as great. For $x=0.1$, y is 0.009, somewhat short of being four times as great as 0.0024. The function behaves like a square near the origin. Analytically, the binomial approximation gives $ax^2(1-x/b)/b = ax^2/b - ax^3/b^2$, and y approaches ax^2/b, as the calculation suggested. The reader can satisfy himself that functions like $ax^2/(b+x^2)$ and $ax/(b+x^2)$ will behave similarly.

To see what $y=1-e^{-ax}$ does in the neighborhood of the origin, use the series expansion for the exponential function ($e^{-ax}=1-ax+\frac{1}{2}a^2x^2 - \frac{1}{6}a^3x^3+\cdots$), and drop all except the first two terms. We get $1-(1-ax) = ax$; that is, the function behaves like a straight line near the origin.

Other functions do the same thing. Take $y=(ax+b)/(x+c)$. This does not vanish at $x=0$, but becomes equal to b/c. To see how it behaves in the neighborhood of the origin, we factor c out of the denominator to get $(ax+b)/c(1+x/c)$, then apply the binomial approximation for the power -1 to the binomial in the denominator, getting $(ax+b)(1-x/c)/c$. Multiplying out, and dropping terms in x^2 since we are interested in very small values of x, we get $(ax-bx/c+b)/c$ as the approximate linear expression for y in the neighborhood of the origin. (See Fig. 1.4.)

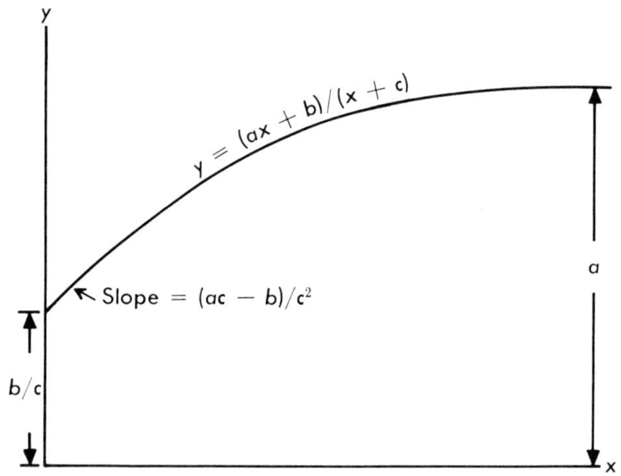

Fig. 1.4. Rectangular Hyperbola, Somewhat More General than Fig. 1.2.

It is of interest to examine the method a little more closely. The first two cases gave ax/b and ax^2/b as approximations for the function near the origin, as if we had obtained the result by simply discarding the term c in the denominator. But the third case, $(ax + b)/(x + c)$, gives a much more elaborate result.

What we have been doing in the binomial approximation is to multiply the numerator by $1 \pm$ a power of x. If there is just one term in the numerator, multiplying by 1 will reproduce it, while multiplying by a power of x will give terms of higher degree in x; and the term of lowest degree is of course the term we keep as x becomes small. But if there are several terms in the numerator, of different degrees in x, multiplying by $1 \pm$ a power of x may give more than one term of a given minimal degree in x.

In particular, it is evident that, with denominator $x + c$, and a numerator of the form $ax + b$, both terms will contribute significantly to the final result. If (as in the next paragraph) the terms are of higher degree or differ by more than 1, only the term of lower degree contributes to the final result.

Thus if we apply the binomial approximation procedure to $y = (ax^2 + b)/(x + c)$, we obtain (again neglecting in the final result all terms of degree in x higher than the first) $(b - bx/c)/c$, a linear expression.

Problems: 1. Ascertain the value at the origin and the behavior near the origin for:

$$y = (ax^2 + bx)/(x + c)$$
$$y = (ax^3 + bx^2)/(x + c)$$
$$y = (ax^2 + bx + d)/(x + c)$$
$$y = (ax^2 + b)/(x^2 + c)$$
$$y = [\ln(a + bx)]/(x + c)$$
$$y = (a + bx)^{1/2}/(x + c)$$
$$y = (a + bx)^{1/2}/(x + c)^{1/2}$$
$$y = (ax + b)(x + c)$$
$$y = (ax^2 + bx)(x^2 + cx)$$

2. Find a rule, analogous to that of the next-to-last paragraph, for denominators $x^2 + c$.

5. Asymptotic Form and Asymptotes. We may ask how a function behaves when the variable x becomes infinitely large, and whether it increases without bound also or approaches some finite value. For $y = ax$ and $y = ax^2$ the answer is obvious: They behave like ax and ax^2, and they increase without bound. But consider again $y = ax/(b + x)$. If x is so large that b is negligible by comparison with x in the denominator (for example, $b = 1$ as before, $x = 1000$), the function becomes practically equal to $ax/x = a$—i.e., it approaches a finite constant.

To examine more carefully how the function behaves, we may again apply the binomial approximation to the denominator. But now, since x is very large, we factor out x rather than the constant, getting $y = ax/x(1 + b/x)$, because as x increases b/x becomes the small quantity that we need for the approximation to be valid. Applying the approximation, we get $a - ab/x$, which in the limit is approximately equal to a. In other words, the function behaves like a constant, not only at the limit, but in the neighborhood of the limit. This is seen in another way if we examine the derivatives. The first derivative dy/dx is equal to $ab/(b + x)^2$, which of course approaches zero as x becomes very large. The same is obviously true of the higher derivatives —they all vanish in the limit. The function is, so to speak, infinitely flat near the limit. (See Fig. 1.2.)

On the other hand, as x approaches infinity, $y = ax^2/(b + x)$ is seen, using the binomial approximation again, to approach $y = ax - ab$, a straight line with positive slope; the function increases without bound.

When the **asymptotic form** of a function is that of a straight line, this means that the curve approaches the straight line in the limit. Such a straight line is an **asymptote** of the curve. But it must be remembered that not all asymptotic forms are linear; for instance, $y = ax^2 + bx + c$ approaches ax^2 in the limit for large x, and e^{ax}, for $a > 0$, increases faster in the limit than any power of x, no matter how high (divide infinite series for e^{ax} by x^n—there is always a term of degree higher than n).

Given a relation between y and x, it is perfectly legitimate to inquire about the asymptotic behavior of x for large values of y! Often this will lead to nothing new: In $y = ax^2 + bx + c$, y approaches infinity with x, and it is equally true that x approaches infinity with y—the asymptotic functions found in the two different ways will be inverses of one another. The cases where x remains finite for infinite y will be treated later under *Poles*, for reasons which will become apparent at that time.

Note that finding the intercepts as well as the slope of an asymptote has practical importance, since one can extrapolate back to the intercept of the asymptote in a graph of experimental data. If the theory predicts a negative intercept, for instance, and the value found is positive or zero, we are getting a broad hint that the model is wrong. Moreover, if the intercept found experimentally behaves as it should with respect to sign, its numerical value will give an estimate of one of the parameters, as will be illustrated frequently in later chapters.

PROBLEMS: Find the asymptotic form of the functions given in the problems at the end of the subsection 3. Skip the problem involving a logarithm—you are not yet equipped for it. (The ambitious reader may consult a handbook for a series that can represent the logarithm for large values, and see what he thinks of the problem then.)

6. Zeroes and Poles. Two other classes of landmarks are the places where a function becomes equal to zero, and those where a function has a zero denominator with a non-zero numerator. For examples of the first kind, consider $y = ax(x - b)$. This is zero both at the origin and at $x = b$, which implies that there must be an extremum between 0 and b, as calculation of dy/dx will confirm. (That the extremum is a minimum follows by way of the second derivative; or by noting that near the origin y **decreases** like $-abx$ (since the higher power of x is negligible compared with the first power for small x). (See Fig. 1.5.)

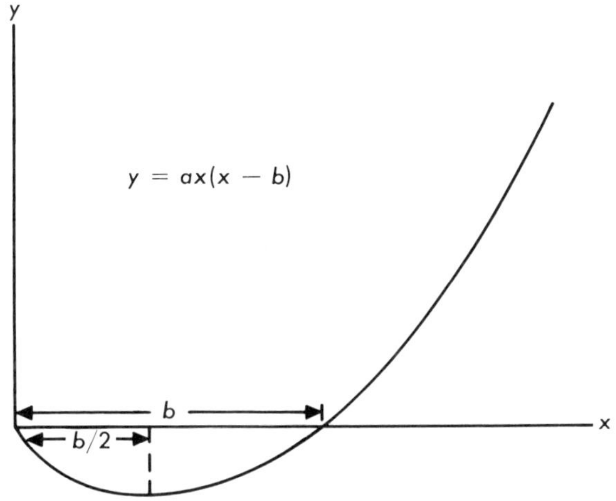

Fig. 1.5. Curve with Minimum and Two Zeroes.

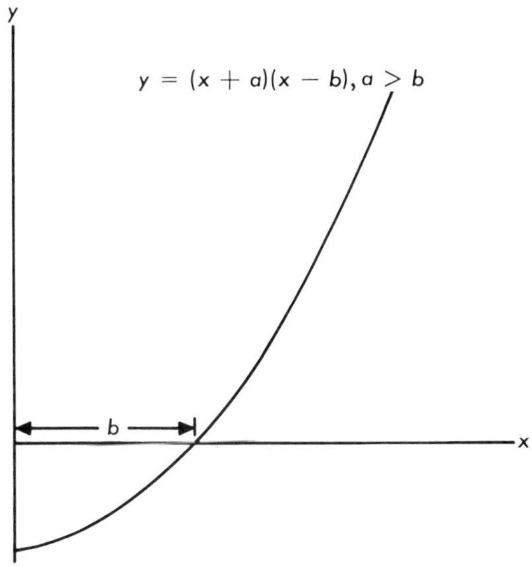

Fig. 1.6. Curve with Zero, Showing Negative Intercept.

A similar function is $y = (x + a)(x - b)$, which still has a zero at $x = b$ (It has another at $x = -a$; but here as elsewhere we ignore negative values, since most of the parameters and variables encountered in this book are intrinsically positive.) Expanding and discarding the square term, we find that the initial slope is $a - b$. (Or find dy/dx and evaluate at $x = 0$.) This may be positive or negative, so the curve may start upward or downward. The intercept is $-ab$, so the curve starts below the x axis. In keeping with the fact that the curve starts downward if $a < b$, the solution of $dy/dx = 0$ is $x_e = \frac{1}{2}(b - a)$, which will be positive and hence appear on our graph only if $b - a$ is positive. (See Fig. 1.6 for $a > b$.)

What about $y = (x - a)(x - b)$? This has two zeroes, a and b. The initial slope is $-(a + b) < 0$, the intercept is $ab > 0$. The curve starts downhill, and eventually rises, since the asymptotic behavior for large x is like $y = x^2$. The minimum lies on that part of the curve that falls between a and b and lies below the x axis. (See Fig. 1.7.)

A simple function with a pole is $y = x/(a - x)$. It is zero at the origin, and approaches x/a near the origin. As x approaches a, the denominator approaches zero and the fraction becomes infinitely great.[5] If $x < a$, the denominator is positive, and so is the whole function. If $x > a$, the function is negative. Thus the function must evidently swing over from positive to negative infinity as x passes through the pole a. The function thus consists of two distinct **branches**, separated by the vertical line $x = a$. The negative branch of the curve eventually flattens out (as in the case of $ax/(b + x)$), and asymptotically approaches $y = -1$. (See Fig. 1.8.)

Note that the vertical line $x = a$ is an asymptote: Instead of saying that y approaches infinity as x approaches a, we could have said that x approaches a as y approaches infinity. The reason for having deferred such cases to this paragraph is that we tend to look at graphs, especially of experimental data, with a built-in bias about the roles of the "independent" variables, even though the distinction is mathematically arbitrary. It seems more natural to say, "When x does this, that happens to y" than the other way around.

[5] If the reader should chance to pick up a rather modern text of "analysis" (a mixture of analytic geometry and calculus), he will find the statement that the function "is not defined" or "does not exist" for values of x that make the denominator vanish. The reason given might be that "infinity is not a number," or that "one cannot divide by zero."

This is perfectly true. The correct way to state what happens is: As x approaches sufficiently close to a polar value (zero of denominator), y becomes and remains greater than any pre-assigned value, no matter how great. (In modern terminology: For every y_0 there exists a $\delta > 0$ such that if $|x - a| < \delta$ then $y > y_0$.) However, if the reader understands this, there is no great point in huffing and puffing about terminology, especially since the modern terminology fails to remind the reader that the function becomes very large in the neighborhood of the point where it is "not defined." I shall therefore use the convenient locutions "approaches infinity" and "approaches minus infinity" throughout the book.

For the reader who was fed rules like "never divide by zero" without any explanation, I recommend a look at p. 46 of Allendoerfer and Oakley (12).

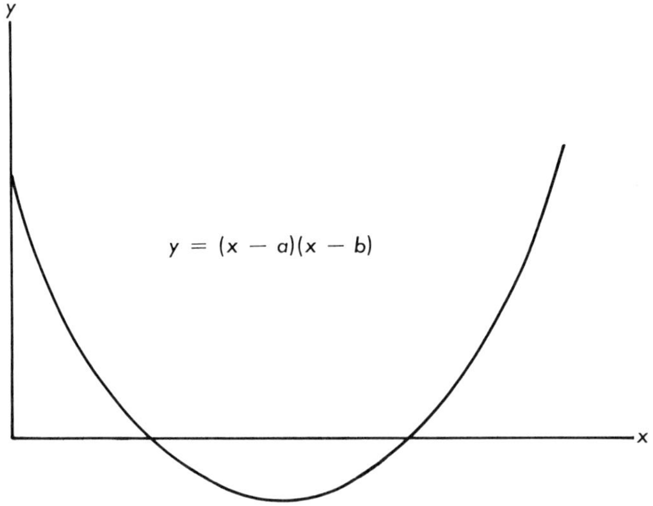

Fig. 1.7. Curve with Minimum and Two Zeroes, Somewhat More General than Fig. 1.5.

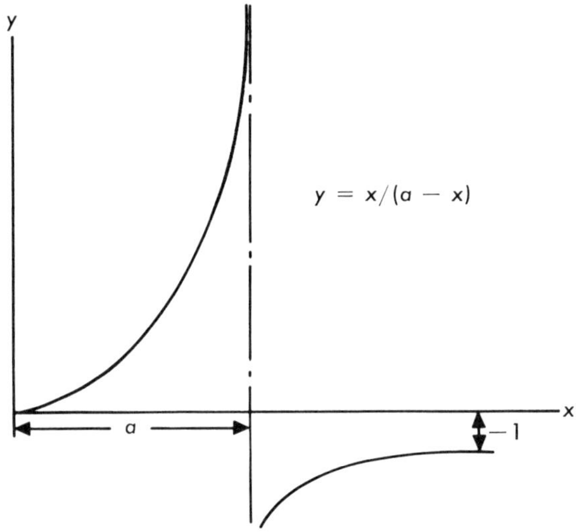

Fig. 1.8. Curve with Pole. Vertical dashed line, $x = a$, is asymptote at pole, Asymptote for large x is $y = -1$.

Suppose $y = x/(x^2 + bx - c)$. The poles must be roots of the quadratic equation $x^2 + bx - c = 0$; they are $\frac{1}{2}[-b \pm (b^2 + 4c)^{1/2}]$. Since $b^2 + 4c > b^2$, the radical is greater than b; hence the solution with the plus sign before the radical is positive and relevant, the solution with the minus sign before the radical will be negative, and will not appear in our graph. The function approaches $-x/c$ near the origin, and decreases until it becomes negatively infinite at the pole, then swings over to a positive branch. It therefore roughly resembles a mirror image of Fig. 1.8. The fact that the function, after starting with negative values, remains negative until it comes to the pole, follows from the fact that it is continuous in this region and has no zeroes except $x = 0$, so it does not cross the positive x axis. The positive branch asymptotically approaches $1/x$, and becomes zero in the limit.

If the denominator were $x^2 - bx + c$, the poles would be given by $\frac{1}{2}[b \pm (b^2 - 4c)^{1/2}]$. These values are real only if the discriminant of the quadratic, $b^2 - 4c$, is positive. If not, the radical and hence the poles are imaginary, and the curve will look somewhat like $ax/(b + x^2)$. If $b^2 - 4c > 0$, both poles will be positive, inasmuch as $b^2 - 4c < b^2$. The curve begins like x/c, has three branches as shown in Fig. 1.9, and vanishes asymptotically

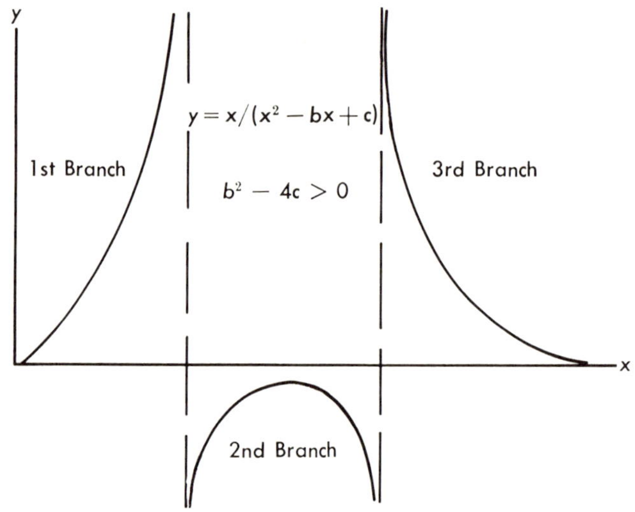

Fig. 1.9. Curve with Two Poles, Showing Asymptotes at Poles.

like $1/x$. There are no zeroes except $x = 0$, so the function cannot cross the x axis except by jumps at poles.

PROBLEMS: 1. Analyze, for behavior near the origin, asymptotic behavior, zeroes, poles, and extrema, the functions:

$$y = (ax^2 + bx + c)/(x + d)$$
$$y = (ax^2 + bx + c)/(x - d)$$
$$y = (ax^2 + bx + c)/(d - x)$$
$$y = (ax^2 - bx + c)/(x - d)$$
$$y = (ax^2 + bx + c)/(x^2 + px + q)$$

2. How would the foregoing problems and the functions of previous subsections behave if we let x and/or the parameters assume negative values?

7. *Symmetry*. The curve of a function is symmetric about the y axis if, for every point at abscissa x, there is a point with the same ordinate at abscissa $-x$ (its mirror image with respect to an imaginary mirror at the y axis). That is, if $y = f(x)$, the graph is symmetric about the y axis if, for all x, $f(x) = f(-x)$. (Example: $y = x^2$.) Such functions are called such as **even functions**, and there are many examples among algebraic functions, those involving even powers of $x[y = (ax^4 + bx^2)^{1/4}]$, or among transcendental functions *of* such functions ($y = e^{-x^2}$). The absolute value of x, $|x|$, plays a similar role to the even powers. But note that some transcendental functions that are not composites of even algebraic functions are nevertheless even—for instance, $\cos x$ (since $\cos x = \cos(-x)$).

For symmetry about a line **parallel** to the y axis, for instance $x = x_0$, the expression $x - x_0$ has the same role as x has for the y axis (which, remember, is the line $x = 0$). Thus $y = x^2 - 2x_0x + a$ is symmetric about $x = x_0$; for $y = x^2 - 2x_0x + x_0^2 + a - x_0^2 = (x - x_0)^2 + a - x_0^2$, an even function of $x - x_0$. (General procedure: Write $x = (x - x_0) + x_0$.)

A function is symmetric about the x axis if for every x it has equal y values above and below the axis, like $y = \pm x^{1/2}$. Clearly, the inverse of such a function will be symmetric about the y axis (the inverse of the square root is the square, for instance); and this may sometime be a convenient way to test the point. (It is the current affectation to say that such **multiple-valued functions**, which have more than one y for a given x, are not functions at all, but merely "relations"—to confine the word "function" to what have always been called **single-valued** functions. For our present purposes, there is nothing significant about this definitional jugglery; but the reader should be aware of it if he should pick up one of the more fashionable textbooks.)

The graph of a function is skew-symmetric or antisymmetric (a recent not-very-suggestive innovation is "symmetric about the origin") if for every pair of values (x, y) on the graph there is a pair $(-x, -y)$ also on the graph—i.e., if $f(x) = -f(-x)$. Such a function is called an **odd function**; and some functions of odd powers of x fall in this category, and so of course does $\sin x$. But it is not true (as one might guess by analogy) that all algebraic or transcendental functions of odd powers are odd functions. For instance, $y = x^3$ is an odd function. But consider $y = x^3 + a$. Changing x to $-x$, we get $(-x)^3 + a = -x^3 + a$. But this is not the same as $-(x^3 + a)$, which would be the negative of the original y. In this case you can see that $y - a = x^3$ is an odd function. The graph is not skew-symmetric with respect to the origin, but the graph you get by dropping it a units downward is; the graph of y is skew-symmetric about the point $(0, a)$. However, not all cases are so easily patched up; and one must examine functions carefully before concluding that they are odd. "Identical" and "different" are exhaustive categories; "identical" and "opposite in sign" are not: $f(-x)$ can be different from $f(x)$ without be equal to $-f(x)$, and so a non-even function need not be an odd function.

Testing for symmetry around lines not parallel to the axes is more complex. If you can transform (linearly) your coordinate system in such a way that the line becomes the new y axis, and the transformed function is symmetric about the new y axis, you can conclude that the original function was symmetric about the original line. (Consult a text of analytic geometry for transformations of coordinate systems.)

One case, however, is simple and rather interesting—symmetry about the line $y=x$. We know that a function and its inverse are symmetrically placed around that line; hence a function symmetric about the line will have to be identical with its inverse. That is, if the coordinates of a point on the graph are (a, b), there must be another point on the graph with coordinates (b, a). To test for this analytically, it is easiest to rearrange the equation $y=f(x)$ into $y-f(x)=0$. Since the left side depends on both x and y, we can call it $F(x, y)$, and write the equation as $F(x, y)=0$. The condition that to every point on the graph there corresponds a point with the coordinates interchanged then becomes $F(y, x)=0$. Taking this and $F(x, y)=0$ together, we can write $F(x, y)=F(y, x)$. In short, if we collect all terms of the original equation on one side of the equal sign, and if we get an equivalent expression by interchanging x and y (possibly after a little rearrangement), the graph is symmetrical about $y=x$. As examples, consider the line $y=a-x$, or the equation of the circle, $x^2+y^2=a^2$, or the rectangular hyperbola, $y=a/x$.

What makes this case interesting is that it has a connection with another use of the word "symmetric". Consider a function of two real variables, say $z=F(x, y)$. Such a function is called symmetric with respect to the variables x and y if $F(x, y)=F(y, x)$ (but not necessarily $=0$). What does this coincidence signify? If z is the third coordinate in a three-dimensional rectangular system, $z=F(x, y)$ is the equation of a **surface**. The condition $F(x, y)=0$ $(z=0)$ defines the curve formed by the intersection of the surface and the x-y coordinate plane (whose equation is $z=0$). In short, if the function defining a surface is symmetric with respect to its variables, the intersection of that surface with the x-y plane is geometrically symmetric with respect to the line $y=x$.

An application of functional symmetry (unrelated to geometric symmetry) will be found in Chapter 4. It is based on this: If $z=F(x, y)$ is symmetric as in the preceding paragraph, and you plot z against x keeping y fixed (say $y=a$), you get exactly the same curve as if you plot z against y

keeping x fixed at $x = a$. (This follows immediately from $F(a, y) = F(y, a)$; assigning all admissible values to x in $F(x, a)$ will give a curve, and assigning all admissible values to y in $F(y, a)$ evidently gives the same curve.)

PROBLEMS: 1. Prove that a function is symmetrical about the line $x = x_0$ if substituting $2x_0 - x$ for x leaves the function unchanged. (Hint: Try to express the function in terms of $x - x_0$; then apply the symmetry condition.)

2. Prove that a function is skew-symmetric about the point (x_0, y_0) if substituting $2x_0 - x$ for x and $2y_0 - y$ for y leaves the function unchanged.

Functions of Integer Variables; Difference Equations

It will pay us to take a fresh look at some familiar types of expression, and to introduce some useful notations. An example of the expressions I have in mind is the factorial, $n!$ in the usual notation. In this expression, "n" refers to an integer; and, for any definite assignment of a numerical value to n, the value of the factorial is uniquely specified, and conversely. The factorial therefore behaves like a function: To every value of an "unspecified" integer it correlates another value, that of the factorial; and the relation is one-to-one. (The graph of $n!$ against n is of course a set of isolated points.)

It is reasonable, therefore, to speak of "n" in such an expression as an **integer variable**, and to classify the factorial as a **function of an integer variable**—a function, that is, of a variable whose admissible values are confined to the integers. With a suitable convention, we could indeed just as well write "$F(n)$" instead of "$n!$". (There is in fact a well-known function of an ordinary continuous variable x, the **gamma function**, which reduces to the factorial when the variable x assumes integer values.) The use of a letter like "n" instead of the more familiar "x" is not essential; but it is a suggestive and well-established convention, by which letters denoting integer-valued variables or parameters are drawn from the middle of the alphabet rather than the end.

Another way of writing functions of integer variables, and one which we shall use frequently in this book, arises when we have a set of numbers, identified by numerical tags or **subscripts**, say $A_1, A_2, A_3, \ldots,$ and we wish to refer to "any of the numbers" or "an unspecified one of the numbers". We then write the standard symbol (A) with a letter subscript instead of a number subscript—thus, A_i, where "i" is in fact an integer variable, which can take on any of the integer values admissible for the original set of subscripts. We speak of "the ith member of the set".

This **variable-subscript** or **variable-index** notation, which is merely another way of writing functions of an integer variable, lends itself to some extremely useful notational devices.

For example, a system of N equations, such as:

$$C_1 = a_1$$
$$C_2 = a_2$$
$$C_3 = a_3$$
$$\cdots$$
$$C_N = a_N$$

may be written simply:

$$C_j = a_j \qquad (j = 1, 2, \ldots, N)$$

Sums and products are readily expressed by means of two symbols. For

$$C_1 + C_2 + C_3 + \cdots + C_N$$

we write

$$\sum_{j=1}^{j=N} C_j$$

The symbol $\sum$ is the Greek capital sigma (which corresponds to English S—for "Sum"). With suitable understanding between writer and reader, the symbolism may be shortened to $\sum_{j=1}^{N} C_j$, or $\sum_{1}^{N} C_j$, or $\sum_{j} C_j$, or sometimes just $\sum C_j$.

Any sum with a definite pattern of integers may be written this way, for example:

$$\sum_{1}^{N} X^j \qquad \text{for } X + X^2 + X^3 + \cdots + X^N$$

The summation symbol obviously has the following properties:

$$\sum aC_j = a \sum C_j$$
$$\sum (C_j + A_j) = \sum C_j + \sum A_j$$
$$\sum_{1}^{N} (C_j + a) = \sum_{1}^{N} C_j + \sum_{1}^{N} a = \sum_{1}^{N} C_j + Na$$

In a similar way, we write for the product

$$C_1 \cdot C_2 \cdot C_3 \cdot \cdots \cdot C_N$$

the expression

$$\prod_{j=1}^{N} C_j$$

$\prod$ being the Greek capital pi (English P, for "Product"). Products of powers or other expressions involving integers may be written as for sums.

If odd or even indices or powers are required, they are readily introduced into the notation by recalling that any even number contains a factor 2—thus, $2j$ is even. Correspondingly, $2j+1$ is odd. Alternating signs in a sum are easily provided for by multiplying the general term by $(-1)^j$ or $(-1)^{j+1}$, depending on which terms are to be negative; the power j will make odd terms negative, the power $j+1$ will make even terms negative.

Since the letter subscripts are variables, they may be changed to other letters, with suitable precautions, in the same way as we may write $y=f(x)$ or $v=f(u)$ at will to designate the same function, provided the assignment of letters is carried out consistently and the new letters are not already in use in another sense in the same expressions. Thus $\sum_j C_j$ equals $\sum_k C_k$ or $\sum_r C_r$. By the same token, we may shift subscripts and limits of sums or products—for instance, $\sum_1^N C_j$ equals $\sum_0^{N-1} C_{j+1}$, as the reader will readily verify.

A necessary precaution arises if one wants to refer to several (still unspecified) distinct members of a set of index-tagged symbols at once. It is then essential to use distinct subscripts, just as in speaking of a function at several distinct but unspecified values of the continuous variable x one does not write "x" every time, but uses pairs like x' and x'' or x_1 and x_2 or x and $x+h$, according to convenience. Thus if we wanted to specify a certain kind of order in the set, we might write "$C_j > C_i$ if $j > i$".

The need for multiple summations occasionally arises. Thus $(\sum C_j)^2$ is $\sum C_j$ multiplied by $\sum C_j$; and this of course is the sum of all possible products among the C expressions, as:

$$C_1C_1 + C_1C_2 + \cdots + C_2C_1 + C_2C_2 + C_2C_3 + \cdots$$

This may be written as a **double summation**, provided the indices are carefully kept distinct:

$$\sum_i \sum_j C_i C_j$$

which is sometimes written $\sum_{i,j} C_i C_j$. The expression "$\sum_j \sum_j C_j C_j$" is really meaningless; but, if one were forced to assign a meaning to it, it would have to mean the same thing as "$\sum_j C_j^2$"—that is, the sum of the squares rather than of all possible products. Thus the "mixing" of two sums, as above, in a double sum demands attention to subscripts.

The double summation obviously has the following properties:

$$\sum_j \sum_k C_j A_k = \sum_k \sum_j C_j A_k$$

$$\sum_j \sum_k C_j A_k = \sum_j C_j (A_1 + A_2 + \cdots) = (\sum_j C_j)(\sum_k A_k)$$

Functions of continuous variables are frequently specified by differential or by functional equations. Thus $y = e^x$ is described by $dy/dx = y$; it is also described by $f(x + z) = f(x)f(z)$. In a similar way, the function of an integer variable, $n!$, is described by:

$$F(n+1) = (n+1)F(n)$$

as the reader may verify by starting with $n = 0$ (and the additional condition $F(0) = 1$) and carrying out a series of substitutions.

Equations of this kind are known as **difference equations**, because the simplest of them is of the form

$$f(n+1) - f(n) = a \qquad (n = 1, 2, \ldots)$$

which describes an arithmetic progression with first term $f(1)$ and common difference a. To use the language we have been using so far in this section, the solution is $f(n) = f(1) + (n - 1)a$.

The simplest equations of this kind, like those just given, may be solved by successive substitutions in an obvious way. They are sometimes referred to as *recursion* equations or recursions; and we will meet them in Chapter 5 and elsewhere in this volume. The topic is an extensive and important one; I will carry it no further here, but merely call attention to its existence and its usefulness, and refer the curious reader to the appropriate texts.

Let me conclude by observing that functions of integer variables, like functions of continuous variables, include functions of **several** variables. Thus in Chapter 5 we will use the binomial coefficients or combinatorial coefficients, which are defined as combinations of factorials. With $F(n)$ for $n!$, we have

$$C_{jn} = F(n)/F(j)F(n-j)$$

a function of the two integer variables j and n.

We may also have mixed cases—continuous and integer variables. Thus in Chapter 5 we will use C_j to stand for the concentration of enzyme molecules with j substrate molecules bound to them. The C_j will therefore depend on a continuous variable, the substrate concentration S. If we wanted to indicate this explicitly, we could write $C_j(S)$ or even $C(j, S)$.

The relevance of these remarks is to the effects of fixing variables (without necessarily specifying them) and of substituting for them. Thus in Chapter 5 we treat the binomial coefficient C_{jn} as a function in which n is fixed though not necessarily specified; the **parameter** n is a constant for a given enzyme, but might be different for another enzyme. The index j, on the other hand, is permitted to vary over all integer values from 1 to n, and behaves as a true variable; and C_{jn} behaves, for purposes of the chapter, as a function of **one** variable (and one parameter).

Again, note that putting $x = a$ (a definite number) in $f(x)$ gives a definite number, $f(a)$; but putting $x = a$ in $g(x, y)$ gives $g(a, y)$, a function of y. Similarly, with a set of numbers $A_1, A_2, \ldots$ given, putting $j = 10$, say, in A_j gives A_{10}, a number. But putting $j = 10$ in $C_j(S)$ gives $C_{10}(S)$, a function of S.

How Much Is Enough?

The material of the foregoing sections would, I can assure the reader, cause any well-bred mathematician to shudder. I have presented a rather scrappy selection of material, with considerable stress on working rules even if with some degree of rationality, and I have trampled carelessly on considerations of rigor and refined terminology. The point of this *gaucherie*, of course, was to bring the average reader to the point where he can get on with enzyme kinetics without constantly getting his feet entangled in elementary manipulations.

Sooner or later, however, any biologist who persists in playing with applied mathematics at the elementary level represented by this book will no longer feel awed by the rudiments of algebra, analysis, and a bit of theory of differential equations, and will indeed be very much at home with such stuff. At this point he may reasonably ask himself two questions (not altogether unrelated): 1. Is there much more mathematics that it would be reasonable or useful for me to learn, or to which I should expose my graduate students? 2. Would I be better off with a more rigorous version even of the elementary kind of mathematics that I have been using?

This is a vexatious and much-debated question. Many physicists, oddly enough, take what I can only consider a vulgarly pragmatic position. "Hurry up and teach our students to do such-and-such," they shout at the cringing mathematician, "and cut out the frills! We want to get them going on some physics!" Biologists, less emboldened by familiarity, tend to ask a more relevant but still secondary question: "How can we fit the necessary mathematics into the curriculum?" Pure mathematicians tend to say, perhaps a little primly, "There is only one kind of mathematics—the right kind—and you either know it or you don't."

Being acutely conscious of everyone's difficulties, I would plead for a compromise; but my position would be closer to that of the pure mathematician than to the others. Let me address myself first to "How much?" and subsequently to "What kind?" If you ask, "Just how much of mathematics will I need if I am going, occasionally or regularly, to use it or to understand its uses in my capacity as a biological scientist?" an attempt at an honest answer would run something like this: "It is impossible to tell. There is an excellent possibility that you might get along happily for a lifetime with the rudiments. If you do need something new, I am almost sure—but almost, not absolutely!—that it is more likely to be something like probability theory or differential difference equations than one of the more exalted and popular contemporary areas like algebraic topology or theory of categories. But the interaction of knowledge and interests is close: The problem you want to solve somewhat dictates the kind of mathematics you have to know to solve it, and the mathematics you know may tend—this is a real danger—to restrict the kinds of problems you will tackle, and even the kinds of papers you will read, no matter how important or fascinating the problems or papers may be. On the whole, I would recommend more rather than less mathematics, with due regard for the exigencies of the curriculum. I would recommend beginning with more of classical mathematics, simply because our experience in applying it to the natural sciences is much more extensive than in the case of hypermodern mathematics. I would stress also that it is important to do enough undisciplined reading—browsing, if you like—to get some sort of picture of mathematics as a whole, a notion of its parts, of what sort of things are discussed under course titles such as 'Lattice Theory' or 'Functions of a Complex Variable'. Then, as you think about a new problem, you will get a notion that a certain branch of mathematics might be just the tool for you; and you can set to work to learn it."

The question "What kind?" is touchy. During the past few decades the spirit as well as the face of mathematics has been radically transformed. Central features of mathematical thinking—the passion for generality, abstraction, and logical precision—have been exaggerated almost into caricatures. Contempt for intuition and an almost neurotic avoidance of concreteness are in the air. The intruder struggles through a thorny thicket of new terminology—some of it indubitably necessary, some apparently replacing adequate older terminology in a spirit of sheer *Schadenfreude*.

I am not trying to make out a case against contemporary mathematical research, even though I find much of its manner unsympathetic. The relevant point is that the change in attitude and content has an influence on teaching. In the first place, capable men in the thick of the living and ongoing part of a discipline naturally enjoy teaching what they know best and are most fascinated by; to see and exhibit the relevance of where you have been to where you are going is a slowly developing talent in a teacher, especially in a culture that has made crass ignorance of history almost into a matter of national pride. It is surely not astonishing that lively young mathematicians would rather talk about the topological properties of the real number system in a classroom than supervise drill in the mechanics of differentiation and integration or the hundred and one ways of testing the convergence of an infinite series. They will argue, moreover, that teaching the most general approach has special virtues, that all sorts of special results, formerly obtained laboriously one by one, fall neatly and effortlessly out of the cornucopia of generality—without realizing that students who know almost nothing of the older methods can scarcely appreciate the contrasting elegance and power of the newer ones.[6]

Then, of course, there are the camp followers—the assiduous writers of textbooks and organizers of committees on undergraduate programs—whose passions are up-to-dateness and fashionableness, who corrupt teaching by a mechanical imitation, at first of living mathematics and eventually of one another.

All of this poses the problem: How do you arrange for students to be taught some real mathematics—something beyond the mere gadgetry of substituting in half-comprehended formulas—without producing people who are scarcely capable of applying mathematics to definite problems and getting definite answers? The answer is not simple, and its discussion is, I fear, beyond the scope of these hasty remarks.

Having been as candid as I can about the practical difficulties, I must come down on the side of mathematical comprehension and depth rather than mathematical technology, for two reasons:

1. The better the mathematics learnt, the greater the eventual practical resourcefulness of the applied mathematician. To take a rudimentary example: Old-fashioned courses in the theory of differential equations consisted almost entirely of teaching standard forms of differential equations, for which someone had once found definite solutions. Even such a straightforward matter as solution by means of infinite series expansions got only a lick and a promise, despite the fact that many of the higher transcendental functions included in such books had in fact been first obtained as infinite series, and could be treated lightly only because someone had once gone through the labor of preparing tables for them. The result was that students so trained could perform ably if confronted by a form that was in the dictionary, or one that could be transformed into such a standard form. If a differential equation that could not be forced into such a mould confronted them, they could only stare at it helplessly, or use numerical approximations—something computers now do better. With a course of somewhat greater generality and sophistication, the student always has some resources; at the very worst, if no closed form can be found for a solution, he knows how to characterize

[6] Many of these people are quite serious about their teaching; and my argument with them must not be misunderstood. Biologists and physicists, who must offer training suitable for their own majors, for each other's majors, for premedical students, and for a miscellany of students fulfilling distributional requirements with varying degrees of reluctance, should understand why mathematicians, confronting an equally complex and often conflicting set of requirements, sometimes fail to find the optimal solution for all of them. Conversely, when scientists, including applied mathematicians, wonder whether the training of a future PhD. in pure mathematics is without qualification the best training for their own majors, it is to be hoped that mathematicians will not consider them merely ignorant and perverse.

and delimit the solution in a variety of ways, to tell whether it is periodic or damped periodic, stable or unstable, etc. In short, there is a vast region of generality and rigor in mathematics, lying somewhere between crude formula-juggling and the cold outer space of contemporary mathematics, knowledge of which confers greater practical power for the needs of the applied mathematician. (And past experience also tells us something else: that what is pure, useless, art-for-art's-sake abstraction today becomes, unpredictably, tomorrow's applicable mathematics.)

2. Real mathematics is incomparably beautiful and exciting. Most people have to grow into it, just as a sense of the marvellous excitement of Bach and Mozart comes, with listening and maturing, to the man who as an adolescent thrilled to Beethoven or even to Wagner, and who found Bach cold and baffling. To take the trouble to learn mathematics because it is a useful tool, to struggle and labor, as we all must, with what Rashevsky once called mathematical dishwashing, and yet never to be rewarded by its gratuitous pleasures—this seems to me infinitely sad and drab. For this wholly unpractical reason, then, as well as for the practical ones, I recommend that anyone who is going to have anything to do with mathematics make an effort to see it as it really is—what Gauss called the Queen of the Sciences (12).

REFERENCES

General Background

To get the feel of mathematics—pure, applied, and personal—the reader can do no better than the late James Newman's admirable collection:

"The World of Mathematics", Simon & Schuster, New York, 1956.

and almost any (or all) of the witty and fascinating books written about mathematics and mathematicians by the algebraist Eric Temple Bell, e.g.:

"The Development of Mathematics", McGraw-Hill, New York, 1940.

For an excellent reference in algebra, I recommend:

"Higher Algebra", by S. Barnard and J. M. Child, Macmillan, London, 1959.

I cannot forbear to mention an old love, probably the first undergraduate algebra text to present the subject in an axiomatic and on the whole systematic fashion, and the first intelligible treatment of the subject I ever saw. It was out of print when I was a student; graduate students in mathematics and physics at Chicago told younger graduate students about it, and were forever blessed. The author knew more about how to teach real mathematics to beginners than the panjandrums of the New Mathematics will ever dream of knowing. I got my copy by writing to the publishers, who found some copies in an attic somewhere; I suggest you do the same:

Henry B. Fine, "College Algebra", Ginn & Co., Boston, 1901.

For all that goes nowadays into the portmanteau word "analysis"—properties of real and complex numbers, essential aspects of analytic geometry, calculus—I cannot do better than recommend:

"A Course of Pure Mathematics", by G. H. Hardy, Cambridge, 1963 (paper).

Hardy was one of the best and liveliest of contemporary mathematicians; his more personal "A Mathematician's Apology" is well worth anyone's time. He writes with clarity and precision, and he deals with all the most useful as well as the most beautiful classical mathematics that can possibly go into a book at that level (it was used for freshmen at Cambridge). Many of the topics dealt with in this chapter are treated far more carefully and extensively. Among other things, the book is a treasury of interesting and useful functions and the analysis of their properties. The prospective reader should be warned of one thing: Hardy, without a word, takes it for granted that his readers are willing to apply a considerable amount of effort and the full concentration of their intelligence to the book; he has nothing in common with the charlatanry of much contemporary mathematics teaching, which seeks to imply that one can learn mathematics without effort or concentration by playing games. Some proofs to which ordinary books would devote a couple of sections are student exercises in this book. Nevertheless, I can think of no other single volume that gives so much, written so admirably, to the reader who is willing to work at it.

For the reader who seeks either a handy reference or a refresher in the purely manipulative aspects of calculus, almost any text of sufficient age will do. The old standby:

Granville *et al.*, "Elements of the Differential and Integral Calculus", Ginn & Co., Boston, 1934

is convenient, and is by no means as mathematically lowbrow as one might fear. An excellent modern book, perhaps the only one not given to making fashionable gestures where they are not required, is:

"Calculus with Analytic Geometry", R. E. Johnson and F. L. Kiokemeister, Allyn & Bacon, Boston, 1957.

For the courageous, I recommend a translation of a classical German text. The pedagogy is not so much bad as nonexistent. But the amazing and useful feature is that, in a not particularly large book, practically any question one might ask is answered clearly, definitely, and correctly, often in one or two lines; and the organization is excellent:

Edmund Landau, "Differential and Integral Calculus", Chelsea, New York, 1906.

Another extraordinarily useful book, much less condensed but less complete, is:

Philip Franklin, "A Treatise on Advanced Calculus", Wiley, New York, 1940.

As a useful elementary introduction to a mixture of topics best described as not-analysis (propositional logic, set theory, etc., and some interesting applications), I recommend:

"Introduction to Finite Mathematics", John G. Kemeny *et al.*, Prentice-Hall, Inc., Englewood Cliffs, N.J., 1956.

Textual References

1. Hempel, C. G., "Aspects of Scientific Explanation", Free Press, Glencoe, 1965.
2. Feigl, H., and Scriven, M. (eds.), "Minnesota Studies in the Philosophy of Science", Vol. 1, Univ. of Minn. Press, 1956.
3. Feller, W., "Introduction to Probability Theory and Its Applications", Wiley, New York, 1950 (Vol. I), 1966 (Vol. II).
4. Simon, H. A., "Models of Man", Wiley, New York, 1957.
5. Kemeny, J. G., and Snell, J. L., "Mathematical Models in the Social Sciences," Ginn & Co., Boston, 1962.
6. Woodger, J. H., "The Axiomatic Method in Biology," Cambridge University Press, Cambridge, 1937.
7. Carnap, R., "Logical Syntax of Language," Harcourt Brace, London, 1937.
8. Polya, G., "How to Solve It," Princeton, 1948.
9. Polya, G., "Mathematic and Plausible Reasoning" (2 vols.), Princeton, 1954.
10. Polya, G., "Mathematical Discovery" (2 vols.), Wiley, New York, 1962.
11. Kuratowski, K., "Introduction to Calculus," Addison-Wesley, Reading, Mass., 1962.
12. Allendoerfer, C. B., and Oakley, C. O., "Principles of Mathematics," McGraw-Hill, New York, 1955.

2

Fundamentals of Enzyme Action

PROPERTIES OF ENZYMES

We begin by summarizing those properties of enzymes which are most relevant to the problems dealt with in this book. In doing so, we shall ignore historically controversial ideas which are now defunct, as well as current unconfirmed speculations. We shall omit the evidence for our statements, and refer the reader to the general texts and handbooks on enzymes (cited in the References) for the details of such evidence.

1. Enzymes are compounds in whose presence certain chemical reactions proceed at higher rates than would be the case in the absence of the enzyme. In many important cases the rate in the absence of enzyme, the **spontaneous** rate, is to all intents and purposes equal to zero. With appropriate experimental precautions, it may be shown that the enzyme is unchanged in amount and properties after the chemical reaction has occurred. This places enzymes in the class of those substances known as catalysts; they are distinguished from other catalysts by the fact that they occur only as products of living organisms.

2. All enzymes about which sufficient information is available are known to be proteins (simple or conjugated). They consequently share all the special properties of proteins. Thus: (a) they are antigenic; (b) they are denatured by such agents as elevated temperature and extreme pH values; (c) their physical state and their catalytic function depend markedly upon a number of physical factors such as pH, temperature, and ionic strength.

3. When enzyme activity (the rate of the catalyzed reaction) is plotted against either pH or temperature, the curve usually has a peak or optimum.

The same is true when enzyme stability is plotted against pH. The region of optimum pH (or temperature) for activity is not necessarily at or near the pH (or temperature) values normal for the living cell from which the enzyme was taken. Nor is the region of optimum pH for activity necessarily at or near the region of optimum pH for stability (and similarly for temperature).

4. All enzymes are functionally specific to varying degrees. This is not the place to discuss all the aspects of this specificity, which enter into various schemes for classifying enzymes. Some examples of various kinds of specificity will suffice:

a. A given enzyme catalyzes only reactions of a certain class, for example hydrolyses: $AB + H_2O \rightarrow AH + BOH$. Usually the range is still narrower, e.g., the hydrolysis of oxygen esters of ortho-phosphoric acid: $ROPO_3H_2 + H_2O \rightarrow ROH + H_3PO_4$. Enzymes of the broader class are known as hydrolases, those of the narrower class as phosphatases.

b. Within classes such as those described in (a), the rate of the catalyzed reaction differs with the substrate (in the case of the phosphatases, with the structure of the group R). Thus there may be a phosphatase for which the rates of hydrolysis of phenyl phosphate and adenosine-3′-phosphate are quite different under otherwise identical experimental conditions.

c. As a consequence of the specificity described in (b), we can distinguish enzymes which are otherwise very much alike but which differ in their **substrate-activity profile**. That is, if we take two such enzymes, and let them act on each of a list of, say, five compounds (e.g., five different phosphate esters in the case of phosphatases), the list of rates for enzyme I acting on substrates A, B, C, D, and E differs from the corresponding list for enzyme II (other conditions being of course the same in all cases).

d. The specificity may be still narrower—the enzyme may act on one and only one substrate. This situation is less common than is usually supposed. At a given date it depends in part on how many compounds have been tested. Thus, for example, there is a class of enzymes which hydrolyze only the terminal phosphate anhydride linkage of adenosine triphosphate (abbreviated ATP), which do not affect any ordinary phosphate esters, and which on the other hand do not split such phosphate anhydride links as those of pyrophosphate or adenosine diphosphate. However, such very specific adenosine triphosphatases do act, though at a much reduced rate, upon inosine triphosphate (ITP), a compound which differs from ATP only in the substitution of —OH for —NH_2 on the number 6 carbon atom of the adenine ring. Some enzymes are more discriminating than others, but it seems fair to say that any enzyme can be fooled if one goes to enough trouble.

e. In addition to specificity with respect to type of bond attacked and chemical structure of compound, most enzymes are highly specific with respect to stereoisomers, and especially with respect to optical isomers. Thus they discriminate, usually quite strictly, between the D- and the L-forms of amino acids or of simple sugars, between the α and β forms of glycosides, and between pyranosides and furanosides.

f. If the same enzyme ("same" in the sense of having exactly the same kind of enzymatic activity) is obtained from two different organisms, the two samples of enzyme always differ antigenically. But, in addition to this, other striking and unexpected differences may be found. Consider, for instance, aldolase, which is defined as the enzyme that (non-hydrolytically) splits 1,6-diphosphofructose into two molecules of triose phosphate (performing essentially the reverse of an aldol condensation). Aldolase obtained from muscle is a simple protein, which is not known to require any added substance other than the substrate for full activity. Aldolase obtained from yeast, on the other hand, is activated by ferrous ion in the natural state, may be activated by zinc, and in the natural state is readily inhibited by iron-complexing substances such as pyrophosphate.

THE PROBLEMS OF A THEORY OF ENZYME BEHAVIOR

The task of theoretical enzymology is to take account of these basic properties of enzymes, to correlate them, and to explain them in terms of fundamental chemical and physical laws. This task may be subdivided into the following problems:

1. To show how the rates of enzyme-catalyzed reactions are related to various environmental factors, such as: (a) the concentration of the enzyme; (b) the concentration of the reacting substances (generally known as the "substrates" of the enzyme); (c) the concentrations of various substances which specifically activate or inhibit the catalysis; and (d) physical factors such as the pH, temperature, and ionic strength of the reaction mixture.
2. To account for the specificities of enzymes in terms of their chemical structure and the chemical structures of their substrates.
3. To explain the occurrence of enzymatic catalysis in physicochemical terms—that is, to show, in terms of the atomic and electronic structures of enzyme and substrate, precisely how the presence of the enzyme raises the rate of chemical reaction above the "spontaneous" rate.

The answer to the first of these three problems will occupy most of this book, for the simple reason that this problem (that of **enzyme kinetics**) has in large measure been satisfactorily solved. The other two problems, for a variety of reasons which include limitations of experimental technique, are only now beginning to be attacked effectively. Thus only the last chapter of the book deals explicitly with the **mechanism of enzyme action**, although it will be seen that certain kinds of kinetic studies contribute to the analysis of mechanism.

THE ASSUMPTIONS OF ENZYME KINETICS

In order to develop a theory of the rates of enzyme-catalyzed reactions, it is necessary to make some concrete assumption about the way in which the enzyme and the substrate molecules interact. For example, one might assume that enzyme molecules increase the energy of substrate molecules by some sort of action at a distance (e.g., electrostatic attractions or repulsions, or electromagnetic radiation). Alternatively, one might assume that energy is transmitted from enzyme to substrate by way of inelastic collisions, with the usual proviso that only sufficiently energized molecules finally react. Again, one might assume that enzyme and substrate form a compound, and that this compound then undergoes internal rearrangement, after which the reaction product and the original enzyme molecule are formed from the rearranged compound.

The notion of action at a distance was entertained at one time (1). It was supported by evidence which has been severely criticized on technical grounds (2), and which has not been confirmed. A quantitative theory based on the idea has not been developed; but it is not difficult to show that it would lead to relations rather different from those found experimentally to date (3).

The idea of energy transmission by inelastic collision was proposed and worked out by Medwedew, and has been elaborated by Hearon and Katzman (4, 5). It leads to a relation between reaction rate and substrate concentration which superficially resembles the relation derived by assuming compound formation between substrate and enzyme; and in a few cases it fits experimental data rather well. However, it will not stand a more critical test, to be described later in this chapter.

The hypothesis of complex or compound formation by enzyme and substrate meets the test of experimental data on reaction rates (with such additional assumptions as may be needed, such as will be described in the course of this book). In recent years, it has received the further support of direct experimental confirmation (6, 7). The case for this model of enzyme action is therefore so strong that it alone will be used as the basis for all the further theoretical developments to be presented.

Before developing the idea in quantitative form, however, it will be worth while to consider in detail some of the fundamental experiments which led up to it.

THE CONCEPT OF THE ENZYME-SUBSTRATE COMPLEX

Just when the notion of a direct stoichiometric combination of enzyme and substrate was first stated is not certain. There is no doubt that it pervades the literature of the late nineteenth and very early twentieth centuries, usually in a fairly mature and clear form. For example, H. E. Armstrong, in a presidential address to the Chemical Society entitled "The Nature of Chemical Change," discussed enzyme action as merely one case among many, and asserted that "complication, not simplification, precedes most, if not all, chemical change, that complex molecular systems are first formed from the interacting substances, and that these, on breakdown, suffer re-arrangement of the parts—such rearrangements taking place in consequence of elements which were previously separated being brought into one common 'sphere of activity' within which it is possible for them to interact. The function of enzymes in promoting hydrolysis (of saccharides), on this hypothesis, consists in bringing water into conjunction with the carbohydrate by combining with both" (8). The idea that the sole function of the enzyme is a passive geometrical one is, of course, no longer entertained; but the picture of the stoichiometric relations of enzyme, enzyme-substrate combination, and products is precisely the one which prevails at present.

The consequences of such a point of view were recognized fairly early, but were not at first fully expressed in algebraic form. Perhaps the earliest extensive quantitative study of enzyme kinetics was performed by O'Sullivan and Tomson (9). Using yeast invertase, these investigators studied methods of extraction and purification, and studied the kinetics of invertase as a function of temperature, pH, substrate concentration, and similar variables. They concluded that the hydrolysis of sucrose by the enzyme followed a "mass action" law, by which they meant that the reaction was first order, a constant **proportion**, or percent, of the substrate being split per unit time.

These conclusions were criticized by Brown (10). Working with the invertase of live yeast, Brown found that a constant **weight** (rather than a constant **proportion**) of the sugar present was split in unit time. Brown then repeated and varied the experiments of O'Sullivan and Tomson, and confirmed his previous results.

In spite of the temptation to be satisfied with this, Brown analyzed the data of both laboratories in a remarkably lucid fashion, employing essentially those concepts of enzyme action which we now hold. He cited evidence from O'Sullivan and Tomson on protection of enzyme against heat by substrate, experiments on compound formation between papain and fibrin (11), and Fischer's concept of configurational specificity of enzymes, to support the view that the invertase must combine with sucrose in the course of its action.

Assuming a mean lifetime of 0.01 sec for this complex before it reacts, he pointed out that in consequence the maximum number of elementary re-actions which one molecule of enzyme could promote per second (what is now called the turnover number of the enzyme) would in such a case be 100.

At low sugar concentrations, the number of complexes formed and able to react per unit time would be proportional to the number of collisions of enzyme and substrate, which would also have to be proportional to the amount of substrate present. If the concentration of sugar is increased, however, till the number of collisions per unit time is greater than the number of conversions the enzyme could perform in that time, the rate could still not exceed 100 conversions per enzyme molecule per second. Under these conditions a constant **weight** of sugar would be split per unit time, this being the maximal possible amount. Brown reasoned that the results of O'Sullivan and Tomson should be valid at low substrate concentrations, performed the experiment, and did indeed get the expected first-order reaction rate.

In the same year E. F. Armstrong investigated a number of disaccharide-hydrolyzing enzymes such as maltases and lactases of various origins (12). The amount of hydrolysis was measured at various times, in the presence of different amounts of enzyme and sugar, and with different enzyme-sugar ratios. In the presence of a large initial **excess** of **sugar**, the actual weight of sugar hydrolyzed after a given time was independent of substrate concentrations (provided it remained in large excess), and the amount split was proportional to the time during which the enzyme was allowed to act. Where a large **excess** of **enzyme** was used, on the other hand, the amount of hydrolysis was directly proportional to the amount of substrate present. Under **all** conditions, the hydrolysis during a given period of time was proportional to the amount of enzyme present.

From these results Armstrong inferred that the outcome of such experiments would be principally determined by the relation between amount of enzyme and amount of substrate. He proposed that the "active mass" of the substrate (the amount which is ready to react at any time) is equal to the amount combined with enzyme.

In a remarkable anticipation of the "steady state" idea which will be presented later, Armstrong described how the amount of this active mass is determined. Enzyme and substrate combine, but then water, the remaining reactant, "competes" with the enzyme for the substrate—that is, the water attacks the active complex and releases the products of hydrolysis. An "equilibrium" is soon reached between the process which forms the active mass and the process which consumes it (the word "equilibrium" is not strictly correct, but it is clear that Armstrong really meant a steady state, in which the reactions going to form the active complex are just balanced by the reactions which remove it.) The level of this "equilibrium" depends on the relation between amount of enzyme and amount of substrate.

If one sets up the experiment with an excess of sugar, the enzyme is, to use current terminology, **saturated with substrate** ("the maximum possible number of effective combinations," in Armstrong's words, is formed). The "active mass" is therefore constant in time, and the amount split is simply proportional to elapsed time. As the amount of unhydrolyzed sugar decreases, the enzyme eventually is no longer saturated, and the active mass becomes a function of the amount of sugar present. If one **begins** the experiment with an "excess" of enzyme (low substrate concentration), the phase of hydrolysis proportional to time is no longer observed, and only the later phase is found.

THE SIMPLEST KINETIC EQUATIONS

At the turn of the century, then, the major notions of enzyme kinetics existed in complete form: the idea of the enzyme-substrate complex, of the turnover number, and of saturation, as well as the concept of specific inhibition of enzymes (e.g., by their reaction products). The mathematical formulation of these ideas is quite straightforward. Let us consider the simple case with which most biochemistry texts begin (and end): that of a reaction involving only one substrate.

The **assumptions** that must be made for such a simple case are:

1. A molecule of enzyme and a molecule of substrate combine reversibly to form a complex or compound. (We will consider in Chapter 6 the situation where more than one substrate molecule unites with one enzyme molecule.)
2. The complex breaks up irreversibly, giving a molecule of product and the original enzyme molecule. (The consequences of reversible action at this stage will be studied in Chapter 3.)
3. The only forms of the enzyme are: free enzyme and enzyme incorporated into complex. The total amount of enzyme is constant throughout the experimental period.
4. The amount of substrate which is bound by enzyme is very small compared with the total amount of substrate present in the mixture. (The result of abandoning this assumption will be explored in Chapter 4.) Consequently, the concentration of free substrate is equal to the concentration of total substrate at any time.

We shall represent the substrate by S, the free enzyme by E, and the E-S complex by C (the reaction product, which will not enter most of the formulae, being denoted by P). To simplify notation, we use the same symbols both in the **stoichiometric chemical equations**, where they denote the **substances**, and in all other equations, where they represent **concentrations** (e.g., in moles per liter); the usual brackets to denote concentrations will be omitted, there being no cases where confusion is likely to occur. The total concentration of enzyme is denoted by E_t. Forward reactions will be numbered, and the corresponding back reactions, if any, will be noted by the same number preceded by a minus (—) sign.

If assumptions 1 and 2 are translated into the symbolism of chemical equations, we get the **stoichiometric scheme** or set of stoichiometric equations:[1]

(2.A) $$E + S \underset{-1}{\overset{1}{\rightleftharpoons}} C \qquad C \overset{2}{\rightarrow} E + P$$

We refer to a set of stoichiometric equations as a **stoichiometric scheme** (or simply a **scheme** for brevity) when they describe completely all of the chemical reactions which are supposed to be going on under our particular assumptions.

There is a separate **kinetic equation** for each compound, reactant or product, pictured in a stoichiometric scheme. Thus, for the scheme 2.A, there will be four kinetic equations, one for each of the four substances E, C, S, and P.

A **kinetic equation** for a given substance expresses the **rate of change** of the concentration of that substance **with time**. In other words, if we are calling the substance X, the kinetic equation for X expresses the quantity dX/dt.

The rate of change for a substance is made up of the **rates of the individual chemical reactions** in which that substance is used or formed (and **only** those reactions). **Each** reaction **step**, forward or backward,

[1] It is possible to condense the way in which two or more equations like 2.A are written. Thus, we can write:

$$E + S \underset{-1}{\overset{1}{\rightleftharpoons}} C \overset{2}{\rightarrow} E + P$$

This can be done only when the equations have a side completely in common. In the case of the example given, the expression C is the common side. To see what can**not** be done under this rule, compare the following:

Complete: $A + B \rightarrow C;\ C \rightarrow X$ Condensed: $A + B \rightarrow C \rightarrow X$

Complete: $A + B \rightarrow C + D;\ C \rightarrow X$ Condensed: $A + B \rightarrow C + D \rightarrow X$

The first condensation is correct; the second is wrong—because it implies that X is formed from C and D together, where in fact X is formed from C only. If the condition of a complete side of an equation in common is not fulfilled, no unbranched condensed version is possible.

The direction of arrows must also be observed. Thus, if we have

$$A \rightarrow X \qquad B \rightarrow X$$

it obviously is not sensible to write either

$$A \rightarrow X \rightarrow B \quad \text{or} \quad B \rightarrow X \rightarrow A$$

According to the two original equations, X is a dead end, and therefore cannot be a link between A and B in a condensed version.

In the same way, any condensed chemical equation can be broken down into a set of simpler equations, each involving only one reaction.

contributes **one term** to the net rate of change. Thus an **irreversible reaction** will contribute just **one term**; a **reversible reaction** will contribute **two**, one for the forward step and one for the backward step.

Thus, if we were confronted with the simple equation $S + E \underset{-1}{\overset{1}{\rightleftharpoons}} C$, we could write three kinetic equations, one each for S, E, and C. Suppose we pick C to begin with: Its rate of change is dC/dt, and it is made up of a term from the forward reaction step (labelled 1) and a term from the reverse reaction step (labelled −1). The first term is positive, because the forward step makes C, and so would tend to make C increase with time; the second term will be negative, because the reverse reaction breaks C down again to S and E, and so would make C decrease with time. Thus we could write the kinetic equation for C in a sort of pseudomathematical way as $dC/dt =$ (Rate 1 forward) − (Rate −1 back). The equation for dE/dt would involve the same terms, but the signs would be opposite; the forward step removes E, and so makes E decrease with time, while the back reaction forms E again, and so makes E increase with time. The pseudomathematical form for dE/dt would be $dE/dt = -$(Rate 1 forward) + (Rate −1 back). The same will be true for dS/dt.

To go from the pseudomathematical equations just indicated to actual usable equations, we must have a way of expressing each of the individual rates mathematically. This is given by the **Law of Mass Action** (13): The **rate** of a reaction is **proportional** to the **product** of appropriate **powers** of the **concentrations** of the **reactants**; the "appropriate **power**" for each reactant is the **number** of **molecules** reacting according to the stoichiometric equation.

Thus, reverting to our illustrative equation, $S + E \underset{-1}{\overset{1}{\rightleftharpoons}} C$, the term which we have referred to as Rate 1 forward is **proportional to** $S \times E$ since one molecule each of S and E reacts. In the same way, Rate −1 back is proportional to C.

Each rate has its own individual constant. To express such constants, we propose to use the letter "k", labelled with a subscript which is the number of the reaction step involved. Thus the constant for Rate 1 forward would be "k_1", and that for Rate −1 back would be "k_{-1}". Such constants are termed "**rate constants**" or "**velocity constants.**"

We are now in a position to write out the rate terms in a true mathematical form. We can, in the case of our illustration, say: Rate 1 forward $= k_1SE$, and Rate −1 back $= k_{-1}C$. Consequently, we can write the net rates as: $dC/dt = k_1SE - k_{-1}C$; $dE/dt = -k_1SE + k_{-1}C$, and so forth.

When we are working, not with a single reversible or irreversible reaction, but with a stoichiometric scheme consisting of more than one reaction, we must take all of the reactions into account. In the scheme 2.A, for instance, reaction 2 contributes a term to dC/dt, and also a term to dE/dt (because C is used up in reaction 2 and E is formed in reaction 2). Thus we would add a term $-k_2C$ to our previous expression for dC/dt, and a term $+k_2C$ to our previous expression for dE/dt. No term will be added to dS/dt, since S is neither made nor used up in reaction 2. But the product P is made in reaction 2, and is not made or used in any other reaction of the scheme. Therefore the rate of change of P will simply be $dP/dt = k_2C$.

Applying this procedure to all four of the variables C, E, S, and P, we get the complete set of kinetic equations:

$$(2.1) \quad dC/dt = k_1ES - k_{-1}C - k_2C \qquad dE/dt = -k_1ES + k_{-1}C + k_2C$$

$$(2.2) \quad dS/dt = -k_1ES + k_{-1}C \qquad dP/dt = k_2C$$

Note that the equation for dE/dt consists of the same set of terms as dC/dt, but with all the signs reversed. (Every reaction step which makes a molecule of E uses up a molecule of C, and every reaction step which uses up a molecule of E makes a molecule of C—as is obvious in just looking at the scheme 2.A.) In short, the two rates of change are related by $dE/dt = -dC/dt$, or, transposing dC/dt to the left side with a change of sign, $dE/dt + dC/dt = 0$, which means that $E + C$ is a constant.

The result just stated also follows from assumption 3 above, which in symbols becomes the **conservation equation**[2] for enzyme (total enzyme consists of free enzyme and bound enzyme):

$$(2.3) \qquad E_t = E + C$$

In view of the relationship 2.3, E is not an independent variable. That is, if numerical values of C and E_t are assigned, the value of E is determined by the equation. Since E_t (total enzyme added in experiment) is supposed to be known, the equation 2.3 can be used to eliminate E (unknown) from the other two equations by solving for $E(E = E_t - C)$, and substituting in 2.1 and 2.2. We obtain:

[2] A **conservation equation** or **equation of material balance** applies to each material component of the system (for example, enzyme, substrate, activator if any). **Conservation of enzyme** is expressed by 2.3. **Conservation of substrate** would read: $S_t = S + C$ (total substrate consists of free and bound substrate, and of no other form, at any given moment). But according to assumption 4, $C \ll S$ (C is very much less than S), and can be neglected in the sum. Therefore, the conservation equation would read $S_t \cong S$ (S_t is **approximately** equal to S). According to this equation, it does not matter whether we write the symbol S or the symbol S_t in the kinetic equations, since they are practically equal; the equation therefore has no effect on any theoretical results. In Chapter 4, the complete substrate balance ($S_t = S + C$) will be used, and then the relationship will have great theoretical importance.

(2.1′) $$dC/dt = k_1(E_t - C)S - (k_{-1} + k_2)C$$

(2.2′) $$dS/dt = -k_1(E_t - C)S + k_{-1}C \qquad dP/dt = k_2C$$

These simultaneous differential equations, expressing the rates at which complex, substrate, and product change with time, express the full content of the assumptions made above. If we could write out the solutions for C, S, and P, as functions of time, they would completely represent the theoretical behavior of the system, Unfortunately, the equations are non-linear (the **product** of C and S appears in them, in addition to the first powers of C and S alone), and no simple expressions for the solutions are obtainable. Chance, with the aid of a computer, obtained numerical solutions, which he applied to data obtained in his laboratory (14).

PROBLEM: Write the kinetic equations for the schemes:

(a) $2E \rightleftharpoons C_1;\quad C_1 + S \rightleftharpoons C_2;\quad C_2 \rightarrow C_1 + P$

(b) $E + S \rightleftharpoons C_1;\quad C_1 + S \rightleftharpoons C_2;\quad C_2 \rightarrow E + P$

(c) $E + S \rightleftharpoons C_1;\quad C_1 \rightleftharpoons C_2;\quad C_2 \rightarrow E + P$

PROPERTIES OF THE KINETIC EQUATIONS

It is possible to learn something about the properties of the functions C, S, and P, without solving the differential equations either analytically or numerically. The key to this sort of study is the role of the derivative as a measure of the slope of a curve, the sign of the derivative as an index of the direction of the curve, and the sign of the second derivative as an index of convexity. To get an intuitive idea of how this works, think of the solution of a differential equation as a graphical process of the following kind: Start at a point where the function is given (initial condition), and calculate the slope from the differential equation. Draw a small line segment from the initial point in the direction of the slope. Now, starting at the new point, recalculate the slope and proceed as before. The curve is approximately rendered by a succession of small line segments whose directions are given by the differential equation.

To proceed with the analysis, let us differentiate the equations 2.1′ and 2.2′ once more (ignoring the dP/dt equation for reasons which will become obvious):

(2.4) $$d^2C/dt^2 = k_1(E_t - C)(dS/dt) - (k_1S + k_{-1} + k_2)(dC/dt)$$

(2.5) $$d^2S/dt^2 = -k_1(E_t - C)(dS/dt) + (k_1S + k_{-1})(dC/dt)$$

Two additional useful relations are obtained by adding 2.1′ and 2.2′, getting:

(2.6) $$dC/dt + dS/dt = -k_2C$$

and by adding 2.4 and 2.5, getting:

$$d^2C/dt^2 + d^2S/dt^2 = -k_2 dC/dt \tag{2.7}$$

When substrate is mixed with enzyme at time $t = 0$, we have $C = 0$ (no complex existed in absence of substrate) and $S = S_0$ (initial value, fixed by experimenter); hence from 2.1′ and 2.2′ we know that $dC/dt > 0$ and $dS/dt < 0$. Using this in 2.4 and 2.5, we find that $d^2C/dt^2 < 0$ and $d^2S/dt^2 > 0$. In words: C is starting to rise along a curve that is concave downward (i.e., with ever-decreasing slope), while S is starting to drop along a curve that is concave upward (i.e., with negative slope ever decreasing in magnitude).

From 2.1′ and 2.2′ it is evident that a continued rise of C and fall of S must eventually lead to the vanishing of the right side of one or both equations. By virtue of 2.6 they cannot vanish simultaneously unless $C = 0$. Since dC/dt has an extra negative term (which is increasing because it is proportional to C), it will reach zero first. Application of 2.4 and 2.5 with $dC/dt = 0$ and $dS/dt < 0$ shows that the signs of the second derivatives are as before. The vanishing of dC/dt must mean either a maximum or an inflection point; the sign and the non-vanishing of d^2C/dt^2 indicate that it is the former—C is now going to decrease.

With both dC/dt and dS/dt negative, the use of 2.4 and 2.5 is ambiguous; the first term of 2.4 is negative and the second is positive, while the reverse is true of 2.5. But since we are starting from a state in which $d^2C/dt^2 < 0$ and $d^2S/dt^2 > 0$, it follows that dC/dt is becoming more negative and dS/dt is becoming less negative. Hence the right side of either 2.4 or 2.5 will eventually vanish. By 2.7, they cannot vanish simultaneously unless $dC/dt = 0$. In view of the extra positive term in 2.4, $-k_2 dC/dt$, which is increasing in view of the second sentence of this paragraph, d^2C/dt^2 will vanish while d^2S/dt^2 is still positive—the C curve has an inflection point.

As we pass through the C inflection point, we arrive at a stage where both second derivatives are positive while both first derivatives are negative. The two curves, in short, are falling and levelling off. Can anything further happen to the signs of the derivatives?

That either C or S should reach a minimum and then begin to increase is physically implausible—it would mean an irreversible process spontaneously reversing itself. But let us look at the mathematics. If $dC/dt = 0$ while $dS/dt \neq 0$, it follows from 2.4 that d^2C/dt^2 would have to become negative (by continuity passing first through zero). But this would describe an inflection point which would turn the C curve downwards rather than upwards. A similar argument shows that a minimum of dS/dt alone is inconsistent with the equations.

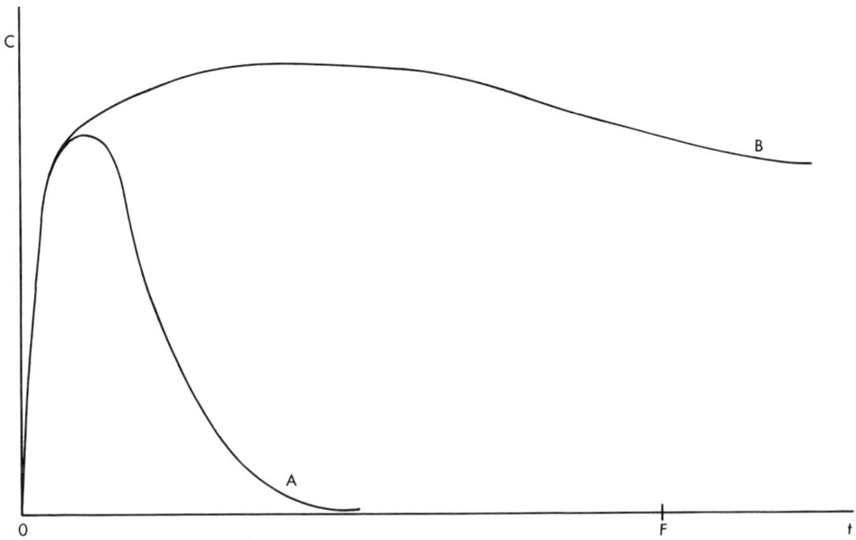

Fig. 2.1. Enzyme-substrate Complex, Variation with Time t during Enzyme Activity. Curve A, rapid reaction; curve B, slow reaction.

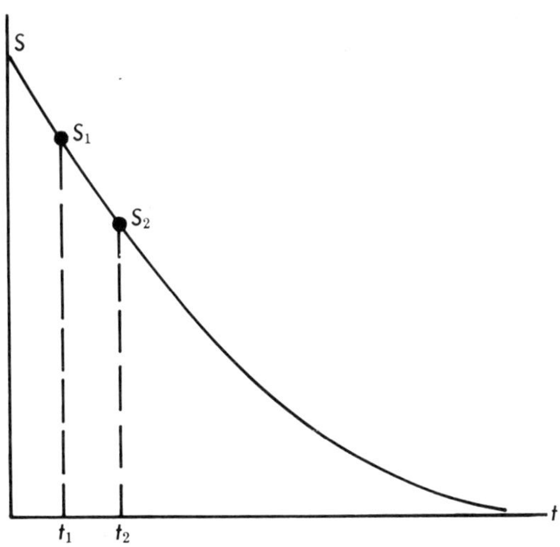

Fig. 2.2. Graph of Substrate Concentration S against Time t during Course of Enzyme-catalyzed reaction.

The only thing that can happen, then, to two curves which are decreasing, with decreasing negative slopes, is to complete the process of levelling off by attaining zero slopes simultaneously. From 2.1′ and 2.2′ it follows that this is possible only for $C = 0$ and $S = 0$. In short, the falling curves approach the t axis, as one would have expected intuitively.

The completed curves are sketched in Figs. 2.1 and 2.2.

The reason for ignoring the P curve is now evident. The function P does not appear in the equations for C and S, in view of the irreversibility of reaction 2. We could thus investigate the fates of C and S, and then predict the behavior of P from the second equation 2.2′: $dP/dt = k_2 C$. According to our discussion, C begins and ends at 0. Thus P (which is initially zero) begins with zero slope, and then rises with increasing slope ($d^2P/dt^2 = k_2\, dC/dt$) until C reaches its maximum. There the P curve goes through an inflection point, and continues to increase, but with decreasing slope, until it levels off at its maximum value when $C = 0$ once more.

THE "STEADY STATE"

It is clear that, except for the asymptote when the reaction terminates, $dC/dt = 0$ at just one instant of time—the maximum point in Fig. 2.1. But the curves A and B in the figure show two possibilities. Let us suppose that the period of the experiment extends from time zero to the time marked by F on the baseline. If, on this time scale, the actual curve of C is the one marked A, C is changing rapidly throughout a large part of the experiment. If the curve is of type B, however, where the declining part of the curve drops very slowly relative to the time scale of the experiment, then **practically** dC/dt can be regarded as zero throughout most of the experimental period. If this is so, a material simplification in the mathematics is possible. If we go back to equation 2.1′, we can put $dC/dt = 0$, and the equation is now a simple algebraic equation relating C and S:

$$k_1 E_t S - (k_1 S + k_{-1} + k_2) C = 0$$

We first divide by k_1, and introduce the definition:[3]

$$(k_{-1} + k_2)/k_1 = K \tag{2.8}$$

Then, solving for C, we get:

$$C = E_t S/(S + K) \tag{2.9}$$

We have now expressed C in terms of S, and of the constants E_t and K.

[3] Observe that a **definition** is for most practical purposes the same thing as an **abbreviation**. It expresses a **decision** to substitute a new (arbitrarily chosen) symbol for a combination of known symbols.

We could now use the conservation equation 2.3 (p. 54) to get E:

$$E = E_t - C = E_t K/(S + K) \tag{2.10}$$

Our aim in all this has been to study the **reaction rate**. In the general system given by equations 2.1′ and 2.2′, there are actually two distinct rates to be considered: the rate at which the substrate is removed, $-dS/dt$, and the rate at which product is formed, dP/dt, which is simply the rate of the irreversible reaction 2, $k_2 C$. These two rates are in general not equal. If the intermediate form C is being built up, then S goes to C faster than C goes to P; that is, $-dS/dt > dP/dt$. If C is in its declining phase, S is going to C slower than C goes to P; that is, $-dS/dt < dP/dt$. But when C is not changing, $dC/dt = 0$, then S is going to C exactly as fast as C goes to P, or $-dS/dt = dP/dt = k_2 C$. That this is true can be confirmed by taking the equation which represents the condition $dC/dt = 0$ and adding it to 2.2′.

In the **practical** or approximate "**steady state**," therefore, we can consider that there is only one rate; we can measure it either in terms of substrate disappearance or product formation, whichever is convenient. We shall denote this rate by the letter "v" (for "velocity"), and we have:

$$v = k_2 C \tag{2.11}$$

Substituting 2.9 into 2.11 in order to eliminate C, we get for the **reaction rate**:

$$v = k_2 E_t S/(S + K) \tag{2.11'}$$

The idea of a reactive intermediate in a steady state is not new to chemical kinetics (15), and has long been used in studies of kinetic mechanisms. The application of this idea to enzyme-catalyzed reactions is due to Briggs and Haldane (16). These authors, however, did not stress (although they recognized) the extremely relative and temporary nature of the "steady" state. The actual situation became obvious when Chance obtained a numerical solution of the kinetic equations 2.1′ and 2.2′ (14). In Chance's ingenious experiments with peroxidase, the length of time which could reasonably be regarded as marking the steady state varied from a fraction of a second to a few seconds at most. It is to be hoped that the situation is more favorable for the majority of enzymes; but experimental evidence bearing on this question has not been obtained in enough cases.

It should be noted that the rapid initial rise of the curves of Fig. 2.1 was observed in all cases by Chance, and that it is essential to the use of the pseudostationary approximation. The approximate "steady state" is bounded on the left by the rise of C from its initial zero value, just as it is bounded on the right by the eventual decline of C to zero again. Only if the rise to the maximum is virtually instantaneous with respect to the time scale of the experiment may one apply the analysis from the beginning of the experiment.

It is of some interest to ask whether the conditions for the **pseudo-stationary condition** are likely to be fulfilled for most enzymes, and what restrictions on the parameters are implied.

To begin with, let us solve 2.1′ approximately by putting $S = S_0$, which amounts to assuming that S does not change significantly while the initial upswing of C is taking place. The approximate solution for the initial **transient** is:

$$C = S_0 E_t[1 - e^{-k_1(S_0+K)t}]/(S_0 + K) \tag{2.12}$$

if we introduce K by means of 2.8. This approaches asymptotically the pseudostationary value 2.9 with $S = S_0$. The question is: How fast?

Suppose we demand that the exponential term drop to the value 0.01 after a time T, so that C is at 99% of its pseudostationary value by T. From a table of the exponential function we find that e^{-x} is about 0.01 when x is about 4.6. We demand, then, that $k_1(S_0 + K)T$ be about 4.6. Taking both S_0 and K to be about $10^{-2}M$, for example, and $T = 0.1$ min, we would get roughly the condition that $k_1 \geqslant 2 \times 10^3$ min^{-1}. Also, since S_0 cannot become negative, $k_1 \leqslant 4.6/KT$, or about 4×10^3 in the example just discussed. Whether these conditions are satisfied can be ascertained if we have a method for measuring C (in Chance's studies (14), for instance, the spectrum of C differed from that of the free enzyme) with sufficient speed; for then we can estimate k_1 from the initial slope of the C-t curve, $k_1 S_0 E_t$. In order that the pseudostationary approximation be reasonably good, as in curve B of Fig. 2.1, it is necessary that C should decrease much less rapidly than S after the maximum of C has been passed. Let us see, at least approximately, what this entails.

In 2.4, factor k_1 out of the right side and introduce K by 2.8:

$$d^2C/dt^2 = k_1[(E_t - C)(dS/dt) - (S + K)(dC/dt)] \tag{2.4′}$$

Since by definition $d(dC/dt)/dt = d^2C/dt^2$, then approximately, during a time θ short enough to justify treating the second derivative as roughly constant,

$$dC/dt = (d^2C/dt^2)\theta \tag{2.13}$$

Substituting 2.4′ into 2.13, and collecting terms in dC/dt:

$$dC/dt = k_1(E_t - C)(dS/dt)\theta/[1 + k_1(S + K)\theta] \tag{2.13′}$$

To get this in a form involving **relative** rates, divide by C on both sides, and multiply the right side by S/S. Noting that $(E_t - C)/C \cong K/S$ if C is given approximately by 2.9 (pseudostationary value):

$$(dC/dt)/C = \{k_1 K\theta/[1 + k_1(S + K)\theta]\}(dS/dt)/S \tag{2.14}$$

In order that the relative rate of change of C, $(dC/dt)/C$, shall be small compared with the relative rate of change of S, $(dS/dt)/S$, the quantity in braces must be small.

If θ is sufficiently large so that we can neglect the 1 in the denominator, the fraction boils down to $K/(S+K)$; this is small if K is small compared with S, hence small compared with S_0.

Suppose θ is only moderately large; and indeed let us suppose, conservatively, that $\theta = 10T$, where T is the time limit on the transient rise of C as before. Then $k_1\theta = 10k_1T \cong 46/(S_0+K)$, and the quantity in braces is approximately $[46K/(S_0+K)]/[1+46(S+K)/(S_0+K)]$. Now there are several cases to be considered. Suppose that K is small compared with S_0. If S has not decreased very much during $\theta = 10T$, the denominator is about 47, and the entire expression is about $K/(S_0+K)$, which is small as desired. If S has dropped to about $\frac{1}{2}S_0$, the denominator is about 24, the expression is about $2K/(S_0+K)$, and this is also small as desired. Now suppose K and S_0 are of the same order of magnitude. If S has not decreased very much during θ, the denominator is about 47 and the numerator about 23, the fraction about $\frac{1}{2}$: the relative rate of change of C is about half that of S. If S has decreased to $\frac{1}{2}S_0$ in θ, the denominator is now about 35 and the numerator about 23: The C rate is about 2/3 the S rate, which seems much too high for comfort.

It would seem, then, that the pseudostationary approximation will be most likely to work fairly well if K is appreciably smaller than S_0, and will be less likely to be satisfied if K is appreciably greater than S_0.

It is important to observe, however, that our demands depend on the nature of the analytical methods we are using. In a great many if not most enzymatic reactions, the analyses can be perfectly well performed if not more than about 10% of the substrate is consumed during the whole experiment. For instance, if $S_0 = 10^{-2}M$, 10% would amount to 1 micromole/ml consumed during the experiment; for most substrates this is a great deal, as far as the requirements of sampling for analysis are concerned. Calculating the relative rates with this requirement, and $K = \frac{1}{2}S_0$, the change of C will only be about 3% while S is consumed to 10%, while if K is even as much as $2S_0$, the change of C during the experiment will only be about 6%. Thus, depending on our analytical powers, the constraints on the parameter K may not be very severe. In the limit for K large compared with S_0, the consumption of C under these conditions will only be about 10%, the same as that of S.

TIME CURVE OF SUBSTRATE

We have seen that the pseudostationary state of an enzyme-catalyzed reaction is not a true steady state—that is, a time-independent plateau—

of the complex C. We have also seen that the approximation which makes use of the so-called steady-state relations is likely to be satisfactory over a fairly wide range of values of the parameters. We shall use the approximation throughout the rest of this volume; and we shall be grateful for it, inasmuch as it is, except for the computer-of-last-resort numerical solutions, all we have.

Before proceeding to look more closely at the properties of the rate equation, let us derive an expression for the time curve of the substrate S. Noting that, when 2.11′ holds, $v = -dS/dt = dP/dt$, we may write:

$$-dS/dt = VS/(K+S) \tag{2.15}$$

where we have introduced the abbreviation

$$V = k_2 E_t \tag{2.16}$$

With the usual techniques of integration, we find:[4]

$$S - S_0 + K \ln(S/S_0) = -Vt \tag{2.17}$$

where S_0 is, as in previous paragraphs, the initial value of S.

The mixture of algebraic and logarithmic terms does not permit explicit solution for S; so, from the perspective introduced in Chapter 1, we ask for the graph of the inverse function , $t(S)$, since it will give us all the information we need. Again following the approach of the first chapter, we begin by asking about the behavior for small t—that is, when S is close to S_0.

To get rid of the logarithm, we first rewrite the argument of the logarithm by adding and subtracting 1, so that we have $\ln[1+(S-S_0)/S_0]$. Since S is by assumption close to S_0, the term $(S-S_0)/S_0$ is small compared to 1. Hence we can use the series expansion of $\ln(1+x)$ given in Chapter 1, and keep only the first term. This converts 2.17 to:

$$(S-S_0)(1+K/S_0) = -Vt \quad \text{or} \quad S - S_0 = -Vt(1+K/S_0) \tag{2.17'}$$

The graph of this is a straight line with intercept S_0 and slope $-V/(1+K/S_0)$. It is represented as the early portion of the curve in Fig. 2.2.

When a good deal of S has been transformed, the relative change $(S-S_0)/S_0$ will no longer be small, and the approximation to the logarithm will no longer be good. But now the logarithm, since its argument is a fraction S/S_0 that becomes ever smaller, becomes increasingly large and negative, while the term $S-S_0$ becomes negative much more gradually, and approaches the finite value $-S_0$ as S vanishes. Fig. 2.3 shows the familiar relation between the logarithm and the straight line. It follows that

[4] This equation was first derived by Henri (17) and has been "rediscovered" a number of times (18, 19).

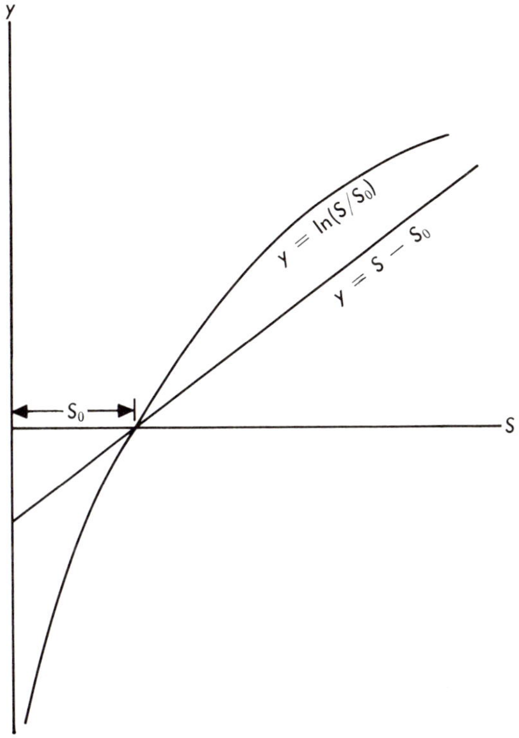

Fig. 2.3. Comparison of Linear and Logarithmic Functions, Showing How Logarithm Dominates for Small Values of the Variable.

we can neglect the linear term compared with the logarithm at the far end of the curve, and rewrite 2.17 as:

$$K \ln(S/S_0) = -Vt \quad \text{or} \quad S = S_0 e^{-(V/K)t} \tag{2.17''}$$

This part of the curve is again shown in Fig. 2.2. It is evident from 2.15 that the curve of $t(S)$ has no extrema, since $dt/dS = 1/(dS/dt)$ does not vanish for any non-negative value of S. It is evident from 2.17 that there are no zerocs, since it follows from $S \leqslant S_0$ that both terms on the left side are negative except for $S = S_0$. The only singularity is at the tail of the curve, where $S = 0$ corresponds to infinite t, as is also obvious from 2.17″. The complete curve is therefore as shown in Fig. 2.2.

Analysis of the Results

The S-t Relation. It is of interest to note some features of the S-t curve which has just been derived. It follows from 2.17′ that $(S - S_0)/t$ is a constant in the initial portion of the curve. That is, **a constant weight of substrate is transformed per unit time.**

The tail of the curve, given by 2.17″, corresponds to small S values. According to 2.15, $-dS/dt = VS/K$ for small S (see Fig. 1.2 and discussion in Chapter 1); and the solution of this approximate differential equation is indeed 2.17″, as the reader can readily verify for himself. But this differential equation may be rewritten as $(-dS/dt)/S = V/K$, which means that **a constant fraction of the substrate is transformed per unit time.**

These two features of the curve correspond to Brown's analysis (10), which was discussed earlier in the chapter. It was these results that led Brown to argue for an enzyme-substrate complex, for a constant turnover number (k_2 in our notation), and for the possibility of "saturating" the enzyme with substrate (a point to which we will come shortly). The mathematical analysis is thus seen to agree with Brown's argument in at least two salient features.

In view of our earlier discussion of the pseudostationary approximation, it is of interest to see how 2.17 can be used to estimate the time required for a given percent conversion of substrate. Taking $S = 0.9S_0$, and using a table to find $\ln(0.9) = -0.1$, we find:

$$t = 0.1(S_0 + K)/V \tag{2.18}$$

Using 2.16 for V, and selecting values of the parameters reasonably well within the practical experimental range, we set $S_0 = K = 10^{-2}M$, $k_2 = 10^4/\text{min}$, and $E_t = 10^{-8}M$ (which corresponds, at a molecular weight of 10^5, to a weight concentration of 1 microgram/ml); and we find $t = 20$ min. With the turnover number k_2 or the enzyme concentration an order of magnitude higher, or S_0 and K an order of magnitude lower, we would get 2 min. The range of t values seems quite reasonable.

Note that we can conveniently generalize 2.18 for any degree of conversion. Setting $S = fS_0$, where f is a proper fraction:

$$t = [(1 - f)S_0 - K\ln(f)]/V \tag{2.18'}$$

The Analysis of Rate Expressions

Let us now consider the rate equation 2.11′. As a function of the parameter E_t, it is simply linear, as pictured in Fig. 2.4.

As a function of S, v is a form of rectangular hyperbola, which was discussed in Chapter 1 and sketched in Fig. 1.2. With the parameters used in 2.11′. the curve appears as Fig. 2.5. It is linear near the origin, behaving like VS/K (using the abbreviation V from 2.16). The asymptote for large S is $k_2 E_t = V$. Since by definition $v = k_2 C$, this means that the asymptotic value of C is E_t; that is, for large S the entire quantity of enzyme is combined with substrate at any given moment. This is in keeping with Brown's argument: For large S, if a molecule of enzyme is freed from substrate by reaction

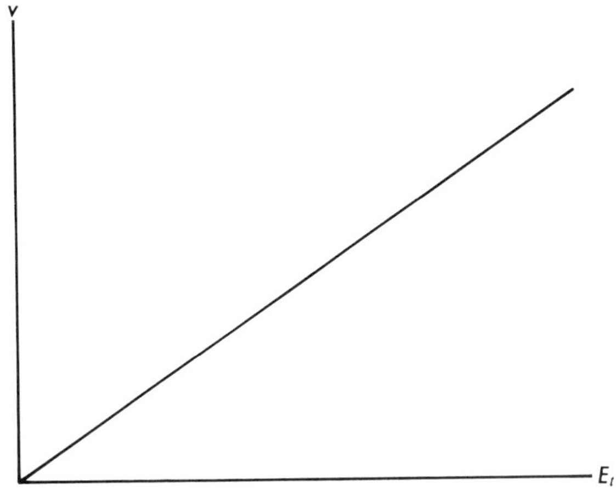

Fig. 2.4. Variation of Rate of Enzyme Activity, v, with Concentration of Enzyme Used, E_t.

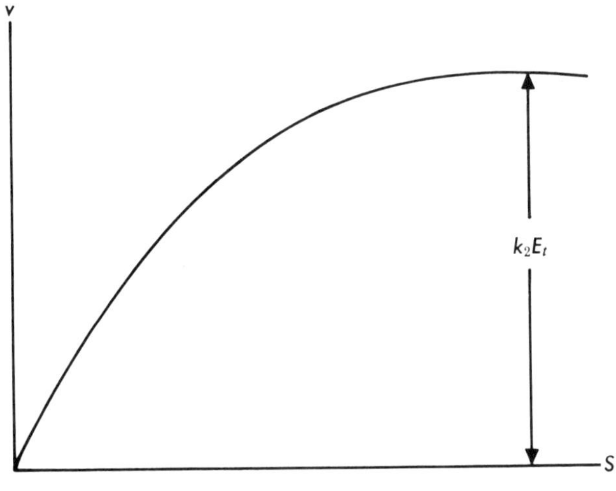

Fig. 2.5. Variation of Rate of Enzyme Activity, v with Concentration of Substrate Added, S.

2, it is instantly recaptured by another molecule of S in reaction 1. We speak of the enzyme in this condition as being **saturated** with substrate.

Fitting data to a hyperbola is feasible but laborious. It is of interest to seek forms of 2.11′ which will be more sensitive both to agreement and to deviation. The best-known of such forms was introduced by Lineweaver and Burk (20). Retaining the abbreviation V, and equating the **inverses** of both sides of 2.11′, we get:

$$1/v = (K + S)/VS = K/VS + 1/V = (K/V)(1/S) + 1/V \tag{2.19}$$

If $1/v$ is considered as a function of $1/S$, 2.19 gives a straight line with intercept $1/V$ on the $1/v$ axis and slope K/V (Fig. 2.6). The measurement of

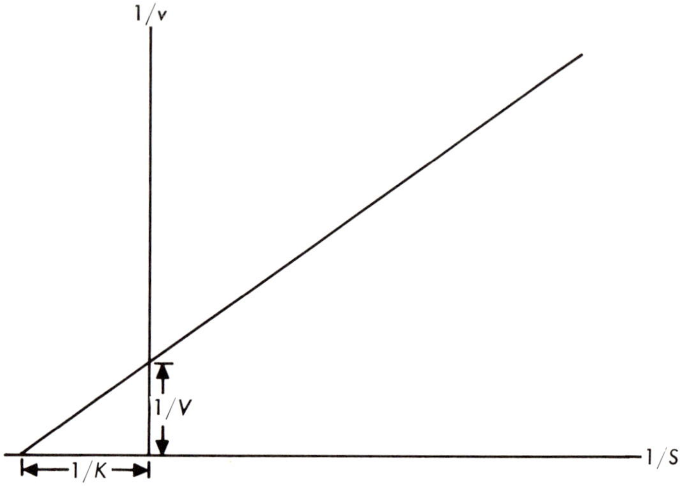

Fig. 2.6. Reciprocal Plot of Fig. 2.2 (Lineweaver-Burk Technique), Showing Graphical Construction for Calculating Kinetic Constants.

intercept and slope on an experimental curve that fits this form therefore its estimation of the parameters K and V. Moreover, the intercept on the (negative) S axis is $-1/K$, giving another estimate.

Multiplying 2.19 through by S, we get another version (21):

$$S/v = K/V + S/V \tag{2.20}$$

Plotting S/v against S (not against $1/S$!), we again get a straight line, with intercept K/V and slope $1/V$, and a second intercept $-K$.

The two forms have advantages under different circumstances. If an experiment is set up with the different substrate concentrations fairly evenly spaced (for example, 10^{-3}, 2×10^{-3}, 3×10^{-3}, etc.), the $1/S$ plot will tend to crowd the points together in one direction (10^3, $10^3/2$, $10^3/3$, etc.), making the extrapolation to the axes less certain. If the concentration series is set up in the common form $10^{-3}M$, 5×10^{-3}, 10^{-2}, etc., once again the reciprocal plot will give very uneven spacing of experimental points. The same will be true if we double at each step, as in viral dilution series. But a concentration series that, for some presumably good and sufficient reason, is crowded towards one neighborhood (for example, 10^{-3}, $10^{-2}/9$, $10^{-2}/8$, etc.) will be spread out rather uniformly by a $1/S$ plot.

Let me mention finally a form which is of no special value for the developments of this chapter, but which will be useful in later chapters. If we introduce the abbreviation $r = v/V$ (**relative rate**, rate as fraction of maximum rate), and solve 2.11′ for S, we get:

$$S = K[r/(1-r)] \tag{2.21}$$

a linear relation between S and $r/(1-r)$, with slope K and zero intercept.

Calculation of Rate from Data. We have been talking about v as if it were a directly measurable quantity, like the reading of an automobile speedometer. Our situation here is more closely comparable to working with an odometer and a stopwatch. What the experiment will give us is a tabular version of the S-t curve, which we may plot as in Fig. 2.2. The rate v at any time t is the slope of this curve.

We may therefore either (a) draw the tangent to the curve at any desired point (for example, T in Fig. 2.7), and obtain the slope of the tangent

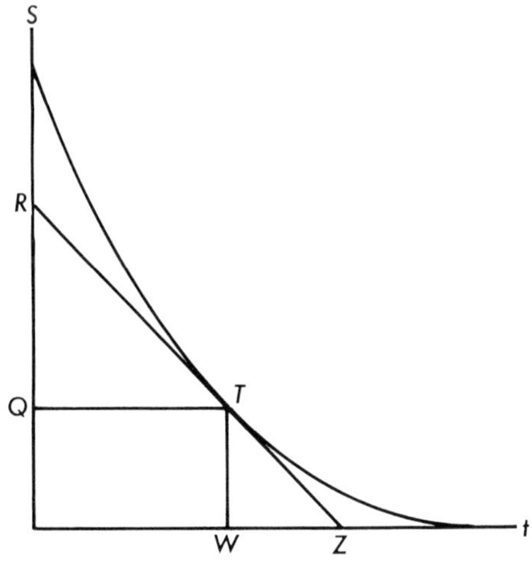

Fig. 2.7. Graphical Construction for Calculating Rate v from S-t Curve. Method (a) of text.

graphically (for example, QR/QT or WT/WZ in the figure), or (b) we may use the data themselves (or the points on a fitted curve if desired) to calculate a divided difference such as $(S_2 - S_1)/(t_2 - t_1)$ (see Fig. 2.2) as an estimate of the slope—an estimate which will be better according as the points selected are closer together.

Unless one is working with a system in which permeability or mixing problems introduce doubts about the significance of the earliest points, it is usually best to calculate **initial** rates—that is, to calculate the slope of the S-t curve at $t = 0$. This method insures a minimum of complications due to such factors as loss of enzyme by denaturation (which will be considered in Chapter 11), or to the fact that most enzyme-catalyzed reactions are not truly irreversible and will have a significant back reaction when P becomes large enough.

It should finally be observed (in case it does not seem obvious) that fitting data to equations like 2.11′ or 2.19 must be carried out with the actual S values corresponding to the point of the S-t curve at which the slope is measured. The admissible S value is S_0 only if the initial slope is used.

Analysis for and of Parameters. We observed earlier that the initial slope of the C-t curve is $k_1 S_0 E_t$. From this and $V = k_2 E_t$ we can get k_1/k_2. We can also get information from the area under the C-t curve, if the latter be extended through the experiment. For since $dP/dt = k_2 C$:

$$P(t) = k_2 \int_0^t C\,dt \tag{2.22}$$

The value $P(t)$ being directly measurable, it is divided by the integral (area under the C-t curve) to give k_2.

If we do not know E_t directly in moles per liter, we can now calculate it from k_2 and $V = k_2 E_t$. Given k_1 and k_2, we can now estimate k_{-1} from $K = (k_{-1} + k_2)/k_1$. Note that $K - k_2/k_1 = k_{-1}/k_1$ is the affinity constant (association constant) measuring the affinity of S for E. It is a common error to use K as the affinity constant,[5] for example in studying series of chemically related but distinct substrates. If this practice leads to anything but nonsense, it will be purely coincidental.

Not every enzyme system is obliging enough to lend itself to the elegant methods by which the C-t curve has been studied in certain cases. If we are limited to measurements in the pseudostationary region, we can simply get K and V from graphs like those of 2.19 and 2.20. If we are working with pure enzyme and know the value of E_t, we can get k_2—and this is the end

[5] This is probably due to the confusion in many textbooks between the present theory and an older one, derived by Henri (17) and incorrectly attributed to Michaelis and Menten (22) (but see 23). In this it is assumed that: (a) reaction 1 (forward and back) is much faster than reaction 2, and is always in true equilibrium, and (b) $v = k_2 C$. If this theory is developed in the same way as the one presented above, the v-S relation is **formally** the same as 2.11′, except that $K_m = k_{-1}/k_1$ replaces K. An immediate difficulty of such a theory is that equilibrium of reaction 1 implies $dS/dt = 0$. The "dynamic equilibrium" argument designed to get around this is that dS/dt is not really zero but very small; it differs from zero because reaction 2 disturbs it at the small rate $k_2 C$. What happens, according to this argument, is that every time a C molecule is removed by reaction 2, there is a rapid readjustment (from S to C) via reaction 1, to maintain the C level. The trouble with this explanation is that it is formally equivalent to saying that $-dS/dt = k_2 C$ (and that both are small); but $-dS/dt = k_2 C$ is obviously equivalent to the $dC/dt = 0$ of the pseudostationary approximation of Briggs and Haldane (see 2.5).

It is high time that this confusion, including the doubly incorrect reference to K as "Michaelis constant", disappeared from the literature.

of the line. With a crude enzyme preparation, where E_t is not known independently, the end of the line comes one stage earlier.

An extremely important aspect of the study of the parameters, the significance of which will become quite obvious in later chapters, is this: The simple system we have been discussing implies not only that K and V are parameters, but that K is independent of E_t while V is proportional to E_t. We shall see that there are enzyme systems that fit equations like 2.19, but whose parameters do not behave in this way—in which, for example, K is a function of E_t, and V is a more complex function of E_t than a simple multiple. Hence every system studied should be analyzed from this point of view as well as that of the v-S relation. Having obtained a v-S curve or its linearized equivalent at one value of E_t, and calculated K and V, one should proceed to repeat the process for other values of E_t. One may then decide whether the apparent K and V values live up to the simple specifications of this chapter. If not, one is immediately alerted to the fact that the enzymatic mechanism is more complex than had been suspected. Since, as was made clear in Chapter 1, the search for mechanisms is the most significant aspect of scientific theorizing, this kind of study is vital to the fruitful utilization of enzyme kinetics.

CONCLUDING REMARKS

The theory developed in this chapter has been a very serviceable one, and as far as it is applicable it has been amply supported by its ability to represent a great variety of enzyme systems as well as by direct experimental substantiation of its fundamental hypothesis (6, 7).

The chain reaction or inelastic collision theory, once a significant rival, has faded out of sight. As developed by Hearon and Katzman (5), it leads to a v-S equation whose graph bears a considerable resemblance to 2.11′, namely:

$$v = V(1 - e^{-S/A}) \tag{2.23}$$

where A is a constant. Analysis along the lines of Chapter 1 shows that it is linear (like VS/A) near the origin, asymptotically approaches V for infinite S, and has no other noteworthy features. This description would certainly seem close to that of Fig. 2.5. However, if we recast 2.23 in the reciprocal form, we find a story that differs from equation 2.19 and Fig. 2.6:

$$1/v = 1/V\{1 - e^{-[1/(A/S)]}\} \tag{2.24}$$

When $1/S$ is large, $1/(A/S)$ is small; hence the exponential series approximation may be used, keeping only the first two terms ($e^{-u} = 1 - u + \ldots$), and we get $1/[V/(A/S)] = (A/V)(1/S)$, which is linear in $1/S$, with zero asymptote. For small values of $1/S$, its reciprocal and hence the exponent in square brackets is large. The negative exponential is therefore small. Accordingly, we can apply the binomial approximation to the denominator in braces, getting $\{1 + e^{-[1/(A/S)]}\}/V$. The exponential term approaches its final value of zero with a horizontal tangent, as may be seen by taking the derivatives and permitting $1/(A/S)$ to become infinite (Hint: u^2e^{-u} vanishes for infinite u, because the exponential e^u increases faster with u than any power of u). The result is therefore as shown in Fig. 2.8, which may be

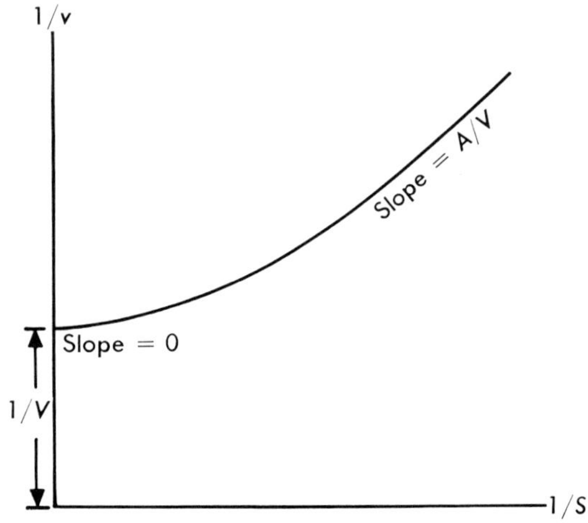

Fig. 2.8. Reciprocal Plot of Medwedew Rate Equation.

compared with Fig. 2.6 as predicted by the intermediate-complex theory.[6]

Despite its honorable history of successes, however, the theory of this chapter is most important for the developments for which it served as a foundation—developments which will be discussed in the chapters that follow. In all of them, the central idea—that everything an enzyme does to substrates and everything that is done to an enzyme by other molecules involves the formation of stoichiometric complexes—is maintained.

REFERENCES

General Background

For an intelligible and readable account of classical enzymology:

Haldane, J. B. S., "Enzymes," Longmans, Green, London, 1930.

[6] Reciprocal plots that level out near the origin have been found experimentally, and are discussed in later chapters; but it is easy to show that they arise from mechanisms quite distinct from that which leads to Fig. 2.8.

For historical perspective:

Bayliss, W. M. "The Nature of Enzyme Action," Longmans, Green, London, 1914.

For a compendious account of enzymes:

Boyer, P., Lardy, H., and Myrbäck, K. (eds.), "The Enzymes," Academic Press, New York, 1959 on.

Textual References

1. Rothen, A., J. Biol Chem. *163*, 345 (1946).
2. Karush, F., and Siegel, B. M., Science *108*, 107 (1948). Iball, J., Science *109*, 18 (1948).
3. Reiner, J. M., Unpublished.
4. Medwedew, G., Enzymologia *2*, 1 (1937).
5. Hearon, J. Z., and Katzman, R., Bull. Math. Biophys. *16*, 259 (1954).
6. Chance, B., in "Advances in Enzymology" (Nord, F. F., ed.), vol. XII, Interscience, New York, 1951.
7. Krimsky, I., and Racker, E., Science *122*, 319 (1955).
8. Armstrong, H., Trans. Chem. Soc. (Lond.) *67*, 1105 (1895).
9. O'Sullivan, C., and Tomson, F. W., J. Chem. Soc. (Lond). *57*, 834 (1890).
10. Brown, A. J., Trans. Chem. Soc. (Lond.) *81*, 373 (1902).
11. Wurtz, A., Comptes rendues *91*, 787 (1880).
12. Armstrong, E. F., Proc. Roy. Soc. Lond. (B) *73*, 500 (1904).
13. a: Guldberg, C. M., and Waage, P., "Etudes sur les affinités chimiques," Christiania, 1867.
    b: Daniels, F., "Outlines of Physical Chemistry," 7th ed., Wiley, New York, 1947.
14. Chance, B., J. Biol. Chem. *151*, 553 (1943).
15. Herzfeld, K. F., Ann. Physik *59*, 635 (1919).
16. Briggs, H. E., and Haldane, J. B. S., Biochem. J. *19*, 338 (1925).
17. Henri. V., Archivio di Fisiologia *1*, 299 (1904).
18. Van Slyke, D. D., and Cullen, G. E., J. Biol. Chem. *19*, 141 (1914).
19. Elkins-Kaufman, E., and Neurath, H., J. Biol. Chem. *175*, 893 (1948).
20. Lineweaver, H., and Burk, D., J.A.C.S. *56*, 658 (1934).
21. Dixon, M., and Webb, E. C., "The Enzymes," Academic Press, New York, 1958.
22. Michaelis, L., and Menten, M., Biochem. Z. *49*, 333 (1913).
23. Oppenheimer, C., "Die Fermente und ihre Wirkungen," vol. I, Leipzig (1924).

3

Reversible Reactions

A SIMPLE MODEL OF REVERSIBILITY

In the assumptions developed in the second chapter, we included the proviso that reaction 2 of scheme 2.A (the reaction from complex C to free enzyme plus product) is irreversible. This effectually made the overall enzyme-catalyzed reaction irreversible, and gave the simplest possible kinetic model capable of dealing at all with the facts of enzyme action. It was recognized to be a simplification, and to limit the application of the resulting formulas.

The fact that enzyme-catalyzed reactions are generally reversible has long been recognized, even in the case of those reactions where the thermodynamic equilibrium of the reaction is far to one side. For example, the hydrolytic reactions such as those of the phosphatases and esterases run virtually to completion if allowed to take their course. Yet it has always been possible, by starting with enormous concentrations of the split products, to induce these systems to run backwards, at least enough for analytical detection of the synthesized materials. In the case of an even more overwhelmingly one-sided reaction like that of hexokinase, it recently proved possible, by the use of isotopic labelling of the products, to demonstrate that a back reaction does occur, even though it has never been detectable by purely analytical methods.

It is almost always feasible to treat a reaction as irreversible by measuring the rate so early that the amount of product formed is too small to give an appreciable back-reaction rate. The only limitation on this device is set by the sensitivity and precision of the analytical methods: if methods for product are not very sensitive, or methods for substrate not very precise, it may be necessary to let the reaction run to an appreciable extent before a reliable rate can be calculated.

Nevertheless, it is of interest to see what the properties of fully reversible reactions will be, and in particular what the equilibrium position of such reactions will be. We shall therefore, in this chapter, study the effect of removing the limitation on reversibility.

The simplest way to do this is to retain the assumptions used in Chapter 2, substituting the word "reversibly" for the word "irreversibly" in assumption 2. This gives the stoichiometric scheme:

$$E + S \underset{-1}{\overset{1}{\rightleftharpoons}} C \qquad C \underset{-2}{\overset{2}{\rightleftharpoons}} E + P \tag{3.A}$$

This differs from scheme 2.A only in the added back reaction step labelled "-2". We retain, of course, the conservation equation for enzyme (2.3) $E_t = E + C$.

Applying the rules for setting up kinetic equations as used in Chapter 2, p. 53, we get (after eliminating E by means of 2.3):

$$dC/dt = k_1(E_t - C)S - (k_{-1} + k_2)C + k_{-2}(E_t - C)P \tag{3.1}$$

$$dS/dt = -k_1(E_t - C)S + k_{-1}C \qquad dP/dt = k_2 C - k_{-2}(E_t - C)P \tag{3.2}$$

The condition for the pseudostationary state is again simply $dC/dt = 0$. Setting the right side of 3.1 equal to zero, we solve for C:

$$C = (k_1 S + k_{-2} P)E_t/(k_1 S + k_{-2} P + k_{-1} + k_2) \tag{3.3}$$

To get the reaction rate, we note that, as in Chapter 2, the two distinct rates—of substrate disappearance (dS/dt) and of product appearance (dP/dt)—now become numerically equal. (Verify by 3.1 and 3.2 that, as in Chapter 2, $dS/dt + dP/dt + dC/dt = 0$. If $dC/dt = 0$, the conclusion follows.) Thus we can calculate the rate, using either $v = -\,dS/dt$ or $v = dP/dt$. These rates are expressed by 3.2 as functions of C. We substitute in either of them the value of C given by 3.3, clear the fraction, and get:

$$v = (k_1 k_2 S - k_{-1} k_{-2} P)E_t/(k_1 S + k_{-2} P + k_{-1} + k_2) \tag{3.4}$$

Observe that this is, by definition, the **net** reaction rate from left to right in scheme 3.A—the net rate at which substrate S goes to product P. The net rate in the backward direction would be the negative of this.

The expression 3.4 applies at any stage of a reversible reaction, S and P being the concentrations of substrate and product at the time of measurement If we now suppose that we are at the beginning of an experiment, in which

S has been added to the enzyme and no P is yet present, we would have a special rate expression, obtained by setting $P=0$ in 3.4.

(3.5) $$v_{if} = k_1 k_2 S E_t / (k_1 S + k_{-1} + k_2)$$

where the symbolism v_{if} means **initial forward** velocity. This expression looks familiar. If we divide boths parts of the fraction by k_1, write V for $k_2 E_t$, and K for $(k_{-1} + k_2)/k_1$ as in Chapter 2, we can rewrite 3.5 as:

(3.5′) $$v_{if} = VS/(S + K)$$

which is precisely the rate expression derived in Chapter 2. This is of course as it should be: If there is no product P present, the added backward reaction step −2 has no effect, and we must come back to the results of the last chapter.

In the same way, we might imagine that we are starting at the opposite extreme, putting in P with no S present ($S=0$). Then, dropping the initial minus sign in equation 3.4, we will get exactly the initial rate of the back reaction:

(3.6) $$v_{ib} = k_{-1} k_{-2} P E_t / (k_{-2} P + k_{-1} + k_2)$$

Here again we can rewrite the expression in a familiar form. Noting that the value of v_{ib} approaches $k_{-1} E_t$ asymptotically with P, we write V' for $k_{-1} E_t$, the maximum possible rate of the back reaction. Analogously, we write K' for $(k_{-1} + k_2)/k_{-2}$. Then, dividing both parts of the fraction by k_{-2}, we have:

(3.6′) $$v_{ib} = V'P/(P + K')$$

This has the same **form** as 3.5′, except that all the constants, as well as P, refer to the back reaction instead of the forward reaction.

These special cases show us how to simplify the complete equation 3.4. If the fraction is divided top and bottom by $k_{-1} + k_2$, and the abbreviations for K and K' are substituted, we have: $(k_2 S/K - k_{-1} P/K') E_t / (S/K + P/K' + 1)$. Introducing the definitions of V and V', and multiplying through by KK' to clear of secondary fractions, we have:

(3.4′) $$v = (VK'S - V'KP)/(K'S + KP + KK')$$

It is easy to see qualitatively how the extra terms affect this relation as compared with 3.5′. The term containing P in the numerator is negative, and decreases the numerator; the term in P in the denominator is positive and increases the denominator, therefore decreases the entire fraction. In short, as common sense would dictate, for all non-vanishing values of product P, the rate will be smaller than in the absence of P. Whether the rate will be reduced appreciably depends on the values of the various constants. For instance, the expression $K'S + KP$ in the denominator of 3.4′ will be practically equal to $K'S$, provided S and P are about the same magnitude but

$K \ll K'$. Similar considerations hold in comparing the S and P terms in the numerator. In the numerator, the maximum rates V and V' are also involved; the P term will be negligible if V' is very small compared with V. And this is another commonsense conclusion: Translating it into words, the effect of the product in slowing up the forward reaction will be negligible provided the maximal rate at which the back reaction can go is very small compared with the maximal rate at which the forward reaction can go.

EQUILIBRIUM IN THE SIMPLE MODEL

The new feature of 3.4′ is of course this: The rate v can be zero without either S or P being zero. In fact, beginning with all S and no P, as the reaction progresses S will grow smaller and P will grow larger, one mole of P being produced for every mole of S removed. Eventually this will cause the first term of the numerator to become small enough and the second term to become large enough so that they are numerically equal; and so the numerator, and hence v itself, will be equal to zero. The same process can be seen to occur if we started with all P and no S, and let the reaction run backward.

The state in which $v = 0$ is, by definition, what is meant by the **equilibrium** of the reaction. In this state, the numerator of 3.4′ is zero, or $VK'S - V'KP = 0$. We can solve for P/S, and get:

$$(3.7) \qquad (P/S)_{equil} = VK'/V'K = K_{ee} = (k_1/k_{-1})(k_2/k_{-2})$$

where we have introduced the symbol K_{ee} to stand for **enzymatic equilibrium constant**, to distinguish it from the ordinary equilibrium constant of a reversible reaction in the absence of enzyme.

The expression on the extreme right of 3.7 is what we get if we substitute for V, V', K, and K' the expressions used to define them. A noteworthy feature of this expression is that it does not contain E_t; this is of course the result of the fact that both V and V' are proportional to E_t, which therefore cancels out of their quotient. This feature—that the equilibrium position of an enzyme-catalyzed reaction is independent of the amount of enzyme added—has always been regarded as one of the characteristic features of such reactions, and indeed of catalyzed reactions in general.

Another feature worthy of attention in K_{ee} is the fact that, as shown by the expression on the extreme right of 3.7, it is simply the product of the true equilibrium constants of the two successive reactions 1 and 2[1]. This is

[1] Reaction 1 is $S + E \underset{-1}{\overset{1}{\rightleftharpoons}} C$. By definition of an equilibrium constant, its equilibrium constant is the value of C/SE at equilibrium. Since at equilibrium $k_1\,SE = k_{-1}C$, this makes the constant equal to k_1/k_{-1}. Applying the same reasoning to reaction 2, $C \underset{-2}{\overset{2}{\rightleftharpoons}} P + E$, its equilibrium constant is k_2/k_{2-}.

the same relationship which is known to hold for the overall equilibrium constant of any set of consecutive reversible reactions; thus, the overall dissociation constant of a dibasic acid is the product of the dissociation constants of the successive stages (3).

A more noteworthy aspect of the equilibrium constant will be realized if we rewrite the expression $(k_1/k_{-1})(k_2/k_{-2})$ as $(k_1/k_{-1})/(k_{-2}/k_2)$. Now the first factor, k_1/k_{-1}, is the reciprocal of the dissociation constant of C in the direction of S and E; in other words, k_1/k_{-1} is a measure of the **affinity** of substrate S for free enzyme E. In the same way, it is obvious that k_{-2}/k_2 measures the affinity of P for free enzyme E. Thus K_{ee} is the affinity of S for E divided by the affinity of P for E. We can therefore say this: At the equilibrium of an enzymatic reaction, product P will predominate over substrate S if S has a greater affinity for enzyme than P. On the other hand, substrate S will predominate if P has a greater affinity for the enzyme than S.

This conclusion, which at first seems delightfully simple and straightforward, appears increasingly odd the more one reflects on it. We appear to be saying that the equilibrium ratio of P to E depends only on the tendency for S to complex with E as against that of P to complex, and has nothing to do with the tendency of each complex to react further, to turn over into its eventual split product.

A MORE REALISTIC REVERSIBLE SYSTEM

Viewing the whole system in the light of what we know about chemical reactions, it is obvious just what is missing. There is no reason to believe that we would get the same complex by the reaction of S with E and by the reaction of P with E—there is every reason to believe that two distinct complexes would be formed. What really would happen, in the light of modern reaction theory, is that S, for example, reacts with E to form a complex; that this complex then undergoes an internal change—a further reaction—which gives a new complex, the same complex as would be formed by P and E reacting; this complex then splits to give P and E. Starting with P and E, we would trace the same path in the reverse order. The correct stoichiometric scheme must accordingly involve two complexes, the S-E complex and the P-E complex, which we will simply call C_1 and C_2 for convenience. The stoichiometric scheme will now be:

$$(3.B) \qquad E + S \underset{-1}{\overset{1}{\rightleftharpoons}} C_1 \qquad C_1 \underset{-2}{\overset{2}{\rightleftharpoons}} C_2 \qquad C_2 \underset{-3}{\overset{3}{\rightleftharpoons}} E + P$$

The kinetic equations are set up as indicated in Chapter 2, and they read (after collecting terms in E, C_1, and C_2):

(3.8) $$dC_1/dt = k_1SE - (k_{-1} + k_2)C_1 + k_{-2}C_2$$

$$dC_2/dt = k_{-3}PE + k_2C_1 - (k_{-2} + k_3)C_2$$

(3.9) $$-dS/dt = k_1SE - k_{-1}C_1 \qquad dP/dt = k_3C_2 - k_{-3}PE$$

We must assume a pseudostationary state for both complexes, and set $dC_1/dt = 0$, $dC_2/dt = 0$. Our conservation equation for enzyme will of course be enlarged to include both complexes:

(3.10) $$E_t = E + C_1 + C_2$$

The relation $dP/dt = -dS/dt$ follows as usual. Either expression can be used for the rate v. We must now solve for the unknowns E, C_1, and C_2, using 3.10 and the pseudostationary forms of 3.8. We then substitute the results in either of the equations 3.9 to get v:

(3.11) $$v = E_t(k_1k_2k_3S - k_{-1}k_{-2}k_{-3}P)/$$
$$[k_1S(k_2 + k_{-2} + k_3) + k_{-3}P(k_{-1} + k_2 + k_{-2}) + k_{-1}k_{-2} + k_{-1}k_3 + k_2k_3]$$

Using the same technique as before (setting $P = 0$ to get the initial forward rate v_{if} and setting $S = 0$ to get the initial backward rate v_{ib}, and analyzing[2] these simpler expressions), we find that the maximum forward and backward rates V and V' and the two Briggs-Haldane constants K and K' are given by:

(3.12) $$V = k_2k_3E_t/(k_2 + k_{-2} + k_3) \qquad V' = k_{-1}k_{-2}E_t/(k_{-1} + k_2 + k_{-2})$$

(3.13) $$K = (k_{-1}k_{-2} + k_{-1}k_3 + k_2k_3)/k_1(k_2 + k_{-2} + k_3)$$

$$K' = (k_{-1}k_{-2} + k_{-1}k_3 + k_2k_3)/k_{-3}(k_{-1} + k_2 + k_{-2})$$

By division:

(3.14) $$V/K = k_1k_2k_3E_t/(k_{-1}k_{-2} + k_{-1}k_3 + k_2k_3)$$

$$V'K' = k_{-1}k_{-2}k_{-3}E_t/(k_{-1}k_{-2} + k_{-1}k_3 + k_2k_3)$$

If we divide the numerator and denominator of 3.11 by $k_{-1}k_{-2} + k_{-1}k_3 + k_2k_3$, and substitute from 3.13 and 3.14, we find (after clearing of fractions by multiplying numerator and denominator by KK'):

(3.11′) $$v = (VK'S - V'KP)/(K'S + KP + KK')$$

which is identical in form with 3.4′, except that the constants are now defined by 3.12 and 3.13.

[2] Proceed essentially as with 3.4 and 3.5, etc. By dropping P, we have $v_{if} = E_tk_1k_2k_3S/[k_1S(k_2 + k_{-2} + k_3) + k_{-1}k_{-2} + k_{-1}k_3 + k_2k_3]$. To put this in the form $VS/(S + K)$, remove the coefficient of S in the denominator by dividing through numerator and denominator with it. This gives $v_{if} = S[k_2k_3E_t/(k_{-2} + k_2 + k_3)]/[S + (k_{-1}k_{-2} + k_{-1}k_3 + k_2k_3)/k_1(k_{-2} + k_2 + k_3)]$. The values of V and K follow at once with this kind of grouping. Getting v_{ib} by dropping S from 3.11, we rearrange in the same way to find V' and K'.

If we make reaction 3 irreversible by setting $k_{-3}=0$, the entire system is irreversible. By virtue of 3.13, K' becomes infinitely large under this condition. This in turn makes the terms in K' dominate in both numerator and denominator of 3.11′. K' being cancelled, the equation reduces to $v=VS/(K+S)$, which is the correct formula for a simple irreversible enzyme reaction, as in equation 3.11′. However, V and K are now given by 3.12 and 3.13.[3]

Still more interesting is the situation if we produce the irreversibility at step 2 instead of step 3, making $k_{-2}=0$. This makes $V'=0$; but all the other constants remain finite and non-zero, though k_{-2} drops out of them. The term in P disappears from the numerator of 3.11′; but the other term in P remains in the denominator. Thus the rate of reaction is reduced by product P, even though the reaction is irreversible. By virtue of the reversible reaction 3, P can compete with substrate S for free enzyme E, and thereby reduce the rate. Such **inhibition by product** has been well known for a long time (1). Further examples of it will be considered in Chapter 7.

EQUILIBRIUM

Since the rate formula is the same as 3.4′, the equilibrium formula will naturally be the same as 3.7, except for the expression of K_{ee} in terms of the various velocity constants. We have:

$$(3.15)\qquad (P/S)_{equil}=VK'/V'K=K_{ee}=(k_2/k_{-2})(k_1/k_{-1})/(k_{-3}/k_3)$$

The interpretation of this result is now not only simple but entirely sensible. The last two factors on the extreme right of 3.15, $(k_1/k_{-1})/(k_{-3}/k_3)$, once again the **ratio of the affinities of S and P** respectively **for the enzyme.** But the first factor, k_2/k_{-2}, is the equilibrium value of the ratio C_2/C_1—it is the ratio of the turnover constants which determine the rates of the critical reactions by which the $S-E$ complex is internally transformed to the $P-E$ complex and vice versa. Thus K_{ee} depends, as common sense tells us it should, not only on the relative affinities of S amd P for enzyme, but on the relative rates at which the two complexes can be converted to one another.

There is still another way to look at 3.15, and it is of special value because it permits an independent test of the validity of the equation. The expression for the equilibrium value of P/S, namely $VK'/V'K$, can also be written as

[3] A look at these expressions will show the reader that the elaborate strategies of Chapter 2 for sneaking up on individual rate constants (not to mention the imbecile attempts to interpret K as an affinity constant, also described in Chapter 2) have suddenly become not very valuable antiques.

$(V/K)/(V'/K')$. For the interpretation of the parts of this quotient, we look for the rate expression which would hold for very low substrate values.

Thus, in 3.5′, if S were small compared with K, the rate would approach the value VS/K. In 3.6′, if P were small compared with K', the rate expression would approach the value $V'P/K'$. In short, if we start an experiment with **only S present**, and a **low concentration of S**, the reaction rate is first-order with respect to the substrate, and the time course of the reaction will be given by equation 3.17″:

$$\ln(S/S_0) = -(V/K)t$$

The practical significance of the equations is this: If we assay substrate at various times during the reaction, and plot the logarithms of the S values against the clock time t, the plot will have the form of a straight line, and the slope of this line will be V/K. (If we plot S against t on semilogarithmic graph paper, the slope will be $(V/K)/2.303$, since the usual semilog paper uses a scale based on common rather than natural logarithms.)

In a similar way, if we start with only P present at low concentration, its rate of change will be first-order, and the time curve of P will be:

$$\ln(P/P_0) = -(V'/K')t$$

A plot of $\ln P$ against t will be a straight line with slope V'/K'.

The ratio $(P/S)_{equil}$ is simply the ratio of the two slopes just mentioned. Thus it is possible to get the numerical value of the equilibrium constant K_{ee} in two independent ways: (a) to let the reaction arrive at equilibrium, and measure $(P/S)_{equil}$ directly; or (b) to study the forward and back reactions at low S and low P values respectively, obtain the semilogarithmic time plots, and divide the slope V/K by the slope V'/K'. The two calculations should give the same numerical value, within experimental error. This calculation has in fact been performed with actual data, with satisfactory agreement (1).

With regard to method (a), a practical comment is in order at this point. As the reaction slows up and grinds to a halt, it will often be difficult to make sure that the true equilibrium has been reached at any given time, since experimental error may tend to obscure a slow but nevertheless real change in one direction. It is therefore advisable to perform the experiment in the two possible ways—starting from S, and starting from P. The same equilibrium state must be reached from both sides. If the two values are in good agreement, the result may be used with confidence. If there is some margin of difference beyond experimental error, it points to the need for running the experiment longer; moreover, the average of the two equilibrium ratios will be a better estimate of the true value than either one alone.

ENZYMATIC COMPARED WITH NONENZYMATIC EQUILIBRIA

An important although somewhat abstract theoretical problem raised by the material of this chapter is the following: Does the equilibrium constant K_{ee} of an enzyme-catalyzed reaction agree with the equilibrium constant K_e of the corresponding spontaneous reaction? According to classical thermodynamics (4), the nature of an equilibrium state is independent of the path by which it is reached. Therefore the same equilibrium ratio of P to S should be reached whether an enzyme intervenes or not, provided that the enzyme is in fact acting as a true catalyst—that is, the enzyme is not changed in amount or otherwise at the end of the reaction.

At equilibrium, the spontaneous reaction $S \rightleftharpoons P$ has the equilibrium ratio or equilibrium constant:

(3.16) $$K_e = e^{-\Delta F/RT}$$

where T is the absolute temperature, R is the ordinary gas law constant (numerical value about 2 calories per mole per degree), and ΔF is the difference in free energy between one mole of P and one mole of S, under standard conditions (4). On the other hand, according to 3.15, $K_{ee} = K_1 K_2 K_3$, where K_1 is the equilibrium constant of reaction 1 of scheme 3.B, $E + S \underset{-1}{\overset{1}{\rightleftharpoons}} C_1$ K_2 is the equilibrium constant of reaction 2, and K_3 that of reaction 3 (for $k_1/k_{-1} = K_1$, and so forth). To each separate reaction we can apply the thermodynamic formula for the equilibrium constant:

(3.17) $$K_1 = e^{-\Delta F_1/RT} \qquad K_2 = e^{-\Delta F_2 RT} \qquad K_3 = e^{-\Delta F_3/RT}$$

where ΔF_1 is the difference in free energy, under standard conditions, between one mole of C_1 and one mole each of E and S, and the other two differences in free energy can similarly be interpreted in terms of reactions 2 and 3 of the scheme 3.B.

We therefore have, using the relations of 3.17:

(3.18) $$\begin{aligned} K_{ee} &= (e^{-\Delta F_1/RT})(e^{-\Delta F_2/RT})(e^{-\Delta R_3/RT}) \\ &= e^{-(\Delta F_1 + \Delta F_2 + \Delta F_3)/RT} \end{aligned}$$

It is another well-known thermodynamic relation that, for a sequence of reactions, the overall standard free energy difference between the final and initial state is equal to the sum of the free energy changes for the individual steps of the sequence. That is, applying the rule to scheme 3.B:

(3.19) $$\Delta F = \Delta F_1 + \Delta F_2 + \Delta F_3$$

Substituting in 3.18, we get finally:

(3.20) $$K_{ee} = e^{-\Delta F/RT}$$

Comparing 3.20 with 3.16, we see that $K_{ee} = K_e$. That is, the theoretical equilibrium ratio for the enzyme-catalyzed reaction is the same as that for the spontaneous reaction. Note that the validity of this depends on the assumption that the enzyme acts as a true catalyst: The actual overall reaction of scheme 3.B is $S + E \rightleftharpoons P + E$, and this is the same as $S \rightleftharpoons P$ only if the enzyme has in no way been changed in the reaction, so that the symbol "E" as used on the two sides of the reaction equation not only means the same kind of molecules, but the same average molecular energy as well. Under this condition, E does not contribute anything to the overall free energy difference, and so we can use ΔF to denote the overall energy change for the enzymatic as well as the spontaneous reaction. The changes which the enzyme may undergo during the sequence of reactions are in no way restricted. The enzyme may be activated energetically (for example, by picking up energy from the solvent), or it may change its configuration, undergo partial denaturation, polymerize, or what you will—but all of the changes must be reversed by the end of the reaction sequence.

Experimental evidence for the theoretical identity $K_{ee} = K_e$ is scanty. The fact that K_{ee} is independent of enzyme concentration has been well established (see, for example, Bayliss (5); and one might suppose that this settles the matter. But the step from a very tiny concentration of enzyme to no enzyme at all is a jump too large to be made by extrapolation. No modern studies with satisfactory preparations of reasonably pure enzymes are known to us. Some early work like that of Dietz (6) on the hydrolysis of ethyl butyrate by pancreatic lipase as compared with hydrolysis by HCl seems to contradict the identity of the equilibria; but the crudity of the preparations and the heterogeneous nature of the incubation mixtures leave the significance of such results open to question.

The technical difficulties of the problem are considerable. Most of the enzyme-catalyzed reactions for which good enzyme preparations are available would be unsuitable because the corresponding spontaneous reactions are so slow as to be below the practical level of detection. The problem is not beyond hope of solution; but up to the present it has certainly not been solved.

REFERENCES

1. Haldane, J. B. S., "Enzymes," Longmans, Green, London, 1930.
2. Gamble, J. L., Jr., and Najjar, V. A., Science *120*, 1023 (1954).
3. Bull, H. B., "Physical Biochemistry," Wiley, New York, 1943.
4. Getman, F. H., and Daniels, F., "Outlines of Physical Chemistry," Wiley, New York, 1943.
5. Bayliss, W. M., "The Nature of Enzyme Action," Longmans, Green, London, 1914.
6. Dietz, W., Z. physiol. Chem. *52*, 279 (1907).

4

The Effect of Substrate Binding

THE RATE EQUATION: DERIVATION AND ANALYSIS

One problem which was ignored in the preceding chapters is the possibility that appreciable amounts of substrate might be tied up in the complex C. We observed in Chapter 2 that there should be a conservation equation for the substrate as well as for the enzyme. It would be:

$$S_t = S + C \tag{4.1}$$

where S is the free substrate, as before, and S_t is the total substrate concentration. In contrast to the behavior of the total enzyme E_t, the quantity S_t does vary during the course of an experiment, since substrate is being converted to product. But the relation 4.1 will hold at all times: the total (unconverted) substrate consists of two parts—one part unbound and capable of reacting with enzyme, and one part bound in the complex C and so not available for reaction with another molecule of enzyme.

Assumption 4 (p. 51) postulated that C was very small compared with S_t, so that S_t was practically the same as S. We therefore used only the symbol "S" in the derivations of Chapter 2. In this chapter we are concerned with exploring the results of using the more general relation 4.1.

No other change in the assumptions will be made here. The stoichiometric scheme is therefore still:

$$S + E \underset{-1}{\overset{1}{\rightleftharpoons}} C \qquad C \overset{2}{\rightarrow} E + P \tag{2.A}$$

The kinetic equations will thus be the same as in the second chapter (p. 54, equations 2.1 and 2.2):

(2.1) $dC/dt = k_1SE - (k_{-1} + k_2)C \qquad dE/dt = -k_1SE + (k_{-1} + k_2)C$

(2.2) $dS/dt = -k_1SE + k_{-1}C \qquad dP/dt = k_2C$

Moreover, taking the derivative of both sides of 4.1:

(4.2) $$dS_t/dt = dS/dt + dC/dt$$

Thus, in the steady state when $dC/dt = 0$, we will have:

(4.2′) $$dS_t/dt = dS/dt$$

It will therefore make no difference, as far as our rate measurements are concerned, whether the analytical method is one which destroys the enzyme and releases all substrate that is bound to it, or assays only free substrate S, or assays product P.

The pseudostationary condition is $dC/dt = 0$ as before. But the resulting algebraic equation, which comes from setting the dC/dt expression from 2.1 equal to zero, is not really the same as 2.9. For we must not only eliminate E by means of 2.3.

(2.3) $$E_t = E + C$$

but also eliminate S by means of 4.1, which we can solve for $S = S_t - C$. Write the pseudostationary condition as:

(4.3) $$k_1SE - (k_{-1} + k_2)C = 0$$

we divide by k_1 and use the definition 2.8, $K = (k_{-1} + k_2)/k_1$. We then eliminate E and S, and get:

(4.3′) $$(S_t - C)(E_t - C) - KC = 0$$

This equation is **quadratic** with respect to C, and its solution is:

(4.4) $$C = \tfrac{1}{2}[(S_t + E_t + K) \pm \sqrt{(S_t + E_t + K)^2 - 4S_tE_t}]$$

Since $v = k_2C$, we immediately have the expression for the reaction rate simply by multiplying 4.4 by k_2. However, the quadratic equation naturally gave two solutions, corresponding to the plus and the minus sign in front of the radical. We must ask ourselves if both these solutions are physically meaningful. Now there can be no complex C (and no reaction) if either substrate or enzyme is missing from the mixture. Setting either $S_t = 0$ or $E_t = 0$ in 4.4, it will be seen that C becomes zero only if we take the minus sign in front of the radical.

The only meaningful rate expression is therefore:

(4.5) $$v = \tfrac{1}{2}k_2[(S_t + E_t + K) - \sqrt{(S_t + E_t + K)^2 - 4S_tE_t}]$$

In Chapter 2, we proceeded to analyze 2.11′ by finding out how it would look if plotted against S and E_t. We can proceed along the same lines here, with regard to S_t and E_t. But first, remembering that the square root of a negative number is imaginary, we must make sure that the expression under the radical sign in 4.5 is positive. Writing out the radicand, we have

$$\begin{aligned}(S_t + E_t + K)^2 - 4S_t E_t &= S_t^2 + E_t^2 + K^2 + 2S_t E_t + 2S_t K + 2E_t K - 4S_t E_t \\ &= S_t^2 + E_t^2 + K^2 - 2S_t E_t + 2S_t K + 2E_t K \\ &= (S_t - E_t)^2 + K^2 + 2K(S_t + E_t) > 0\end{aligned}$$

It is clear that v is positive everywhere except for $S_t = 0$ or $E_t = 0$, since $(S_t + E_t + K)^2 - 4S_t E_t < (S_t + E_t + K)^2$. To get the behavior of v, considered as a function of S_t, near the origin, we write the radicand as $S_t^2 + (E_t + K)^2 + 2S_t(K - E_t)$, and factor out the term independent of S_t, $(E_t + K)^2$, to get a suitable expression for the application of the binomial approximation (Chapter 1). We have $(E_t + K)^2\{1 + [2S_t(K - E_t) + S_t^2]/(E_t + K)^2\}$, wherein the fractional term will evidently become small as S_t does so. The radical, by the binomial method, will therefore be approximately $(E_t + K)\{1 + [2S_t(K - E_t) + S_t^2]/2(E_t + K)^2\}$. Multiplying out, and neglecting the quadratic term compared with the first power as S_t diminishes, we find that v behaves like $k_2 E_t S_t/(K + E_t)$ near the S_t origin.

The asymptotic form for large S_t is readily found in a similar way (see Chapter 1). Writing the radicand as above, we now factor out S_t^2, recalling the rule that one factors out the term which is going to be the greatest in the limit. Following through with the binomial approximation again, and eventually dropping $\frac{1}{2}(E_t + K)^2/S_t$ in the limit for large S_t, we find that v approaches $k_2 E_t$.

Finally, $dv/dS_t = 0$ has no solution—there are no extrema. Despite the superficial dissimilarity, then, the behavior of 4.5 with respect to S_t is very much like that of 2.11′ with respect to S (Fig. 4.1).

However, v is not linear in E_t. Indeed, noting that S_t and E_t appear symmetrically in v, and recalling the consequences of such behavior from Chapter 1, we can at once conclude that the plot of v against E_t for constant S_t is of the same form as the plot of v against S_t for constant E_t; and if the constant value selected were the same in the two cases the graphs would actually be identical. (See Fig. 4.2.)

The symmetry between S_t and E_t is a natural result of the fact that we have dropped assumption 4 and added the conservation equation 4.1. In Chapter 2 we were saying, in effect, that the substrate can tie up the enzyme but the enzyme cannot tie up the substrate. In the present chapter we are recognizing that it works both ways. It is therefore reasonable that the two symbols should enter the formulas in the same way.

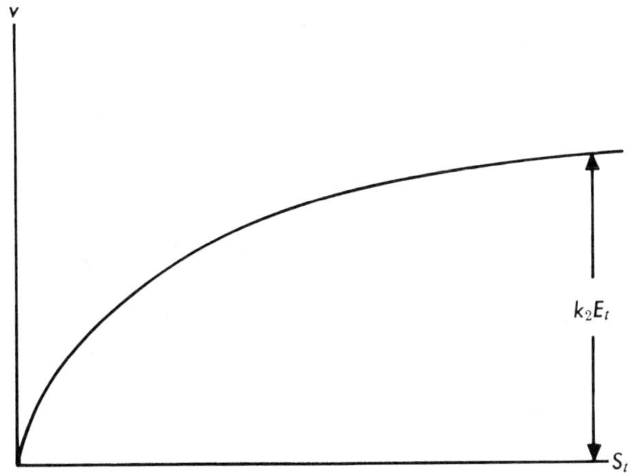

Fig. 4.1. Rate v as Function of Total Substrate Concentration S_t, When Substrate Bound by Enzyme is Appreciable Fraction of S_t.

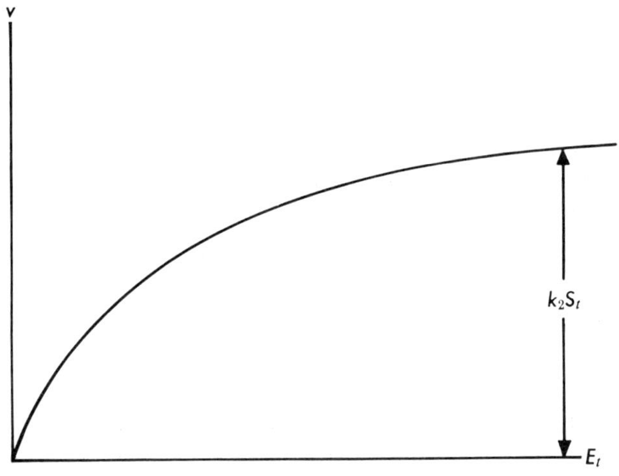

Fig. 4.2. Rate v as Function of Enzyme Concentration E_t for Same Case as Fig. 4.1—saturation with enzyme.

SATURATION WITH ENZYME

The fact that v is not simply proportional to E_t, but levels off as E_t becomes sufficiently large, is an important new result of adding 4.1 to our system of equations. In Chapter 2 we found that it is possible to "saturate" the enzyme with substrate. In this chapter we have shown that it is at least theoretically possible to saturate the substrate with enzyme. Whether this can always be realized in practice is another matter; we know of only two published cases in which the data are clear (1,2). To reach the saturation level requires, as we have just seen, that E_t should be large compared with either S_t or K. This can probably be achieved only with highly purified enzymes, where high enzyme concentrations become a practical possibility. At the same time, it may prove necessary to work with low substrate concentrations—and this will be limited by the sensitivity of the analytical methods. The fact that enzyme saturation is not usually observed is explained by the fact that enzyme concentrations are in general practice very tiny compared with substrate concentrations, even when the latter are quite low. When purified enzyme is available—the most favorable situation for getting high E_t—the investigator who has toiled hard to get ten or twenty milligrams of fairly pure protein is understandably reluctant to squander his product: he generally uses only as much enzyme as is needed to give conveniently measurable reaction rates.

A DIAGNOSTIC PLOT OF THE RATE EQUATION

We pointed out in Chapter 2 that the direct plot of v against S does not lend itself readily to estimation of the constants V and K. It was for this reason that the reciprocal plot of Lineweaver and Burk (3) was introduced. With the more complex form of 4.5, the need for a simpler form to aid in analyzing the relationships is even more urgent.

We return to equation 4.3′:

$$(S_t - C)(E_t - C) = KC \tag{4.3′}$$

which can be rewritten as:

$$(E_t - C) = KC/(S_t - C) \tag{4.3″}$$

If we let S_t become very large, the right side becomes very small, and approaches zero. The left side must therefore also approach zero, so that C approaches E_t, as we found in the preceding section.

It is now convenient to introduce the **relative rate** r, defined by:

$$r = v/V = k_2C/k_2E_t = C/E_t \tag{4.6}$$

We solve this equation for C ($C = rE_t$), and substitute in 4.3′. After cancelling the common factor E_t on both sides of the equation, dividing by $1 - r$, and transposing rE_t to the right side, we have:

$$S_t = rE_t + Kr/(1 - r) \tag{4.7}$$

If we had wished to express the relation in terms of v rather than r, we could have solved $v = k_2 C$ for C, getting $C = v/k_2$, and after substituting in 4.3′ and rearranging we would have had:

$$S_t = v/k_2 + Kv/(V - v) \qquad (4.8)$$

Let us consider how we may plot 4.7 (obviously the graph of 4.8, with its very similar form, would follow along similar lines). The first term of course represents a straight line through the origin with slope E_t on a graph of S_t against r. The second term starts out as almost a straight line; when r is very small compared to 1, it is practically equal to Kr. As r increases, the denominator $1 - r$ becomes smaller, the numerator r becomes larger, and so the whole term increases. As r approaches 1, the denominator approaches zero, so that the whole term becomes infinitely large: The curve rises almost vertically. The composite curve is sketched in Fig. 4.3. No value of r beyond 1 is considered; the rate of reaction (v) cannot be greater than the highest possible rate (V), so the relative rate cannot exceed 1.

Figure 4.3 is essentially Fig. 4.1 sideways. The variable r differs from v only by the constant factor $1/V$; plotting against r means plotting against v on a scale where the largest value of v, namely V, is taken as 1. In Fig. 4.1 we graphed v against S_t; in Fig. 4.3 we are graphing S_t against (relative) v. Thus the relation of S_t and v can be inferred without solving the quadratic equation 4.3′. (Recall the discussion of inverse functions and their graphs in Chapter 1.)

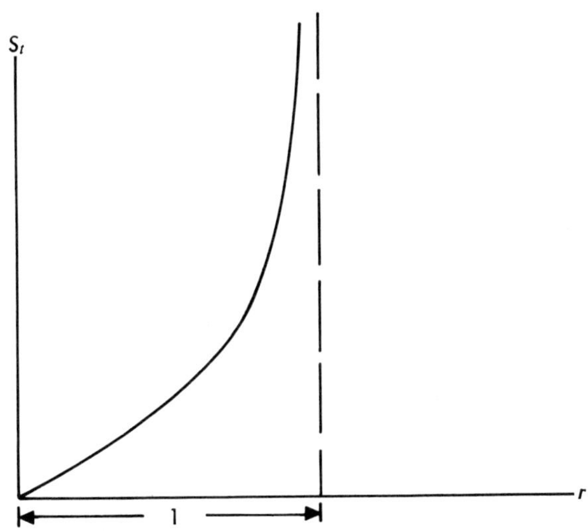

Fig. 4.3. Relation of S_t to Relative Rate r.

The fact that 4.7 contains the combination $r/(1-r)$ suggests plotting S_t against that combination rather than against r itself. Let us introduce a symbol for the combination: $u = r/(1-r)$. We can solve this backwards for r in terms of u: $r = u/(1+u)$. Equation 4.7 can now be rewritten as:

$$S_t = E_t u/(1+u) + Ku \tag{4.7'}$$

Let us sketch 4.7′ by the usual methods. Near the origin (u small) the first term varies as $E_t u$. The combined curve for small u values is therefore the sum of the two straight lines $E_t u$ and Ku, which is a straight line passing through the origin, with slope $E_t + K$. For large u, the first term of 4.7′ approaches E_t asymptotically. The combined curve therefore approaches the asymptote $S_t = E_t + Ku$, a straight line with slope K and intercept E_t. The whole curve is plotted schematically in Fig. 4.4.

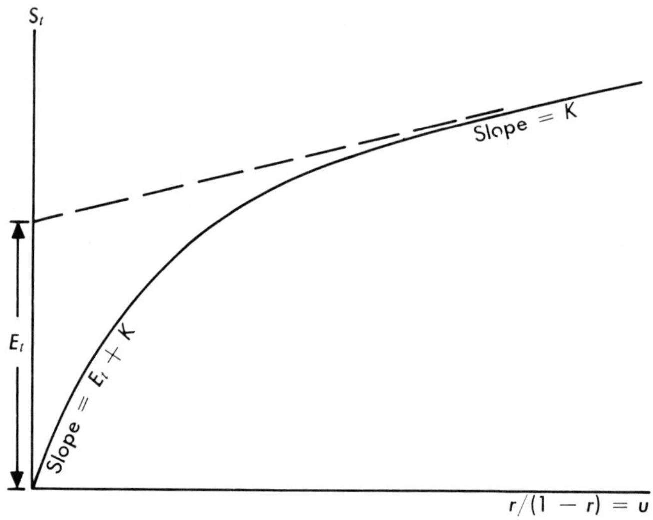

Fig. 4.4. Relation of S_t to $r/(1-r)$. Note intercept E_t.

The parameters can be estimated from this graph, as with the reciprocal plot in Chapter 2. The initial and final parts of the curve are practically straight, so their slopes can be obtained graphically, giving $E_t + K$ and K. The value of E_t obtained by difference can be checked with the value of E_t given by the intercept of the asymptote (the dotted line in Fig. 4.4).

A special feature of this graph is the fact that it gives the value of E_t by itself. The reader will recall that in Chapter 2 it was possible to get from ordinary experimental data only the value of the product $k_2 E_t = V$, but not the value of either factor alone. By adding the conservation equation 4.1 to our set of assumptions, we are able to find the value of E_t. Knowing V (for example, from a graph like Fig. 4.1), we can calculate k_2 by division ($k_2 = V/E_t$ by definition of V).

This value can also be checked independently if enough enzyme can be used to saturate the system, as in Fig. 4.2. The rate at saturation with enzyme is $k_2 S_t$. This gives the relation $v = -dS/dt = -dS_t/dt = k_2 S_t$. The solution of this is $\ln S_t - \ln S_{t0} = -k_2 t$ (S_{t0} is the value of S_t at the beginning of the experiment, the time $t = 0$). This is a straight line on a semilogarithmic graph, and its slope is $-k_2$ 2.303. This value of k_2 can be checked against the value obtained as described in the last paragraph.

One limitation on the process of getting E_t from Fig. 4.4 is this: If $E_t \ll K$, then $E_t + K$ will not differ much from K. The curve will hardly be distinguishable from a simple straight line. In fact, looking at equation 4.7, the reader can see that, if E_t is extremely small, the first term practically drops out, and the equation has the form $S_t = Kr/(1 - r)$. This equation, it will be remembered, was introduced in Chapter 2 (p. 67), as an alternative form of equation 2.11′. In other words, if $E_t \ll K$, the rate relations of this chapter become practically the same as those of Chapter 2. This is not surprising, since the novelty in Chapter 4 is to keep account of the amount of substrate bound by enzyme: if the amount of enzyme is very tiny, then the amount of substrate bound by it is necessarily tiny also, and therefore does not materially affect the relation between v and S_t.

The Lineweaver-Burk reciprocal plot ($1/S_t$ against $1/v$) can also be used; but the equation is appreciably more complex than 4.7 or 4.8. We include the graph for reference (Fig. 4.5).

ENZYME EQUIVALENT WEIGHT

The importance of the new aspect of this chapter—the possibility of extimating E_t from experimental data—is considerable. If we are using a purified enzyme, so that most or all of the protein added is enzyme, we can then make use of the well-known relation that grams divided by moles equals molecular weight. Thus, if the weight concentration of enzyme is W grams per liter, then the molecular weight is $M = W/E_t$. (E_t will come out in moles per liter provided S_t is expressed in those units.)

This statement must be qualified in the light of the next chapter, where it will appear that, under certain conditions, there may be several independent enzymatically active groups on a single molecule of enzyme. In such a case, the quantity which we are here calling E_t will actually be the number of active groups multiplied by the true molar concentration of enzyme, or in other words the **equivalent concentration of active groups**. (The language is chosen by an exact analogy to polyvalent acids and bases. With a strong dibasic acid like H_2SO_4, if the concentration is 1 mole per liter, the **equivalent concentration or normality**—the concentration of hydrogen ions—is 2 equivalents per liter.)

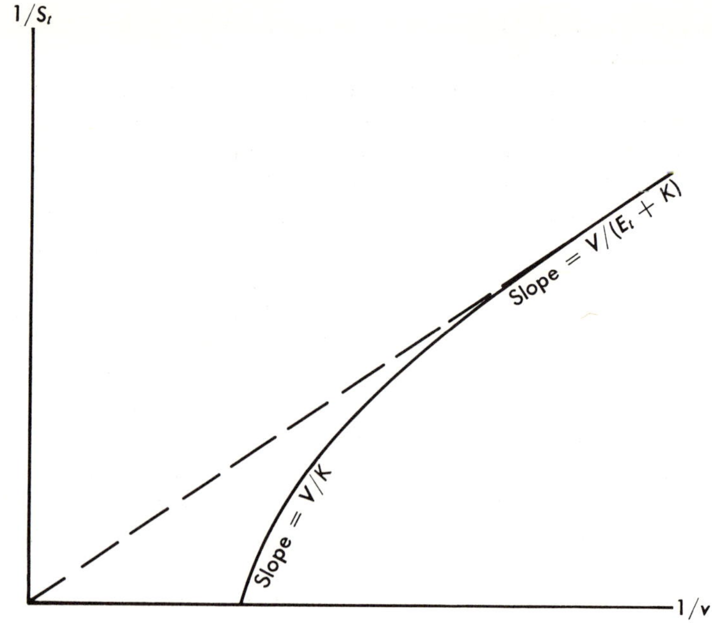

Fig. 4.5. Reciprocal Plot of Fig. 4.1. Note that Asymptote (dashed line) passes through origin.

If there are n active groups per enzyme molecule (and if they are independent of one another, as considered in Chapter 5), and the true molar concentration of enzyme is E_t, then the apparent molar concentration, or equivalent concentration, as obtained by the methods of the present chapter, will be nE_t. Dividing into W, we would have $W/nE_t = M_e$, the **equivalent weight**. (Again we follow the standard chemical usage—if a molecule has n equivalents, the equivalent weight is related to the actual molecular weight by $M_e = M/n$.) If the molecular weight has been determined by an independent method (one of the standard methods for molecular weights (4) of proteins, for instance), we can get the number of equivalents n by $n = M/M_e$. In other words, the experimentally determined molecular weight, divided by the weight concentration of enzyme in grams per liter, and multiplied by apparent molar concentration of enzyme, gives the number of active groups.

REFERENCES

1. Stadtman, E. R., Novelli, G. D., and Lipmann, F., J. Biol. Chem. *191*, 365 (1951).
2. Ochoa, S., Stern, J. R., and Schneider, M. C., J. Biol. Chem. *193*, 691 (1951).
3. Lineweaver, H., and Burk, D., J.A.C.S. *56*, 658 (1934).
4. Edsall, J. T., in "The Proteins" (Neurath, H., and Bailey, K., eds.), vol. I, pt. B, Academic Press, New York, 1953.

5

Multivalent Enzymes

Suppose, using the methods described in Chapter 4, we were to calculate that we had, in a certain experimental mixture, 10,000,000 enzyme molecules per cc. A natural question would be: Could we not just as well have 1,000,000 molecules, each containing 10 enzymatically active sites or groupings? The protein molecules which constitute enzymes are so large that there is certainly no *a priori* reason why such a thing could not happen.

In this chapter we shall consider how an enzyme with more than one active site per molecule would differ from the kind of one-place enzyme molecule which we have considered up to now. In particular, we shall inquire whether there is any situation in which one could not tell the difference on the basis of kinetic measurements alone.

With regard to each individual active site, let us make the same assumptions that we used in Chapter 2. In other words, we shall assume that each site is capable of combining reversibly with one molecule of substrate, and that the complex gives rise irreversibly to **free site** and product.

What shall we assume about the various active sites on the same enzyme molecule? It is reasonable (although not compulsory) to suppose that all the sites are inherently identical—that is, looking at their chemical structure or at the structure of their immediate surroundings on the enzyme, we would not be able to tell them apart unless we had arbitrarily put tags (or numbers) on them.

We could make the most complex assumption: that each site has its own individuality, if only by virtue of the chemical groups surrounding it on the enzyme molecule. But the increase in mathematical complexity which would result from this would serve no purpose; for at present the detailed topography of most protein molecules is so little known that such a theory could hardly be implemented. In fact, most of what is known of this aspect of protein structure has come from the study of enzyme reactions and their relation to substrate structure (as we shall see in later chapters).

What is the consequence of assuming the various active sites to be basically identical? First of all, it means that, if none of them is occupied by substrate, there is an equally good chance for a substrate molecule to become attached to any one of them. Therefore the chance that a substrate molecule will become attached to that particular enzyme molecule **somewhere** is simply proportional to the number of sites.

INTERACTION OF ENZYME SITES

Suppose some sites are full and some are empty. If another substrate molecule comes along, its chance of attaching to the enzyme will again be proportional to the number of empty sites. But there will have to be another factor which measures the **effect** of the **occupied groups**.

What does this mean exactly? We know that reaction between two chemical compounds depends on various features of their structure, such as the presence of charged or polar groups, availability of electrons or of electron-accepting groups, etc. This applies to the reaction of enzyme site and substrate as well as to other chemical compounds.

When one site has been filled by substrate, the factors that determine binding and reaction at another site may be modified—and *a priori* the modification may either help or hinder matters.

At this point we could again bring in the detailed topography of the sites. Since most electrical forces become smaller with distance between the attracting or repelling groups, obviously a given site will influence those which are nearest to it more than those which are further away. Sites which are centrally located among the others (if any) will be nearer to more other sites, and will therefore be more influential. The course of events on any enzyme molecule would therefore depend on the past history at every step—on which sites had happened to become filled earliest.

To illustrate, suppose a linear arrangement of sites, ..., and suppose an end site is filled first (*x* . . . . . .). If the extra force due to occupancy is attractive, it will increase the chance of occupancy of the sites nearest to it (*x x* . . . . . or *x* . *x* . . . .). The effect will now be increased as two attracting centers are available, and so the molecule will be filled up, with one end being filled first. The tendency to form completely filled enzymes will be high.

If the force is repulsive, the next molecule will have the best chance of attacking at the farther end (*x* . . . . . *x*). The next one will attack (but with still slower rate, because there are now two repelling centers) with greatest probability near the middle ($x \ldots x \ldots x$ or $x \,.\, x \ldots x$). The chances of filling the rest of the gaps will rapidly decrease.

The same reasoning could be applied if the sites were arranged in a circle, or a square, or any other regular or random arrangement. The numerical results would obviously be different in the various cases, since these results would depend on the exact spacing of the sites. The statistical problem could indeed be worked out; it is an example of what in probability theory is called a Markov process (1)), or in statistical mechanics a cooperative process (2).

However, it is to be expected that the selection of an initial site to be filled would be a random event, varying from one molecule to another in the assembly of enzyme molecules; and thus the various possible "histories" of filling are likely to be represented in the assembly.

We shall therefore make an assumption which is equivalent to a sort of averaging over the various orders in which the sites of an enzyme could be filled up. We shall simply assume that all vacant sites on vacant enzymes are formed at a characteristic rate and broken up at another characteristic rate. We shall further assume that all vacant sites on singly occupied enzymes are formed and broken up at characteristic rates, but that these are in general different from the rates for the vacant enzyme molecules, and so on.

To illustrate the ideas, we will first present the simplest possible case—that of an enzyme with two active sites.

THE TWO-PLACE ENZYME

Terminology: As before, we use S for substrate, P for product, E for free enzyme, and E_t for total enzyme (actual number of enzyme molecules). We will somewhat change the reaction step numbering system, however, as well as the symbols for rate constants. We denote the singly occupied enzyme by C_1, and the doubly occupied (completely occupied) enzyme by C_2.

The stoichiometric scheme is:

$$S + E \underset{-1}{\overset{1}{\rightleftharpoons}} C_1 \qquad C_1 \overset{1a}{\rightarrow} E + P \tag{5.A}$$

$$S + C_1 \underset{-2}{\overset{2}{\rightleftharpoons}} C_2 \qquad C_2 \overset{2a}{\rightarrow} C_1 + P$$

Instead of calling the reaction from C_1 to $E + P$ reaction 2, we are calling it reaction $1a$, and so on. The purpose of this is to have the numbers 1, 2, and so on refer to the successive **stages** of occupancy of the enzyme; this will be particularly helpful in keeping a simple notation when we pass from the two-place enzyme to the general many-place enzyme.

Our assumptions are embodied in the fact that corresponding reactions for the two stages of enzyme occupancy are distinguished and assigned different numbers and different rate constants. Thus we assume that k_1 is not equal to k_2, and that they differ by more than the probability factor which we discussed above (equal to 2 in the first case and 1 in the second, since these are the numbers of vacant sites available for reaction at the two stages). The same is true for the other constants; and there are six distinct constants in all.

The kinetic equations for the system are:

$$(5.1) \qquad \begin{aligned} dC_1/dt &= k_1SE - (k_{-1} + k_{1a})C_1 - (k_2SC_1 - k_{-2}C_2) + k_{2a}C_2 \\ dC_2/dt &= k_2SC_1 - (k_{-2} + k_{2a})C_2 \\ dS/dt &= -k_1SE + k_{-1}C_1 - k_2SC_1 + k_{-2}C_2 \\ dP/dt &= k_{1a}C_1 + k_{2a}C_2 \end{aligned}$$

The conservation equation for enzyme is:

$$(5.2) \qquad E_t = E + C_1 + C_2$$

Introducing the pseudostationary conditions, $dC_1/dt = 0$, $dC_2/dt = 0$, into 5.1, we solve for C_1 and C_2 (strategy hint: solve the second equation for C_2 first, then substitute into the first equation and get C_1):

$$(5.3) \qquad C_1 = SE/K_1 \qquad C_2 = SC_1/K_2 = S^2E/K_1K_2$$

where, following the pattern of previous chapters, we define:

$$(5.4) \qquad K_1 = (k_{-1} + k_{1a})/k_1 \qquad K_2 = (k_{-2} + k_{2a})/k_2$$

We put 5.3 into 5.2, and solve for E:

$$(5.5) \qquad E = E_t/(1 + S/K_1 + S^2/K_1K_2)$$

Since total substrate in the system, $S + P + C_1 + C_2$, is fixed, putting $dC_1/dt = 0$ and $dC_2/dt = 0$ it follows that $dS/dt + dP/dt = 0$, so that once again we have $v = -dS/dt = dP/dt$. Using the form $v = dP/dt$, with dP/dt given by 5.1, we substitute for C_1 and C_2 by means of 5.3, and eliminate E from the result by means of 5.5. The rate is thus:

$$(5.6) \qquad v = (k_{1a}S/K_1 + k_{2a}S^2/K_1K_2)E_t/(1 + S/K_1 + S^2/K_1K_2)$$

This is a new form of rate equation, for it contains the second power of S as well as the first. The rate is zero at the origin, and varies as $k_{1a}E_tS/K_1$ near the origin. When S becomes very large, v asymptotically approaches $k_{2a}E_t$. That is, the rate at saturation with substrate is the rate of turnover of doubly occupied enzyme. This agrees with the commonsense intuition that, at saturation, all enzyme would have both places filled with substrate.

Autoinhibition

If we now proceed to test the relation of v to S for a maximum value, a novel situation emerges. We have:

$$(5.7)\quad dv/dS = E_t[k_{1a}/K_1 + 2k_{2a}S/K_1K_2 + (k_{2a} - k_{1a})S^2/K_1{}^2K_2]/(1 + S/K_1 + S^2/K_1K_2)^2$$

The familiar condition for a maximum is $dv/dS = 0$. One way for this to occur is that the numerator of dv/dS should become zero, while the denominator remains finite. We accordingly set the numerator of the derivative equal to zero (after clearing of fractions):

$$(5.8)\qquad k_{1a}K_1K_2 + 2k_{2a}K_1S + (k_{2a} - k_{1a})S^2 = 0$$

This is a quadratic equation for S, and can be solved as usual. The significance of the solution is, of course, that this is the value of S for which v has its maximum value. If no physically meaningful solution exists (that is, a real and positive number, since S is a concentration), then v does not have a maximum when plotted against S.

It is immediately evident that there can be no solution if $k_{2a} - k_{1a} \geqslant 0$. For in that case each of the terms would be positive as long as S is a real and positive number. But the sum of three positive terms cannot possibly add up to zero. There is no way for the first two terms to be anything but positive. The only way they can be cancelled, therefore, is for the third term to be negative—that is, $k_{2a} - k_{1a} < 0$. This condition also guarantees that the solution is real, since it makes the discriminant of the quadratic 5.8 positive.

This result is novel, then, If $k_{2a} < k_{1a}$, the graph of v against S will reach a maximum, then decrease again, and finally arrive at a nonzero asymptotic value. The physical meaning of this result is quite simple. The condition $k_{2a} < k_{1a}$ says that the turnover rate of a doubly occupied enzyme is less than that of a singly occupied enzyme. As the amount of substrate is increased, starting from low values, the amount of singly occupied enzyme will at first predominate, and the rate will tend to approach the intrinsic saturation value for an active site, $k_{1a}E_t$. But in the meantime the second site on some of the enzyme molecules is being filled up, and these filled enzymes yield product at a slower rate. As the filled enzymes become more numerous, therefore, and the singly occupied enzymes become scarce, the overall rate decreases, until all enzymes are fully saturated with substrate on both sites. At that point the rate will be the saturation rate for doubly occupied enzyme. We have here an odd case of **autoinhibition**—the inhibition of enzyme by its own substrate. This case is illustrated in Fig. 5.1, curve C.

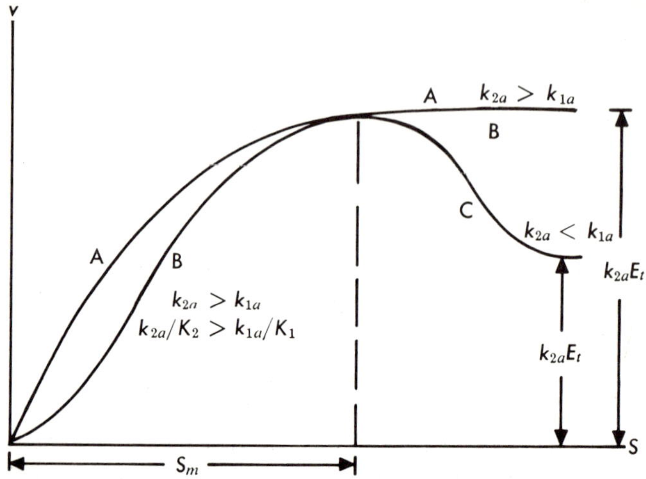

Fig. 5.1. Rate v as Function of Substrate Concentration S, for Enzyme with Two Active Sites per Enzyme Molecule. Curves A and B, turnover on second site is greater when first site is filled. Curve C, turnover on second site is less when first is filled **(auto-inhibition).**

If $k_{2a} - k_{1a} > 0$, on the other hand, no peak can occur, and the curve of v against S would seem to be fairly conventional, at least qualitatively—it simply rises from zero, and eventually flattens out. However, calculation of the second derivative indicates that there will be an inflection point, provided $k_{2a}/K_2 > k_1/K_1$. This case is sketched in B of Fig. 5.1, the simpler case in A. Is there a reliable way to distinguish such a case from the case of Chapter 2, where there is only one site per enzyme molecule?

The Diagnostic Plot

The Lineweaver-Burk plot of $1/v$ against $1/S$ furnishes such a way. If we take the reciprocal of 5.6, and divide numerator and denominator by S^2, so as to express the right side entirely in terms of $1/S$ instead of S, we have:

$$(5.9) \quad 1/v = [(1/S)^2 + (1/K_1)(1/S) + 1/K_1K_2]/[(k_{1a}/K_1)(1/S) + k_{2a}/K_1K_2]E_t$$

This relation may be analyzed by the methods of Chapter 1, and it is found to be linear near the origin and to exhibit a linear asymptotic form. The initial slope is $K_2(k_{2a} - k_{1a})/k_{2a}E_t$; the slope of the asymptote is $K_1/k_{1a}E_t$. The intercept of the asymptote is $(1 - k_{2a}K_1/k_{1a}K_2)/k_{1a}E_t$.

If $k_{2a} < k_{1a}$, the initial slope is negative; the curve is like Fig. 5.2. The minimum corresponds to the maximum of Fig. 5.1B.

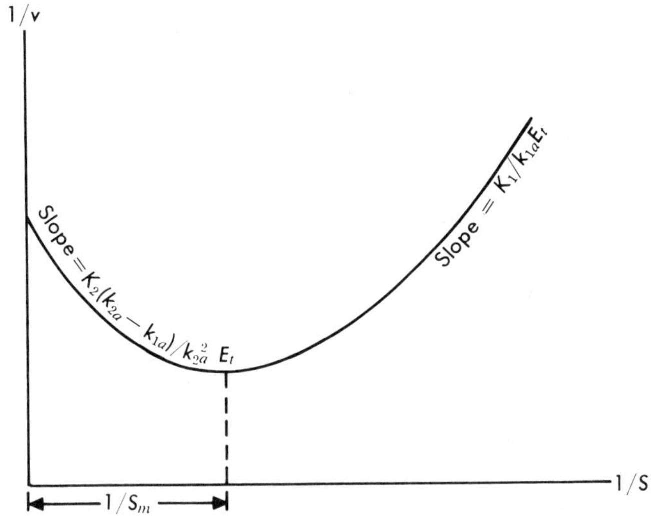

Fig. 5.2. Reciprocal Plot of Curve B, Fig. 5.1.

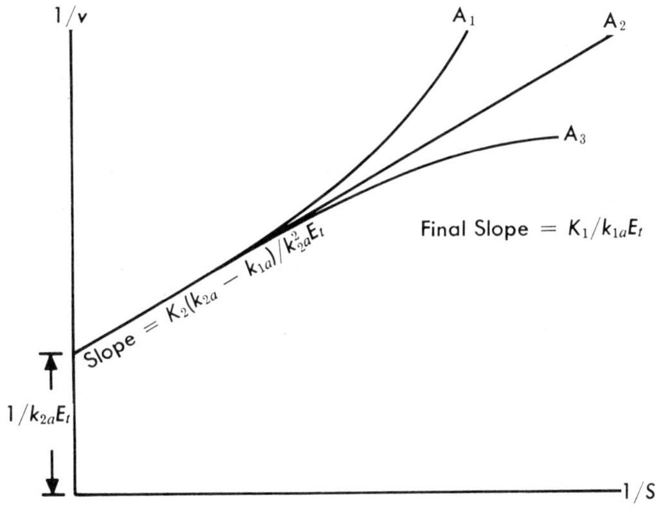

Fig. 5.3. Reciprocal Plots of Curve A, Fig. 5.1. Showing How Reciprocal Plot Can Distinguish Cases Which Seem Identical in Direct Plot.

On the other hand, if the initial slope of the reciprocal plot is positive ($k_{2a} > k_{1a}$), there are three cases, according as the final slope is greater than, equal to, or less than the initial slope. This situation is sketched in curves A_1, A_2, and A_3 of Fig. 5.3, corresponding to curve A of Fig. 5.1. Once again we see that the reciprocal plot is more discriminating than the direct plot of v against S. The form of Fig. 5.1A makes it almost impossible to tell, from experimental data, that there are three distinct cases. These cases show up clearly in Fig. 5.3, because of the ease with which we can detect departures from the simple linear form.

It is readily shown that the equality of initial and final slopes is enough to ensure the linearity of the reciprocal plot (drawn as A_2 in Fig. 5.3). The condition for equality of slopes, which is $K_2(k_{2a} - k_{1a})/k_{2a}^2 E_t = K_1/k_{1a} E_t$, can be rewritten (after cancelling the common factor E_t, solving for K_2, and inverting the fraction), as $1/K_2 = (1/K_1)k_{1a}(k_{2a} - k_{1a})/k_{2a}$. This relation can be used to eliminate $1/K_2$ from 5.9 whereupon the numerator of 5.9 factors readily into $1/S + (1/K_1)(k_{2a} - k_{1a})/k_{2a}$ and $1/S + (1/K_1)k_{1a}/k_{2a}$, while the denominator can obviously be rewritten as $(k_{1a}/K_1)[1/S + (1/K_1)(k_{2a} - k_{1a})/k_{2a}]E_t$. The denominator thus cancels a factor of the numerator, leaving only a single linear factor that contains $1/S$. The resulting relation (writing V for $k_{2a} E_t$) is:

$$1/v = (K_1/V)(1/S) + (1/V) \tag{5.10}$$

This is formally identical with 2.19 (p. 66), and its graph is linear.

In short, then, the reciprocal plot for a two-place enzyme is a straight line only if a special relation holds among the constants of the system (namely, the relation for equality of slopes). Except for one special situation (**Independent Sites**, below), this relation would hold only by a lucky accident. Therefore, even when $k_{2a} - k_{1a} > 0$, the two-place enzyme system should be distinguishable from the simple case of Chapter 2, where the reciprocal plot is always linear.

To distinguish the present case from that of Chapter 4, where the reciprocal plot is also non-linear, note that in Fig. 4.5 the two limiting slopes will become more and more nearly equal as one uses less and less enzyme; for one slope is V/K and the other is $V/(E_t + K)$. In the present case, on the other hand, both slopes will change exactly in proportion to $1/E_t$, and will maintain a constant ratio to each other as the total amount of enzyme is varied.

Independent Sites

Let us now consider the constants more closely—in particular, let us ask what would happen if the two active sites on the enzyme had no influence on each other at all. To do this, we must first inquire how an influence would show up in the rate constants.

To begin with, consider an empty enzyme site, completely isolated from other influences, but exposed to substrate at concentration S. According to the Mass Action Law, the rate at which such a site would react with substrate would be proportional to the concentration of substrate S. Let us call the rate constant for this basic reaction m_1; the rate is m_1S. If the number of molecules with such sites were E, the total rate would be E times as great: m_1SE. But if there are two such sites on each empty enzyme molecule, there are twice as many opportunities for combination, and the total rate will be twice as great again: $2m_1SE$. This is precisely the rate which was written in 5.1 as k_1SE. Since m_1 expresses a property of the site in isolation, we can call it an **intrinsic rate constant.** The **statistical factor** of the actual rate is 2, and the effective rate constant $k_1 = 2m_1$.

Consider now that a substrate molecule combines with one of the two sites, leaving the other one empty. The intrinsic rate of the back reaction (complex to free substrate plus free site) is simply proportional to the number of occupied sites. For one site it is, let us say, m_{-1}. Then for C_1 molecules with one site occupied on each, it is $m_{-1}C_1$. Since we are talking about enzyme molecules with just two places, and we are asking about the back reaction when just **one** of these is occupied, the statistical factor is 1: the total rate is $m_{-1}C_1$—which we wrote in 5.1 as $k_{-1}C_1$, and $k_{-1} = m_{-1}$.

In the same way, the rate of breakup of such sites to give product will be, say, $m_{1a}C_1$—which we wrote in 5.1 as $k_{1a}C_1$.

Now, with one site filled, consider the same three reaction steps at the second site. If the second site is influenced by the first, its intrinsic rate of reaction will be different when the first site is filled. This fact will be expressed by labelling the intrinsic rate constants differently.

Thus, the second site reacts with substrate at the rate m_2S rather than m_1S. If there are C_1 molecules with sites of this kind, they will react at total rate m_2C_1S. There is just one empty site available on each of them. The statistical factor to multiply m_2C_1S is therefore 1, and the overall rate at which once-filled enzyme goes to twice-filled enzyme is then m_2C_1S, which we wrote in 5.1 as k_2C_1S.

In the same way, the rate at which the back reaction occurs on twice-filled enzymes is $2m_{-2}C_2$—because there are C_2 of these molecules, and there are two filled sites capable of reaction on each one (and the constant is m_{-2} instead of m_{-1} because of the influence exerted by the filled sites on each other). This is the rate which we wrote in 5.1 as $k_{-2}C_2$.

Finally, by similar reasoning, the overall rate of formation of product from twice-filled enzyme is $2m_{2a}C_2$—which was termed $k_{2a}C_2$ in 5.1.

We are therefore in a position to write the rate constants in a way which separates the statistical factors (how many groups available to react) from the intrinsic reaction rates of the sites. These intrinsic rates (the m's) again are separated into two classes: those where the rate is influenced only by the site itself, by its surroundings in the enzyme molecule, and by **another empty site** (the constants with subscript 1), and those where the rate is influenced by the site itself, its surroundings, and **another full site** (the constants with subscript 2). We have:

$$k_1 = 2m_1 \qquad k_{-1} = m_{-1} \qquad k_{1a} = m_{1a} \tag{5.11}$$

$$k_2 = m_2 \qquad k_{-2} = 2m_{-2} \qquad k_{2a} = 2m_{2a}$$

We have, up to this point, deliberately used symbols like m_1 and m_2, to emphasize that the presence of filled sites may affect the properties of other sites. Suppose we were now to drop this assumption. We postulate that filling a site on an enzyme molecule has no effect on the properties of another site on the same molecule—either because the complex is so much like the free site, from an electronic point of view, that no significant change of the effective force fields occurs, or because the sites are far enough apart so that the fields due to one site are too weak at the other site to be influential.

In that case there will be no difference between the intrinsic reaction rate of a site when the other site is empty or when the other site is full. That is:

$$m_1 = m_2 \qquad m_{-1} = m_{-2} \qquad m_{1a} = m_{2a} \tag{5.12}$$

What will be the effect of this simplification on the effective rate constants, and on rate formulas like 5.7 and 5.9? Using 5.11 and 5.12, we get:

$$k_1 = 2k_2 \qquad k_{-1} = \tfrac{1}{2}k_{-2} \qquad k_{1a} = \tfrac{1}{2}k_{2a} \tag{5.13}$$

In addition, if we use the definitions 5.4 of K_1 and K_2, and substitute into them from 5.13, it will be seen that $K_1 = (1/4)K_2$.

If these relations are substituted into the condition for the reciprocal plot 5.9 to be a straight line $(K_2(k_{2a} - k_{1a})/k_{2a} = K_1/k_{1a})$, it is seen that this condition is **identically fulfilled** (that is, after cancelling everything that can be cancelled, we get $1 = 1$). That is, an enzyme with independent sites can only have a linear reciprocal plot, and never the other types.

Similarly, eliminating k_{2a} and K_2 from 5.6, we find that the numerator contains a factor $1 + S/(2K_1)$, while the denominator is simply the square of that expression. So, after simplifying, we find that:

$$v = k_{1a} E_t S/(K_1 + \tfrac{1}{2}S) \tag{5.6$'$}$$

This would be the same as 2.11′, except for the factor 1/2. The oddity is easily handled, however. We must remember that the **intrinsic** constants for the sites themselves are the m's. We can easily express everything in terms of them. Let us call the common value of m_1 and m_2 (see 5.12) simply m; similarly we use m_- for the common value of m_{-1} and m_{-2}, and m_a for the common value of m_{1a} and m_{2a}. Defining the symbol K as $(m_- + m_a)/m$ by analogy with the usual definitions (K has the same relation to the individual sites of the two-place enzymes that K in Chapter 2 had to the individual one-place enzymes), we can see by using 5.4 and 5.11 that $K_1 = \frac{1}{2}K$, while $k_{1a} = m_a$. Putting these results into 5.6′ and simplifying, we have:

$$v = 2m_a E_t S/(K+S) = m_a(2E_t)S/(K+S) \tag{5.6″}$$

This equation is in the familiar form of 2.11′. The second way of writing the right side of the equation answers one of the questions with which this chapter began. The quantity $2E_t$ is simply the total number of sites. We may therefore say: If the sites on a two-place enzyme are independent of each other, they have a rate formula which is just the same as if each site were on a separate enzyme molecule. If the sites are independent, in short, we cannot tell 2,000,000 enzymes with one site each from 1,000,000 enzymes with two sites each.

THE GENERAL CASE OF THE *n*-PLACE ENZYME

It seems reasonable to suppose that a good part of the analysis of the two-place enzyme should apply when there are more than two active sites on each enzyme molecule. However, there is no reason to take the point entirely for granted. It is possible with a little additional effort to get a good deal of information with regard to the more general case. In particular it is very easy to show that it behaves, in the special case of independent sites, precisely the same way as the case of the two-place enzyme.

At this point some preliminary remarks on the symbols to be used are needed. First, we shall use "n" to stand for the number of active sites on an enzyme molecule. Next, we shall number the various types of complexes, as we did in the last section: Complexes with one site occupied have concentration C_1, those with two places occupied have concentration C_2, and so forth, up to those complexes which have all n places occupied, with concentration C_n. For the sake of uniformity of notation, we will use C_0 to refer to free enzyme, which we have usually denoted by E.

Each of the stages from 1 to n will be subject to rather similar conditions—each one is formed from a lower stage by combination with substrate, each one can release substrate again, each one can turn over to give product, and so on. We will therefore use the variable-subscript notation introduced in Chapter 1 and define C_j to mean "the concentration of the complex that has j places filled."

Using the variable-subscript notation, it is easy to write the stoichiometric scheme of the system in a single line:

$$S + C_{j-1} \underset{-j}{\overset{j}{\rightleftharpoons}} C_j \qquad C_j \xrightarrow{ja} C_{j-1} + P \qquad (j = 1, 2, \ldots, n) \tag{5.B}$$

For the rate constants, we follow the symbolism of the preceding section, merely extending it by the variable-subscript device. Thus the rate constants corresponding to the three chemical steps in the general form of 5.B above (formation of j-fold complex, back reaction of j-fold complex, turnover of j-fold complex to give product) would be written k_j, k_{-j}, and k_{ja}. In the same way, we will write K_j for the quantity $(k_{-j} + k_{ja})/k_j$.

The splitting of the constants into the "statistical" and 'intrinsic" factors follows the same line of reasoning as for two-place enzymes. When a complex with $j - 1$ places filled reacts with substrate to give j-fold complex, it can react at any one of $n - (j - 1) = n - j + 1$ empty places. When a j-fold complex reacts backwards or turns over to product, it can react at any one of its $\boldsymbol{j}$ **full** places. Expressing these ideas in variable-subscript notation:

$$(5.14) \qquad k_j = (n - j + 1)m_j \qquad k_{-j} = jm_{-j} \qquad k_{ja} = jm_{ja} \qquad (j = 1, 2, \ldots, n)$$

which is analogous to 5.11 for the two-place enzyme.

We substitute these relations into the definition of K_j. Denoting the combination $(m_{-j} + m_{ja})/m_j$ by M_j, we can write:

$$(5.15) \qquad K_j = M_j/(n - j + 1) \qquad (j = 1, 2, \ldots, n)$$

The kinetic equations corresponding to scheme 5.B are (remembering that C_j not only comes from C_{j-1}, but also from C_{j+1}):

$$(5.16) \qquad dC_j/dt = (k_j SC_{j-1} - k_{-j} C_j - k_{ja} C_j) - (k_{j+1} SC_j - k_{-(j+1)} C_{j+1} - k_{(j+1)a} C_{j+1}) \qquad (j = 1, 2, \ldots, n - 1)$$

$$dC_n/dt = k_n SC_{n-1} - k_{-n} C_n - k_{na} C_n$$

The general equation does not apply to C_n, because there is no form with $n + 1$ places to which C_n can go; hence a separate equation is written for this form.

The pseudostationary conditions are of course simply $dC_j/dt = 0$ for $j = 1, 2, \ldots, n$. The conservation equation for enzyme is:

$$(5.17) \qquad E_t = \sum_0^n C_j$$

The easy way to solve the pseudostationary equations is to start with the last equation—the one for C_n—first. If we set $dC_n/dt = 0$ in 5.16, and solve for C_n, we get $C_n = SC_{n-1}/K_n$ (using the definition of K_n as $(k_{-n} + k_{na})/k_n$).

If we now take the first equation of 5.16 in pseudostationary form, and put $j = n - 1$, this gives

$$0 = (k_{n-1}SC_{n-2} - k_{-(n-1)} C_{n-1} - k_{(n-1)a} C_{n-1}) - (k_n SC_{n-1} - k_{-n} C_n - k_{na} C_n).$$

Note that the last three terms are the same as the three terms of dC_n/dt, which we just used, and thus are automatically equal to zero in the steady state. Dropping them, and using the remaining three terms, we solve for C_{n-1} to get $C_{n-1} = SC_{n-2}/K_{n-1}$, using the definition $K_{n-1} = (k_{-(n-1)} + k_{(n-1)a})/k_{n-1}$. Continuing this process (putting $j = n - 2$ in 5.16, etc.), we can see that the solutions fall into a definite and regular pattern, in which each concentration equals the next lower concentration multiplied by S and divided by a constant of the Briggs-Haldane type. Thus generally:

$$C_j = SC_{j-1}/K_j \qquad (j = 1, 2, \ldots, n) \tag{5.18}$$

This is a system of n linear equations in n unknowns, and can be solved by elimination. In this particular case the process of elimination is specially simple. Each unknown (C_j) depends only on unknowns below it in the list (as C_{j-1}). The elimination in such a case is called a **recursion** (Chapter 1). Write the relation for C_{j-1}, namely: $C_{j-1} = SC_{j-2}/K_{j-1}$. Substitute in 5.18, and get $C_j = S^2C_{j-2}/K_{j-1}K_j$. Now write out the expression for C_{j-2}, and substitute back in the relation just obtained. The result is $C_j = S^3C_{j-3}/K_{j-2}K_{j-1}K_j$. Again a pattern is evident: Each substitution adds a factor S and another Briggs-Haldane constant, and puts a lower-stage concentration in. Continuing the process for j steps, we finally have:

$$C_j = S^jC_0/K_1K_2 \ldots K_j \qquad (j = 1, 2, \ldots, n) \tag{5.18$'$}$$

We have thus expressed every C_j in terms of C_0. We now substitute from 5.18′ into 5.17. With the abbreviation p_j for the product $K_1K_2 \ldots K_j$, we have

$$E_t = C_0 + SC_0/p_1 + S^2C_0/p_2 + \ldots + S^nC_0/p_n$$

Solving for C_0 and using the symbol $\sum$ to abbreviate the sum (Chapter 1):

$$\begin{aligned} C_0 &= E_t/[1 + (S/p_1 + S^2/p_2 + \ldots + S^n/p_n)] \\ &= E_t/[1 + \sum_1^n (S^j/p_j)] \end{aligned} \tag{5.19}$$

Putting this value of C_0 back into 5.18′, we have:

$$C_j = (S^jE_t/p_j)/[1 + \sum_1^n (S^j/p_j)] \qquad (j = 1, 2, \ldots, n) \tag{5.18$''$}$$

The turnover step at **each stage** in scheme 5.B gives rise to product. Therefore the reaction rate (v) in the steady state is the sum of all the turnover steps ($v = k_{1a}C_1 + k_{2a}C_2 + \ldots$). (It is left as an exercise for the reader, using scheme 5.B, to write the equations for dS/dt and dP/dt, and to show that, as usual, in the steady state $dS/dt = -dP/dt$.) Thus $v = \sum_1^n k_{ja} C_j$. Using 5.18″ for C_j:

(5.20) $$v = \sum_1^n (k_{ja} E_t S^j / p_j) / [1 + \sum_1^n (S^j / p_j)].$$

This equation has also been derived by Botts and Morales (2).

The general properties of v as a function of S are the same as those of v for a two-place enzyme, given by 5.7. Thus v is zero when S is zero, and approaches $k_{na} E_t$ asymptotically as S increases without bound.

The problem of the existence of a maximum in the v-S curve (or possibly more than one maximum) closely resembles the problem in the two-place case ($n = 2$), despite the apparent great increase in the complexity of the formula (5.20 compared with 5.7).

To approach this by way of the usual $dv/dS = 0$ leads, even after appreciable transformation and simplification, to somewhat complicated expressions. A simpler approach is based on an elementary argument: The function passes through the origin with a positive slope, it has an asymptote $v = k_{na} E_t$, and it is continuous. Hence, if it intersects the asymptote for any finite positive value of S, it must have at least one extremum (and if only one, it will be a maximum).

The condition that v, as given by 5.20, intersects the line $v = k_{na} E_t$, leads (after multiplication to remove the v denominator) to cancellation of the terms in S^n. We collect like powers of S, and divide through by the coefficient of S^{n-1}, getting a polynomial equation in the standard form where the coefficient of the highest power is 1:

(5.21) $$\sum_j^{n-1} (k_{ja} - k_{na}) p_{n-1} S^j / p_j (k_{n-1,\,a} - k_{na}) - k_{na} p_{n-1} / (k_{n-1,\,a} - k_{na}) = 0$$

We now fall back on some rules of the theory of equations (Note to investigative reader: Look up an ancient and compendious college algebra; the theory of equations has become an unfashionable topic of late). One such rule is this: If such an equation has a negative constant term, it has at least one positive root. Thus if $k_{n-1,\,a} - k_{na} > 0$, there is at least one solution of 5.21, hence at least one maximum of 5.20.

If not, the next criterion available is Descartes' Rule of Signs, according to which the number of positive roots is at most equal to the number of changes of sign between adjacent terms of the polynomial. It is obvious from 5.21 that, with the constant term positive, we will arrive at a negative term (hence a change of sign) just as soon as one of the k_{ja} exceeds k_{na}. If $k_{ja} \leqslant k_{na}$ for all j, the curve fails to cross the asymptote, and there is no maximum above the asymptote.

This would not rule out extrema so low as to fall below the asymptote. We cannot eliminate this possibility rigorously; but we can indicate that it is improbable. Differentiating v (denoting the numerator by "N" and the denominator by "D"):

$$dv/dS = [D(dN/dS) - N(dD/dS)]/D^2 \tag{5.22}$$

Setting this expression equal to zero:

$$N/D = v = (dN/dS)/(dD/dS) \tag{5.23}$$

Carrying out the indicated differentiations of numerator and denominator and substituting into 5.23:

$$v = [(\sum jk_{ja} S^{j-1}/p_j)/(\sum jS^{j-1}/p^j)]E_t \tag{5.23'}$$

This is the value v would have if S corresponded to an extremum. If we write the condition for this expression to be less than the asymptotic value $k_{na} E_t$, we can cross-multiply and cancel the nth degree terms as before, collect terms, and get:

$$\sum_{j}^{n-1} j(k_{ja} - 1)S^{j-1}/p_j < 0 \tag{5.24}$$

This condition can be satisfied only if at least one $k_{ja} < 1$, so that at least one term becomes negative—and one negative term probably would not do it. Now we cannot logically exclude the possibility that such constants be less than 1; but to my knowledge no such turnover constants have ever been recorded, and such an enzyme would be so slow as to be totally uninteresting.

The interpretation of our results is similar to that of the two-place case. The critical role of the relation between the various k_{ja} and k_n is intelligible in terms of the balance between the role of added substrate in producing **more** active complexes and its eventual possible role in producing high-order complexes that turn over slowly.

If we now return to the special case where occupation of some places on an enzyme molecule does not affect the properties of the remaining places, the formulas simplify considerably. As before, we express the condition of noninteraction among enzyme sites by saying that the fundamental kinetic constants do not depend on j, the number of occupied sites. Thus:

$$m_j = m \qquad m_{-j} = m_- \qquad m_{ja} = m_a \text{ for all values of } j \tag{5.25}$$

Since we have defined $M_j = (m_{-j} + m_{ja})/m_j$ (p. 102, following equation 5.14), we now get (substituting into M_j from 5.25) $M_j = (m_- + m_a)/m$. This means that M_j also is independent of j in this special case; and we will therefore now write it simply as M. The relation between M_j and K_j (equation 5.15, p. 102) now reads simply $K_j = jM/(n - j + 1)$—that is, K_j depends on j only through the **statistical** factors.

From these considerations, and the definition $p_j = K_1K_2 \ldots K_j$, it is now possible to write:

$$p_j = [M.1/(n - 1 + 1)][M.2/(n - 2 + 1)] \ldots [M.j/(n - j + 1)]$$

We can rearrange this by factoring out all of the M's, which occur j times, giving M^j. We also multiply together the numerators and denominators of the statistical factor fractions:

$$p_j = M^j(1)(2) \ldots (j)/(n)(n - 1) \ldots (n - j + 1) \tag{5.26}$$

The statistical part of 5.26 is a familiar expression (3). The formula for the number of independent combinations that can be made by selecting j things out of n is $(n)!/(j)!(n - j)!$. Such a formula is often called a **combinatorial coefficient**. Denoting it by C_{jn}, we have:

$$\begin{aligned} C_{jn} &= (n)!/(j)!(n - j)! = (n)(n - 1) \ldots (3)(2)(1)/(j)(j - 1) \ldots \\ &\quad \times (3)(2)(1)(n - j)(n - j - 1) \ldots (3)(2)(1) \\ &= (n)(n - 1) \ldots (n - j + 1)/(1)(2) \ldots (j) \end{aligned} \tag{5.27}$$

The notation C_{jn} has been adopted for typographical simplicity. Usages to be found in the mathematical literature include ${}_nC_j$, C^n_j, $C(n, j)$ and $\binom{n}{j}$.

Now, comparing 5.27 with the statistical part of 5.26, we can write:

$$p_j = M^j/C_{jn} \tag{5.26'}$$

We now rewrite the rate expression 5.20, p. 104, for our special case. Since by 5.14 (p. 102) $k_{ja} = jm_{ja}$, 5.25 gives $k_{ja} = jm_a$. Also, by 5.26′, the expression $S^j/p_j = S^j/(M^j/C_{jn}) = S^jC_{jn}/M^j = C_{jn}S^j/M^j = C_{jn}(S/M)^j$. If we use these results in 5.20, we have:

$$v = \sum_1^n jm_a E_t(S/M)^jC_{jn}/[1 + \sum_1^n (S/M)^jC_{jn}] \tag{5.20'}$$

The final step is to simplify 5.20′ if possible. The difficulty with simplifying a quotient of two sums is obvious. If we have $(a + b + c + d)/(e + f + g + h)$, the only way to simplify is to see whether the numerator and denominator have a common factor (other than 1) which can be cancelled. Our best hope for 5.20′, therefore, is to see if each part can be written as a product of factors.

The denominator is easy. The combinatorial coefficients C_{jn} remind us of the well-known formula (3) for the expansion of a binomial:[1] $(1+x)^n = 1 + \sum_1^n x^j C_{jn}$. We therefore recognize that the denominator of 5.20′ is simply $(1+S/M)^n$.

The numerator is not quite as obvious: It is not immediately recognizable as a binomial expansion. If we factor out from each term the expression $m_a E_t(S/M)$ which does not depend on j, the remaining factor is $\sum_1^n j(S/M)^{j-1}C_{jn}$. We recognize this as the derivative, with respect to (S/M), of the denominator of 5.20′. Since we have shown that the denominator is $(1+S/M)^n$, the summation in the numerator is $n(1+S/M)^{n-1}$, and the entire numerator is $m_a E_t(S/M)n(1+S/M)^{n-1}$.

We can now write the complete simplified version of 5.20′:

$$(5.20'') \qquad \begin{aligned} v &= m_a E_t n(S/M)(1+S/M)^{n-1}/(1+S/M)^n \\ &= m_a E_t n(S/M)/(1+S/M) \\ &= m_a(nE_t)S/(M+S) \end{aligned}$$

In writing the last form of this, we have grouped the factor n along with E_t, since nE_t is exactly the total number of sites on E_t molecules with n sites each. We see that, when the sites do not interact, we have a simple expression of the Briggs-Haldane type for the rate, with the total number of active sites in place of the total number of enzyme molecules. As in the case of the two-place enzyme, we see that the general n-place enzyme behaves, in the absence of interaction among the sites, as if each site were on a separate molecule.

REFERENCES

1. Feller, W., "An Introduction to Probability Theory and Its Applications," Wiley New York, 1950.
2. Botts, J., and Morales, M., Trans, Far. Soc. *49*, 696 (1953).
3. Rosenbach, J. B., Whitman, E. A., Meserve, B. E., and Whitman, P. M., "College Algebra" (4th ed.), Ginn and Company, Boston, 1958.

[1] The reader will recall how the combinatorial expression arises. The nth power of the binomial, $(1+x)^n$, is simply $(1+x)(1+x)\ \ldots$ taken n times. If we multiply out n factors, each containing two terms, we will get altogether 2^n terms. Some of these will consist of x to the power j. These will be the terms one would get by taking as a factor the term x out of j of the n factors $(1+x)$, and taking the term 1 out of the remaining $n-j$ factors. There will be exactly C_{jn} such terms x^j, since C_{jn} is just the number of ways in which one can select j objects out of a total of n objects. If we combine all of the terms x^j, therefore, we will have a term $C_{jn}x^j$. The same reasoning holds for every power of x in the expansion.

6

The Analysis of Complex Biological Systems

The analysis of complex biological systems is one of the major tasks of enzyme kinetics. When enzymes and their substrates are at hand in pure form, many questions can be answered directly, without resort to inferences from kinetic data. But many of the problems confronting biological scientists require the study of complex and incompletely analyzed systems, ranging all the way from extracts of the soluble portions of living cells up to the cells themselves (and sometimes beyond cells to tissues). In such instances we may often reach probable if not absolutely certain conclusions on the basis of kinetic data.

Some of the aspects of complex systems will be dealt with later in the book. For example, the results of linking a number of enzymes in chains or networks will be treated separately in Chapter 9. Much material from other chapters will be relevant to the problem; for instance, the manipulation of pH, temperature, and inhibitors is important. In this chapter, we shall formulate and attempt to solve a number of the elementary and everyday problems which attend the study of enzymes in complex systems.

THE OBJECTIVES OF ENZYME ASSAY

Most of the problems arise from asking, in one form or another, the simple question: How much of enzyme X is there in system Y? This is the problem of enzyme "assay." It is a loaded question, and some of the difficulties to which it leads are semantic as well as technical. Indeed, the semantic problems had better be cleared away before the technical ones are even considered.

Consider a fairly typical situation. The proposed research is to study the correlation between limb development in the chick embryo and the alkaline phosphatase of the limb bud. The problem is to decide upon the best method of assay for the enzyme. In connection with each aspect of the problem, several alternatives arise, and a decision must be made among them.

1. The "limb-bud content of phosphatase" might mean one of at least two distinct though related things:

   a. The activity of the enzyme—the number of moles of substrate converted by the enzyme per hour per milligram of tissue mass.
   b. The number of moles of enzyme protein per unit of tissue.

2. If the activity of an enzyme is to be studied, there are at least two obvious sets of conditions which can be employed:

   a. The pH, ionic strength, substrate concentration, etc., reproduce those found in the intact tissue—as nearly as possible.
   b. The conditions are those for the exhibition of maximal enzyme activity.

3. Most enzymes will attack more than one substrate, including artificial compounds as well as those found in nature. The measurement of enzyme activity may be carried out under one of at least three conditions in this respect:

   a. Using one or all of the natural substrates only.
   b. Using the substrate which gives the highest activity under the chosen conditions, even if it is not a natural substrate.
   c. Using a representative set of all available substrates.

As to **aspect 1**, we can **eliminate alternative 1b** (moles of enzyme per unit of tissue), for practical purposes, with some simple considerations. To get 1b, we must know how many activity units a mole of enzyme will have under the experimental conditions chosen for assay; with such knowledge, we could calculate moles of enzyme from measured tissue activity. To get the information, we would have to purify limb-bud alkaline phosphatase, ascertain its molecular weight, and assay it under the chosen conditions (which includes observing how the other contents of a limb-bud extract or homogenate would affect the activity as compared with a simple salt-buffer mixture).

This would be a considerable technical problem, particularly in the fictitious example we have chosen. The technical problem of getting enough limb-bud material for a single assay is by no means a minor one, and the same would hold for many medical research problems in which the amount of tissue is limited to what can be safely obtained with needle biopsy. Assuming the requisite amount of tissue to be made available by heroic efforts, the purification and physical characterization of the enzyme might consume a year or two. Perhaps one could find a sufficiently pure preparation of alkaline phosphatase from more plentiful material (for example, alkaline phosphatase of intestinal mucosa, for which fairly pure starting material is even commercially obtainable). But this would be far from satisfactory, in view of the radical differences which are found between "identical" enzymes from different tissues (for example, aldolase of muscle and aldolase of yeast (1)).

Unless there is something about the proposed application of the information that demands moles of enzyme rather than activity, we are therefore reasonably likely to pick alternative 1a.

Turning to **aspect 2, alternative a** (duplicating intracellular conditions), it is soon obvious that it is easier to make the choice than to carry out that decision. We could prepare the tissue extract or homogenate in a medium which attempts to duplicate the important features of the intact cell interior, provided we knew enough about them. We can think of such factors as pH, ionic strength, number and distribution of specific ions, and so on, about some of which we know very little, and of that little even less that is reliable. Most people think of a mixture of inorganic ions like the alkali metal and alkaline earth cations, ammonium ion, and anions such as chloride, sulphate, and phosphate, and perhaps traces of iron, with pH somewhat above 7, as a balanced and natural medium, assuming the concentrations are adjusted to be "isotonic" (that is, to have the same osmotic pressure as an approximately $0.14M$ solution of NaCl). But some tissues have high concentrations of amino acids, which contribute appreciably to osmotic balance as well as to the specific ion pool. Some tissues, like muscle, are rich in nucleotides. All tissues contain polyelectrolytes such as the nucleic acids, whose effects on ionic balance will be considerable, and whose specific influences on enzyme activity are almost unexplored. As to the intracellular pH, it is largely a matter of speculation: The study of this problem is fraught with technical difficulties, and all of the investigations which have been made are open to criticism. As to whether the internal osmotic pressure of most cells is the same as that which is usually termed "isotonic", not enough tissues have been studied to support a generalization. The idea is certainly untrue for bacterial protoplasts, which lyse almost as readily in an isotonic saline medium as in distilled water.

If we turn to tissue slices in order to avoid the difficulties just cited, the problems of permeability arise to plague us. For many substrates, penetration from outside to inside of the cell occurs slowly, if at all, under conditions which will permit the cells to remain in good condition. In the case of yeasts and some bacteria, it is frequently possible to circumvent this by pH shifts (since often uncharged forms penetrate more readily than charged forms of substrates) or by the use of extremely high external concentrations. When we deal with the cells of most vertebrate tissues, however, the range of variation permitted for pH and concentration is generally quite narrow. Moreover, it is a fact (though infrequently realized) that a tissue slice, by virtue of the way in which it is prepared, is an ill-defined mixture of uninjured and injured cells, with the former in a precarious majority; and it would be difficult and laborious to make sure how much of the measurements, in any given instance, is due to the uninjured cells.

Tissue slices have their place, of course, when one wants to study factors that demand retention of tissue structure, as in the elegant studies of biochemical response to stimulation of brain tissue by McIlwain and his collaborators (2). They may serve also for studies that require intact structure at the cellular level, as in studies of the problems of permeability (3). But for studies of enzymes or assays of enzyme activity they are almost worthless.

To measure enzyme activities under the conditions that prevail in the intact cell interior is a worthwhile objective, but not necessarily related to what one wants to accomplish in the usual enzyme assay. It is best achieved with the use of isotopic tracers in an intact organism; but this technique in turn presents special problems. In particular, the isotope method for measuring rates is useful chiefly when a vast amount of detailed knowledge of metabolic pathways is already available (4); in addition, it requires the use of preparative and analytical techniques of special rigor and refinement. The kinetic aspects of these problems are to be considered in detail in Chapter 9.

Let us agree, then, that we are most likely to be interested in measuring enzyme activity under optimal conditions, however natural or unnatural these may be. (Recall, for example, the absurdly high pH optima of most alkaline phosphatases.) Practically, this will mean that we will go through the usual preliminaries of measuring activity in some kind of tissue preparation over a reasonably wide range of **pH** values and of **substrate concentrations**, and perhaps even over a range of **temperatures**. The **interactions** of these factors, if any, will be detected by studying, for instance, a range of possible **combinations** of **pH** values and **substrate concentrations.** In the same way, requirements for activators or cofactors (see Chapters 7 and 8) will be explored, together with the optimal concentrations of these.

Considering **aspect 3** (choice of substrate or set of substrates), it is clear that the choice among the alternatives depends on what one expects to do with the information. If one simply wants a **convenient comparative assay**—for instance, to compare activity of limb-bud enzyme in four-day embryos with that in five-day embryos, and so forth—**alternative b** is the method of choice. On the other hand, if one is concerned, for instance, with the phosphatase activities of both limb-buds and metanephros, and would like to know whether these are really due to the same enzyme, then alternative c is recommended. **A "substrate profile" is a powerful aid to the identification of an enzyme**: It was with this technique that Tsuboi (5) neatly demonstrated that the high acid phosphatase which appears in human serum in advanced cases of prostatic cancer originates from prostatic tissue, and is distinct from the much lower level of acid phosphatase detectable in normal serum. Alternative a has nothing to recommend it, except perhaps for studies in biochemical evolution: The occurrence—more precisely, the detectability—of certain compounds (for example, phosphate esters) in a tissue depends upon a number of accidental factors, including the current state of refinement of our analytical techniques.

Having thus boiled the problem of assay down to two neat packages—comparative assay with most active substrate under optimal conditions, or identification by substrate profile—we must now qualify and generalize somewhat. As we said in a more limited context in the preceding paragraph: How one proceeds depends on what use is to be made of experimental information. Thus, it is important to set up a study on enzyme activities, of the kind we have just been discussing, along one of two possible lines:

1. Have a clear idea, to begin with, what it will mean if this or that experimental result is found; or

2. Plan the experiments in such a way that one is least likely to miss any interesting or significant phenomena.

For example, the study of limb-bud phosphatase may be, as so many such studies are, a sort of fishing expedition, in which one hopes that some suggestive or startling correlation may be found without having a clear initial idea of what it might be like. Then one would particularly recommend two variations of the simpler assay procedure: a. Test several substrates, including at least one natural substrate. b. Run a fairly complete concentration curve on each substrate, with at least enough values to permit calculating K values. (The additional labor is much less than it sounds. Remember that the preliminary experiments in which optimal conditions are found have already involved concentration curves and the testing of a range of substrates; the laboratory worker will therefore already be somewhat proficient with the techniques, and probably have all the solutions at hand.)

The reasoning is simple: The developmental process, instead of involving changes in the **amount of enzyme activity**, may instead involve changes in the **properties of the enzyme**—changes which would be reflected perhaps in the substrate profile or in the K values. This may result from replacement of one type of enzyme protein by another protein which is still an alkaline phosphatase; or it may result from modification of an existing enzyme protein without actual breakdown and replacement. A simple comparative assay might miss such a development altogether. (The reader will readily think of similar possibilities in other experimental situations.)

We may conclude this section with two remarks:

1. Remember that there is nothing sacred about mass statistics. A technically adequate study of a sample of six may be of great use; a technically unsatisfactory and poorly planned study of 100 cases is almost certain to be a waste of time.

2. In the absence of ideas, an enzyme assay has no magical superiority over a look through a microscope, and 25 enzyme assays will merely be 25 times as pointless.

THE PROBLEMS AND PRINCIPLES OF ENZYME ASSAY

One usually wants, in doing an enzyme assay, to make a comparison among assay values obtained at different times, or with samples differing according to some definite criterion (for example, normal and cancerous tissue, or tissue from fed and starved animals, and so forth). Let us exclude, for the moment, such a problem as the identity or nonidentity of two samples of enzyme which catalyze the same reaction (for example, serum and prostatic acid phosphatases). Does the kinetic theory of the preceding chapters in any way suggest restrictions on the manner in which such data are to be obtained?

It is sometimes stated (6) that an enzyme assay, to be valid, must satisfy at least three conditions: Activity must be proportional to the amount of enzyme source added (that is, to the volume of homogenate or extract added to a given total volume of incubation mixture (Chapter 2)); activity must be constant during the time period of the assay; and assay must be carried out at saturating substrate concentration.

The first requirement reflects an uncritical acceptance of the simple kinetic system of Chapter 2. According to the theory of that chapter, v is always simply proportional to E_t, other factors being unchanged. According to Chapter 4, however, it is possible for the relation between v and E_t to be non-linear. Is Potter's first requirement then grossly incorrect? Not at all. The discussion of Chapter 4, particularly on p. 86, indicates that the non-linear v-E_t relation is likely to be an unusual occurrence, especially with crude preparations. To get a level of enzyme high enough for "enzyme saturation," one will usually have to use a highly purified preparation. The amount of most enzymes which one will actually find in most cells or tissues will be sufficiently low so that the substrate binding factor will be negligible, and the kinetic assumptions of Chapter 2 (perhaps with appropriate modification for other complications) will hold. A conservative statement would be this: If assay of an enzyme in a crude preparation shows that v and E_t are not approximately proportional to one another, one should suspect the presence of an activator or inhibitor of the enzyme in the preparation, or of some similar complication of a type which one might well term an artefact. How an activator or inhibitor would influence the v-E_t relation will be considered in Chapter 8. Certain other factors which might lead to non-linear v-E_t curves will be considered in later sections of this chapter.

The second and third conditions are closely linked. When enzyme is saturated with substrate, we have $v = -dS/dt = V$, which is a constant independent of time t. This means that $S = S_0 - Vt$, so that the plot of S against t is a straight line. Is this condition really essential for a valid enzyme assay?

Suppose one sample contains enzyme at level E_t; the enzyme has the constants k_2 and K. A second sample contains enzyme at level E_t'; the constants are of course the same. We add substrate at concentration S, not necessarily saturating. The first sample has a rate $v = k_2 E_t S/(S + K)$; the second sample has a rate $v' = k_2 E_t' S/(S + K)$. By simple division, $v/v' = E_t/E_t'$. That is, the observed rates are in the same proportion as the enzyme concentrations (as must be true if v is proportional to E_t). Comparing the rates permits us to compare the enzyme concentrations; if the rate in one sample is twice that in the other, the enzyme concentration is twice as high. This is the kind of comparability we need for an assay. Why, then, should a further restriction be placed on the conditions of assay?

The answer is that the restriction is really a practical rather than a theoretical one, although the theory shows the reason for it. First, let us recall again that a constant rate means a straight-line relation between S and time, or between P and time. The rate is simply the slope of the curve. In fact, recalling the discussion of Chapter 2. (p. 67), we note that we do not measure rate directly, but always calculate it as the slope of an S vs. t or P vs. t curve. The practical advantage of the straight-line case is obvious. It is easiest to fit well with a curve, and it is easiest for the graphical (or even numerical) estimation of slope.

There are two other aspects of the situation which are most readily seen from the curve of v against S (Fig. 6.1). First of all, the rate is maximal when the enzyme is saturated with substrate; the rate at point a is about a third of the rate at the point c. Hence, if a point like c is always chosen for assay, the **sensitivity** of the assay is increased to its highest value—the assay will detect reliably a lower level of enzyme if one measures the enzyme at its greatest level of activity. The **accuracy** of the assay is also increased in relation to fixed errors of measurement. For example, suppose the rate at point a in the

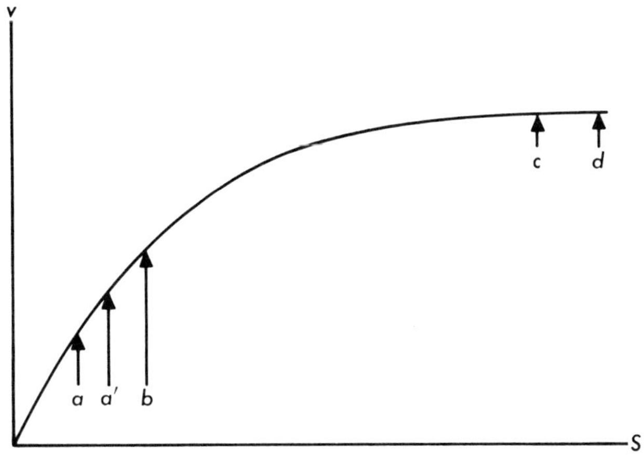

Fig. 6.1. Rate v of Simple Enzyme System, Plotted against Substrate Concentration S. Labelled to illustrate the effect of working at various substrate concentration on the accuracy of an assay.

figure is 30 units, at point c 90 units. Suppose the uncertainty in an instrument (e.g., the uncertainty in the colorimeter reading, as determined from the spread of values on repeated readings of the same standard solution) is equivalent to 3 velocity units. Then the irreducible instrumental error in the case of point a will be 10% of the reading (3 out of 30), while at point c it is about 3% (3 out of 90).

The figure brings out another practical consideration with regard to accuracy. Contributing to the errors in comparative assay will be the usual gravimetric and volumetric errors, and this holds in particular for the preparation and dispensing of the substrate solution. What this means for the results can be seen by considering what happens to the measured rate if the amount of substrate in the incubation mixture is really at the point b instead of a. Compare this with what happens at saturation if the substrate is really at point d instead of c. The error in the rate is extremely small in this latter case. In the former it is quite large, and it is still appreciable even if the error displaces the substrate point only to a'.

We may therefore say this: If practical considerations of experimental error and sensitivity of analytical methods were excluded, one could in principle assay an enzyme at any substrate concentration, provided conditions were always maintained the same for all samples. In practice, the maximal ease of calculation, maximal sensitivity, and maximal accuracy and reproducibility are obtainable when the enzyme is assayed under the condition of substrate saturation.

The foregoing considerations apply more or less without qualification to the problem of a truly routine assay, such as might be desired in a clinical laboratory: All that such an assay demands is a sensitive, accurate, reproducible method. Other considerations enter if the assay is to serve a purpose in a research problem, as was hinted in the preceding section. If the aim is to study the effect of disease, diet, or drug upon the enzymatic activity of a tissue or organism, a broader if more laborious survey is desirable. The observed condition or experimental manipulation may alter the **amount** of enzyme; but it might equally well alter only its turnover rate or K value for one or more substrates, or its sensitivity to some inhibitor, or indeed, without altering the enzyme in any way, it might change the tissue level of some cofactor or coenzyme or endogenous inhibitor. From the research point of view, any such alteration will affect the function of the enzyme in the living cell, and hence will be of interest. In problems of this kind, therefore, it will be advisable to obtain a complete curve of v as a function of S, and to calculate K as well as V. If a cofactor or coenzyme is known to be required, a study of the rate-concentration curve of that factor is also advisable. If the outcome of such experiments is negative, but the worker still cherishes a hunch that this particular enzyme is somehow important in this particular connection, he may pursue the study into the realm of response to inhibitors or response to pH variations.

In short, it must be recognized that the term "assay" is at least partly a misnomer in connection with any research project on an enzymatic activity. What such a project may demand is truly a "study" of the enzyme in any or all of its various aspects, even when purification of the enzyme is not contemplated.

ENDOGENOUS SUBSTRATES

A complication that occurs frequently in the study of crude tissue preparations is the presence of substrate in the tissue before any substrate is added. Such a substrate is commonly referred to as "endogenous," while added substrate is called "exogenous." For example, when a particular kind of phosphatase is being assayed, phosphate esters present in the tissue may be acted on just as well as the phosphate ester which is added for assay purposes. The questions which arise are then, roughly stated: a. How will this affect a comparative assay? For instance, if two samples of tissue have different amounts of endogenous substrate, will this give an incorrect ratio of activities when the two are compared? b. Can we find out whether the endogenous substrate is the same as the exogenous? c. If a correction for the endogenous material is required, how can one ascertain its magnitude experimentally and apply it? d. How will endogenous substrate affect the calculation of a K value for the enzyme, assuming that we want such information?

Let us suppose that we are dealing with a simple enzyme system of the kind described in Chapter 2. If the total amount of one substrate, exogenous and endogenous together, is S_t, the rate is expressed as usual by (sce 2.11′, p. 59):

(6.1) $$v = k_2 E_t S_t/(K + S_t)$$

If the exogenous (added) substrate is denoted by S as before, while the endogenous substrate concentration is denoted by S_e, we have obviously:

(6.2) $$S_t = S + S_e$$

Substituting in 6.1:

(6.3′) $$v = k_2 E_t(S + S_e)/(K + S + S_e)$$

To complete the picture, we must take account of the fact that the endogenous supply is introduced along with the enzyme. If we were to double the volume of homogenate or extract used, we would thereby double the amount of enzyme and the amount S_e of endogenous substrate. If the homogenate volume is H, we may therefore write:

(6.3) $$E_t = aH \qquad S_e = bH$$

where a and b are constants. If we divide the second equation of 6.3 by the first, we get:

(6.4) $$S_e/E_t = b/a = f \qquad \text{or} \qquad S_e = fE_t$$

where f is introduced as an abbreviation for the expression b/a. Putting the relation 6.4 into 6.1′, we have finally:

(6.1″) $$v = k_2 E_t(S + fE_t)/(K + S + fE_t)$$

Let us first consider that S is fixed, and is not large enough to saturate the enzyme; and let us vary E_t. When E_t is small, fE_t is small compared to S, and may be neglected. The rate is approximately $k_2 E_t S/(K + S)$ (the classical Briggs-Haldane form). If E_t is made sufficiently large, the rate asymptotically approaches $k_2 E_t$. Both of these limiting expressions are proportional to E_t; the graph of v against E_t will therefore approximate a straight line at both ends. The initial slope is smaller than the final slope, however, since it contains the factor $S/(K + S)$, which is less than 1.

Thus the plot of v against E_t for moderate values of S will give an indication of the presence of endogenous substrate if the plot has the shape of Fig. 6.2. This non-linear effect of concentration of the enzyme preparation we will refer to as the **first dilution effect** ("first," because we will meet in

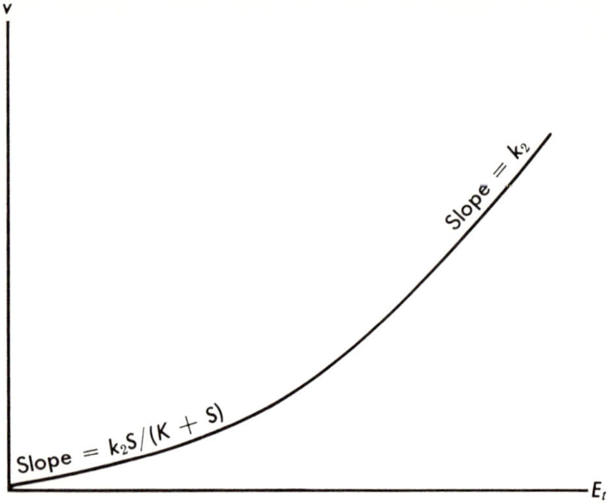

Fig. 6.2. Effect of Endogenous Substrate Present in a Crude Enzyme System on the Relation of Rate v to Enzyme Concentration E_t, Illustrating First Dilution Effect. (Endogeneous substrate same as added substrate.)

this book with other dilution effects due to different mechanisms; "dilution" because going backwards along the E_t axis may be thought of as diluting the enzyme).

Note that, if f is a very small number, the term fE_t will be negligible even for the largest attainable values of E_t (obviously we are limited by the point at which the incubation mixture is practically solid). In that case, the rate will always have the classical form $k_2 E_t S/(K+S)$. But if f is a very small number, this simply means that the amount of endogenous substrate is very small. In short, the first dilution effect will be observable only if the amount of endogenous substrate is significant compared to the exogenous substrate concentration S.

This brings us to the variation of S. The foregoing remarks applied to moderate S values, not large enough to saturate the enzyme present. If S is large enough, the rate approaches $k_2 E_t$—the familiar V.

The answer to the first question on p. 116—How will endogenous substrate affect an assay?—seems to be essentially this: A comparative assay may be vitiated by varying amounts of endogenous substrate if the test is conducted with suboptimal amount of exogenous substrate. The problem can be eliminated simply by making the value of S high enough for saturation. The rate found under this condition is V, and is independent of endogenous substrate. This way, however, we sacrifice a good deal of information.

As to question c—correction for endogenous metabolism—such a correction, as traditionally applied, is barred by the fact that v is not a linear function of S_t. If one lets S become 0 in 6.1″, one gets for the endogenous rate:

$$v_e = k_2 f E_t / (K + f E_t) \qquad (6.5)$$

Subtracting this from the overall rate expression 6.1″, however, obviously does not give the usual kind of rate expression, $k_2 E_t S / (K + S)$, which would hold in the absence of endogenous material. Thus the long-standing custom of measuring the "endogenous rate" (no added substrate) and then subtracting it from the overall rate in the presence of added substrate will not lead to a theoretically meaningful result when applied to a single enzymatic reaction. (This is *a fortiori* the case when a metabolic pathway involving more than one enzyme is concerned. Experimental evidence for this has been obtained by the author and his colleagues in one particular case (7).)

The reciprocal plot of 6.1″ is readily obtained. Taking the reciprocals of both sides, putting V for $k_2 E_t$, carrying out the indicated division by $S + fE_t$, and finally factoring out S from the denominator $S + fE_t$ to get it in the form of $1/S$:

$$(6.6) \qquad 1/v = (1/V)[K/(S + fE_t) + 1] = (1/V)[K/S(1 + fE_t/S) + 1]$$
$$= (1/V)[K(1/S)/(1 + fE_t(1/S)) + 1]$$

The intercept on the $1/v$ axis is $1/V$ as usual. The graph is obviously not a straight line, however. It is linear for small values of $1/S$; the initial slope is K/V as in the classical case. When $1/S$ becomes large, the fraction approaches the asymptote K/fE_t, which is independent of the variabie $1/S$. The shape of the curve is illustrated in Fig. 6.3.

Question d—effect of endogenous substrate on calculation of K value—can be answered as follows in the light of the preceding paragraph: If the variation of substrate concentration S is pushed far enough so that high S values and hence low $1/S$ values are reached, the initial more or less linear portion of Fig. 6.3 can be drawn; we can get V and K from its intercept and slope in the usual way (p. 66, Chapter 2).

The limiting value of $1/v$ for large values of $1/S$ (as indicated on Fig. 6.3) is $(1/V)(K/fE_t + 1)$. This gives an opportunity for checking the estimation of the constants—for example, K. Since $fE_t = S_e$, it can be ascertained by a direct analysis of the homogenate. Assuming that the estimation of V is reasonably trustworthy (since it is obtained by the intercept method with Fig. 6.3), we can regard K as the only remaining unknown. If we denote the experimentally found asymptotic value of $1/v$ by x, we have $x = (1/V)(K/fE_t + 1)$, and we can solve for $K = (Vx - 1)fE_t = (Vx - 1)S_e$. This estimate of K can be used to check the estimate from the initial slope of Fig. 6.3.

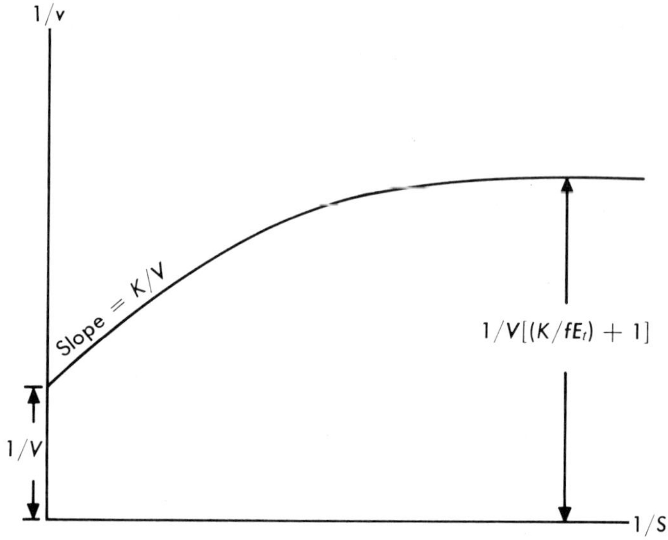

Fig. 6.3. Effect of Endogenous Substrate on the Reciprocal Plot of a Simple Enzyme System. (Endogenous substrate same as added substrate.)

MIXTURES OF TWO SUBSTRATES

Question b, p. 116—Is the endogenous substrate the same as the exogenous? —requires a consideration of the behavior of substrate **mixtures**. We begin with the simple scheme 6.A:

(6.A)
$$E + S_1 \underset{-1}{\overset{1}{\rightleftharpoons}} C_1 \xrightarrow{1a} E + \textit{products}_1$$
$$E + S_2 \underset{-2}{\overset{2}{\rightleftharpoons}} C_2 \xrightarrow{2a} E + \textit{products}_2$$

where the symbolism is self-explanatory. The conservation relation for enzyme is:

(6.7)
$$E_t = E + C_1 + C_2$$

The pseudostationary equations for C_1 and C_2 are of course:

(6.8)
$$k_1 S_1 E - (k_{-1} + k_{1a})C_1 = 0 \qquad k_2 S_2 E - (k_{-2} + k_{2a})C_2 = 0$$

Solving the first of these for C_1 and the second for C_2:

(6.9)
$$C_1 = S_1 E/K_1 \qquad C_2 = S_2 E/K_2$$

where the constants K_1 and K_2 are abbreviations of the usual sort:

(6.10)
$$K_1 = (k_{-1} + k_{1a})/k_1 \qquad K_2 = (k_{-2} + k_{2a})/k_2$$

Substituting from 6.9 into 6.7, and solving for E:

(6.11) $$E = E_t/(1 + S_1/K_1 + S_2/K_2)$$

and hence, putting 6.11 into 6.9:

(6.9′) $$C_1 = (S_1/K_1)E_t/(1 + S_1/K_1 + S_2/K_2)$$

$$C_2 = (S_2/K_2)E_t/(1 + S_1/K_1 + S_2/K_2)$$

Now two situations may arise, depending on whether the analytical procedures with which the rates are measured are specific or not. To make the point concrete, suppose we are studying a phosphatase, and that the two substrates are fructose diphosphate and β-glycerophosphate. If we follow the reaction by analyzing for inorganic phosphate, a product common to the hydrolysis of all phosphate esters, we will be measuring the sum total of the splitting of the two substrates or formation of the two products, $dP_1/dt + dP_2/dt$. The rate equation would be:

(6.12) $$v = k_{1a}C_1 + k_{2a}C_2 = E_t(k_{1a}S_1/K_1 + k_{2a}S_2/K_2)/(1 + S_1/K_1 + S_2/K_2)$$

This expression can now be rearranged by first factoring k_{1a} out from the numerator, and then by multiplying numerator and denominator through by K_1. Thus:

(6.12) $$v = k_{1a}E_t[S_1 + (k_{2a}/k_{1a})(K_1/K_2)S_2]/[K_1 + S_1 + (K_1/K_2)S_2]$$

Suppose that the second substrate (glycerophosphate in our example) is endogenous. Then a relation similar to 6.4 will hold:

(6.13) $$S_2 = S_e = fE_t$$

since, with every portion of enzyme-containing homogenate, a corresponding portion of endogenous substrate is being added.

If we substitute 6.13 into 6.12′, and use the abbreviations $g = k_{1a}/k_{2a}$, $G = K_1/K_2$, and the familiar $V = k_{1a}E_t$, we now have:

(6.12″) $$v = V(S_1 + fE_tG/g)/(K_1 + S_1 + fE_tG)$$

This expression is somewhat similar to 6.1″, and can be analyzed as that equation was analyzed on p. 117 ff. Considering first the variation of E_t (changing amount of homogenate, with added substrate always the same), we see that for very small values of E_t the rate behaves like $VS_1/(K_1 + S_1)$—that is, it has the classical Briggs-Haldane form. For very large values of E_t, the rate approaches $V/g = k_{1a}E_t/(k_{1a}/k_{2a}) = k_{2a}E_t$. We see that the initial and final regions of a graph of v against E_t will be linear, but with different slopes. In this case, however, the final slope is not necessarily greater than the initial slope, even though the factor $S_1/(K_1 + S_1)$ in the initial slope is again less than 1. The final slope may be either less than or greater than 1, depending on whether $k_{1a}S_1(K_1+S_1)$ is less than or greater than k_{2a}. The curve may thus bend either upwards as in Fig. 6.2, or downwards, depending on whether $S_1/(K_1+S_1)$ is less than or greater than $k_{2a}/k_{1a}=1/g$. If the two turnover rates are the same, then of course $g=1$, and the initial and final slopes of the v-E_t curve will be the same as they would be if the two substrates were identical. (Fig. 6.2).

Keeping E_t constant—and varying S_1, we find that the rate is linear near the origin, with intercept $VfE_t(G/g)/(K_1+fE_tG)$, which is simply the endogenous rate V_e. For very large values of S_1 the rate approaches V. It is thus apparent, as in the preceding section, that all complexities—and much information—can be eliminated by working at the saturation level of the exogenous substrate.

It is still not clear that we can infallibly answer question b—that we can tell whether the substrates are the same or different. A Lineweaver-Burk reciprocal plot of $1/v$ against $1/S_1$ is again curved instead of straight, just as in the preceding section. The equation is:

(6.14) $$1/v=(1/V)[(K_1+fE_tG)(1/S_1)+1]/[(fE_tG/g)(1/S_1)+1]$$

The intercept is $1/V$ as before. The initial slope is $(1/V)[K_1+fE_tG(1-1/g)]$. For $1/S_1$ very large, $1/v$ approaches $(1/V)(K_1+fE_tG)/(fE_tG/g)$ instead of $(1/V)(K/fE_t+1)$ as in the case of equation 6.6.

If $g=1$ (that is, $k_{1a}=k_{2a}$), the initial slope reduces to K_1/V as in the preceding section. The final value of $1/v$ reduces to $(1/V)(K_1+fE_tG)/fE_tG$, compared to $(1/V)(K+fE_t)/fE_t$ in the preceding section. The possibility of answering question b reduces to the possibility of distinguishing **experimentally** between f and fG. It is obvious that this cannot be done using the kinetic equations alone without knowledge of f. The only possibility then lies in an analysis of the homogenate, by which we attempt to identify the phosphate ester which decreases during incubation in accordance with the kinetic equations.[1]

In case $g\neq 1$, the problem is much easier, since the initial slopes of the reciprocal plots behave quite differently with respect to E_t, We will denote by y the initial slope of the reciprocal plot, considered as a function of E_t: We have $y=(1/k_{1a}E_t)[K_1+fE_tG(1-1/g)]$ (as shown previously, discussion immediately following 6.14). The plot of y against $1/E_t$ is a straight line with an intercept that is positive or negative according as $g>1$ or $g<1$; if $g=1$ it would pass through the origin.

[1] Specific analysis of fresh homogenate will then give the value of $S_2=fE_t$. We already know experimentally the value V for the rate at S_1 saturation. Since the initial slope of the Lineweaver-Burk plot is K_1/V, we can thus calculate K_1. Knowing this, we can calculate fE_tG from the final level of $1/v$, which, we recall, is $(1/V)$ $(K_1+fE_tG)/fE_tG$. From this, knowing fE_t, we can get G. And from this, knowing K_1, we can calculate K_2. Our information about the kinetic behavior of both substrates is then complete.

Note that if g is sufficiently small, y may actually be negative. Instead of **rising** from $1/V$ to the final value (as for example in Fig. 6.3), the Lineweaver-Burk plot may **fall** from $1/V$ to a final value which under this condition ($g \ll 1$) is **smaller** than $1/V$ (Fig. 6.4). (Recall the formula for the

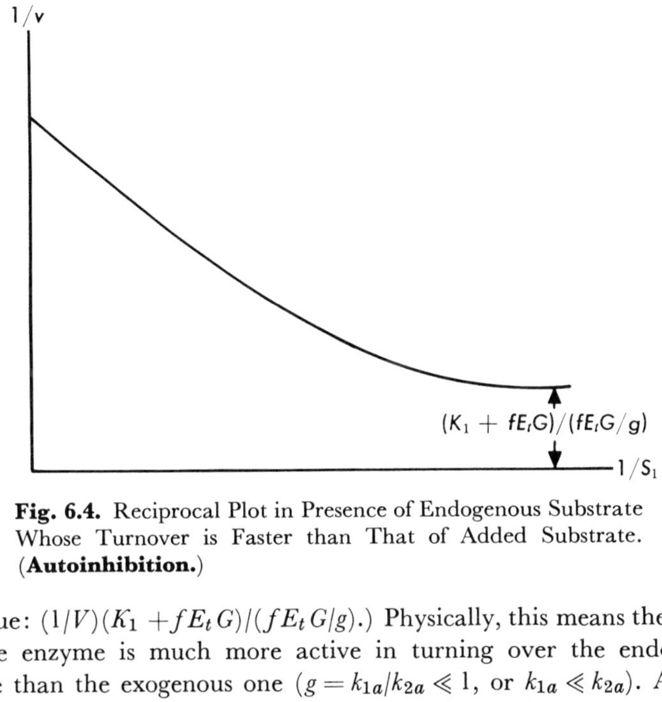

Fig. 6.4. Reciprocal Plot in Presence of Endogenous Substrate Whose Turnover is Faster than That of Added Substrate. (**Autoinhibition.**)

final value: $(1/V)(K_1 + fE_t G)/(fE_t G/g)$.) Physically, this means the following: The enzyme is much more active in turning over the endogenous substrate than the exogenous one ($g = k_{1a}/k_{2a} \ll 1$, or $k_{1a} \ll k_{2a}$). As more exogenous substrate is added to the incubation mixture, it competes with the endogenous substrate for the enzyme, crowds out the more "active" substrate, and thus **reduces** the rate instead of **increasing** it.

The peculiarity of this case will show up also in the direct v-S_1 plot 6.12″. The initial slope dv/dS_1 comes out proportional to $K_1 + fE_t G(1 - 1/g)$, which, as we have just seen, could be negative if $g \ll 1$. Thus the rate (Fig. 6.5) instead of rising as S_1 increases, would actually **sink** from the endogenous value $V(fE_t G/g)/(K_1 + fE_t G)$ to the value V (p. 122).

Under certain conditions, then, particularly in crude enzyme systems, **a substrate may behave as if it were an inhibitor for itself**. (Other cases are treated in various chapters—for example in Chapter 5 and Chapter 11; the mechanisms are different in those cases.) The importance of this case, aside from any intrinsic significance it may prove to possess, is that it emphasizes how kinetic data obtained with a crude system may be extremely misleading and how vital it is to interpret such data cautiously and with the aid

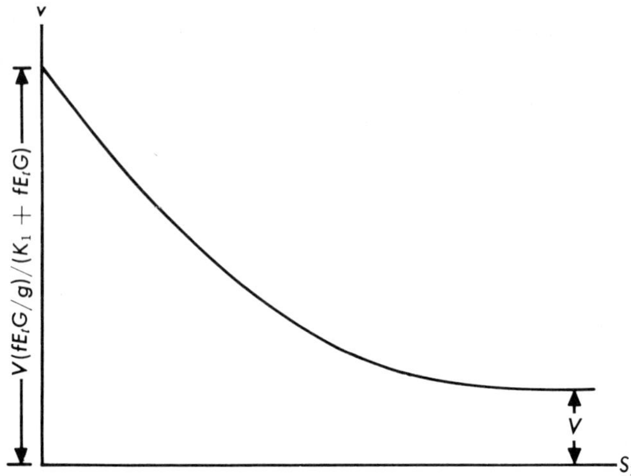

Fig. 6.5. Direct v-S Plot of Same Case as in Fig. 6.4. (**Auto-inhibition.**)

of a painstaking analysis. For obviously, this is not a true case of autoinhibition at all—it is a disguised case of inhibition by a competitive substrate, such as we treat in Chapter 10.

The condition $K_1 + fE_t G(1 - 1/g) < 0$ depends on other constants than g. We can rewrite it as $E_t > K_1/fG(1/g - 1)$, which shows that the occurrence of a freak curve like 6.6 is dependent on E_t—that is, experimentally speaking, on the amount of homogenate used and its richness in the enzyme being assayed. When $E_t < K_1/fG(1/g - 1)$, the v-S_1 curve will be normal. The critical level of E_t is reached at $E_t = K_1/fG(1/g - 1)$ where the slope of the v-S_1 curve is zero for all values of S_1, and we get a simple horizontal line. Then, as E_t is increased, we get the inverted v-S_1 curve, as in Fig. 6.5.

Up to the present, our analysis in this section has been based on the assumption that the analytical procedure used in the experiment is nonspecific—thus, in our example, we supposed that fructose diphosphate and glycerophosphate were substrates 1 and 2, and that their hydrolysis was measured as the release of inorganic phosphate. If a more specific procedure is adopted, the theory will be modified. If we analyze for free fructose, for example, we measure only the rate of hydrolysis of fructose diphosphate. Then 6.12 behaves as if $k_{2a} = 0$, although it is not (since phosphate is still being released from both compounds) and:

(6.15) $$v = k_{1a} C_1 = k_{1a} E_t (S_1/K_1)/(1 + S_1/K_1 + S_2/K_2)$$

Using again the abbreviations G for K_1/K_2 and V for $k_{1a} E_t$, and rearranging, we get, instead of 6.12″:

(6.15′) $$v = VS_1/(K_1 + S_1 + fE_t G)$$

The equation for the reciprocal plot becomes, instead of 6.14:

$$1/v = (1/V))[(K_1 + fE_t G)(1/S_1) + 1] \quad (6.16)$$

This equation differs radically from 6.6 and 6.14. Plotting $1/v$ against $1/S_1$ now gives a **straight line** instead of a curve. This situation resembles the simple case shown in Chapter 2, equation 2.19, p. 66. But there is one important distinction. In 2.19 the coefficient of $1/S$ is simply K/V, where K is a constant that does not depend on E_t. In 6.16, the coefficient of $1/S_1$ is $(K_1 + fE_t G)/V$, a fraction whose numerator does depend on E_t, and in fact is a linear function of E_t. Experimentally, the situation is as follows: One can carry out the Lineweaver-Burk procedure as outlined in Chapter 2, find the rate at different substrate concentrations, and graph the results as the reciprocal plot, from which the values of V and k can be obtained. This can then be repeated with varying amounts of enzyme E_t. In the case of 2.19, the values of V will change (being simply proportional to E_t), but the values of K will remain the same. (Another way to say this is that, if we plot V/v against $1/S$, instead of plotting $1/v$, the various straight lines will be superimposed on each other.) On the other hand, in the case of 6.16, the **"apparent" K** values will **increase as E_t is increased**, since the apparent K value is the expression $K_1 + fE_t G$.

We now also have a straightforward answer to question b—can we tell whether an endogenous substrate is the same as the exogenous substrate? If the variation of rate v with substrate concentration S_1 can be studied with a non-specific and a specific analytical procedure, let the reciprocal plots for the two experimental series be drawn. If the non-specific analysis gives a curve and the specific analysis a straight line, the two substrates are distinct. (A still simpler procedure, when a specific analytical method is available, is to measure the endogenous rate ($S_1 = 0$). If the endogenous substrate is distinct from the exogenous, the specific method will not detect it, and the apparent endogenous rate will be zero. However, such a negative result is not quite as conclusive as the more complex procedure.)

The three cases discussed in the last two sections, based on 6.1″, 6.12″, and 6.15′, can be compared by the method introduced in Chapter 2 and further used in Chapter 4, of plotting S_1 against $r/(1-r)$. It will be remembered that the definition of r is $r = v/V$. Starting with the basic rate equations, we compute r and $1-r$, and finally $r/(1-r)$. We have:

(6.17) For 6.1″: $r = v/V = (S + fE_t)/(K + S + fE_t)$

$$r/(1-r) = (S + fE_t)/K \qquad \text{or} \qquad S = -fE_t + Kr/(1-r)$$

(6.18) For 6.12″: $r = v/V = (S_1 + fE_t G/g)/(K_1 + S_1 + fE_t G)$

$$r/(1-r) = (S_1 + fE_t G/g)/[K_1 + fE_t G(1 - 1/g)] \qquad \text{or}$$

$$S_1 = -fE_t G/g + [K_1 + fE_t G(1 - 1/g)]r/(1-r)$$

(6.19) For 6.15′: $r = v/V = S_1/K_1 + S_1 + fE_t G)$

$$r/(1-r) = S_1/(K_1 + fE_t G) \qquad \text{or} \qquad S_1 = (K_1 + fE_t G)r/(1-r).$$

According to all three of these equations, when S_1 is plotted against $r/(1-r)$, a straight line should result. In this respect they resemble 2.11, and differ from 4.7. However, all of them differ from 2.21 in essential respects.

The curve of 2.21 (simple enzyme system; no endogenous substrate, p. 66) passes through the origin, and has a constant slope K, as in line A of Fig. 6.6, The curve of 6.17 (endogenous substrate same as exogenous) is like curve B of Fig. 6.6—it has constant slope K, but does not pass through the origin,

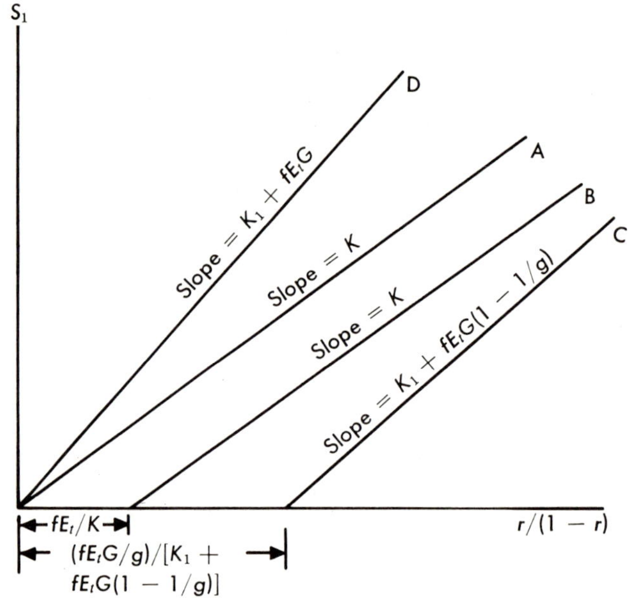

Fig. 6.6. Added Substrate Concentration s_1 vs. $r/(1-r)$. Curve A, no endogenous substrate. Curve B, endogenous substrate identical with added substrate. Curve C, endogenous substrate distinct from added substrate, non-specific analytical method. Curve D, same as case of curve C, but specific analysis for added substrate. If turnover of substrates is equal, case C reduces to B form, but specific analysis still gives D.

and its intercept is a function of E_t. The curve of 6.18 (endogenous substrate possibly different from exogenous; non-specific analysis), line C, has a slope greater than K_1 (which we take to be the same as K in the figure), and both slope and intercept are functions of E_t. The curve of 6.19 (endogenous substrate possibly different from exogenous; specific analysis), curve D, again passes through the origin, but the slope is a function of E_t. If an experimental study takes into account the slope and intercept of the curve of S against $r/(1-r)$, and the way in which these vary with E_t, it will be possible to assign the results to one of these four cases. In the case $g = 1$ (endogenous and exogenous substrates have same turnover rate), the slope of curve C (6.18) reduces to K_1, which is independent of E_t, so that this case would behave just like curve B (6.17). However, the technique of specific analysis will handle this, since the experimental results will fit curve D if the two substrates are distinct.

TWO SIMILAR ENZYMES

We have considered the problems created by the presence of more than one substrate during assay of an enzyme in a crude system. Another problem which may frequently arise in the crude system is the presence of more than one enzyme performing the same general type of function and acting on the same range of substrates. Thus, a tissue homogenate may contain two or three or more phosphatases, proteases, and the like.

The eventual answer to such a problem lies in the techniques of preparative enzymology. If two phosphatases can be isolated from a tissue and shown to be distinct proteins, this is conclusive. But if it proves possible to isolate only one, what then? Is there really only one, or is there perhaps a second one which is extremely unstable as it becomes purified? There are always answers to such questions sooner or later, but they are not always readily obtained.

If the two enzymes are obliging enough to have radically different pH ranges for their activity, this may answer the question. It is certainly true that if there are two enzymes, and the plot of activity against pH for each enzyme has a peak, and if the peaks of the two curves are at quite different pH values, then a mixture of the two enzymes will give a pH curve with two peaks, and a valley between them. But the converse is not necessarily true—that is, that a double-peaked pH curve necessarily means two enzymes (see Chapter 10). (A particularly odd case comes to mind in this connection—some studies by Mommaerts (9) on the ATPase activity of myosin, in which he found that a two-peaked pH curve was obtained with one kind of buffer, but only a single peak appeared with another kind of buffer.)

Not to belabor the point, suffice it to say that the problem of whether an enzymatic activity in a tissue is due to one enzyme or to more than one may not yield rapidly to the obvious fundamental lines of attack. We therefore turn to kinetic methods in search of the answer.

The stoichiometric relations will obviously be the same as for Chapter 2, except that they are repeated for two enzymes with a common substrate.

(6.B)
$$E_1 + S \underset{-1}{\overset{1}{\rightleftharpoons}} C_1 \qquad C_1 \overset{1a}{\rightarrow} E_1 + P$$
$$E_2 + S \underset{-2}{\overset{2}{\rightleftharpoons}} C_2 \qquad C_2 \overset{2a}{\rightarrow} E_2 + P$$

The conservation equations for the two enzymes are:

(6.20)
$$E_{t1} = E_1 + C_1 \qquad E_{t2} = E_2 + C_2$$

We assume that no conservation equation for S is required. In other words, in the range through which S may be varied, the amount of S bound by the enzymes is still quite small compared to the amount of free substrate. To put it still another way, the two enzymes, while using the same substrate, are not "competing" for it.

We can apply the usual steady-state methods to each enzyme. The observed activity will be the sum of the separate activities. We thus get:

(6.21)
$$v = k_{1a} C_1 + k_{2a} C_2 = V_1 S/(K_1 + S) + V_2 S/(K_2 + S)$$

where we have introduced the obvious abbreviations $V_1 = k_{1a} E_{t1}$, $V_2 = k_{2a} E_{t2}$.

Remembering our previous results, we can see that each of the terms of 6.21 will have the familiar hyperbolic form when plotted against S. The sum of the two hyperbolae is not a hyperbola. The question is: How readily can the composite curve be distinguished from a hyperbola? A rough idea of the answer can be obtained if we very crudely approximate each hyperbola by putting together a slanting straight line and a horizontal straight line, joining them by a sharp bend. With this device, the composite is easy to draw. Two cases have been illustrated in Figs. 6.7 and 6.8. In each case we see that the composite of the two curves A and B contains an extra bend. Now imagine that we soften and blur these sets of lines—we give the straight lines some appropriate degree of curvature, and round the sharp bends. It is immediately obvious that the composite in 6.7 might still look not quite like a simple hyperbola; but the composite in 6.8 would surely be indistinguishable from a hyperbola on a simple visual inspection. Since 6.7 was selected to illustrate a rather extreme case, it seems clear that inspection of a v-S graph will seldom if ever suffice to tell the two-enzyme case from the one-enzyme case. This is all the more true since the experimentally obtained curves will be further blurred by experimental error.

Since it is easier to plot a straight line from experimental data than to plot a hyperbola, and easier to detect deviations from straightness, it is natural to try the Lineweaver-Burk reciprocal plot next. Starting with 6.21, one unites

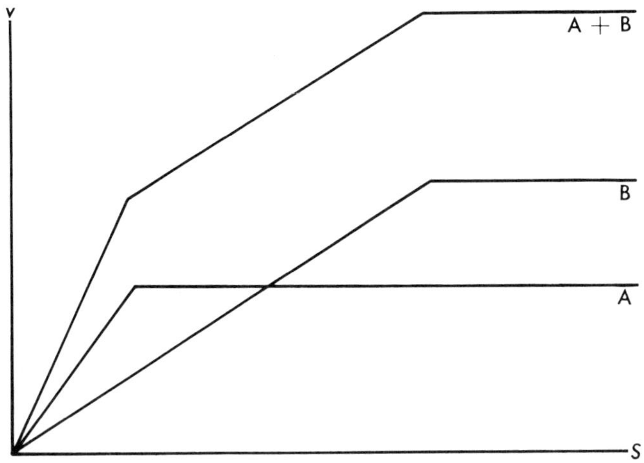

Fig. 6.7. Rate against Substrate Concentration for Two Similar Enzymes. Hyperbola for each rate approximated by broken line to show how total rate appears. Curves A and B, curves for the separate rates of the two enzymes. Curve A + B, total rate (sum.)

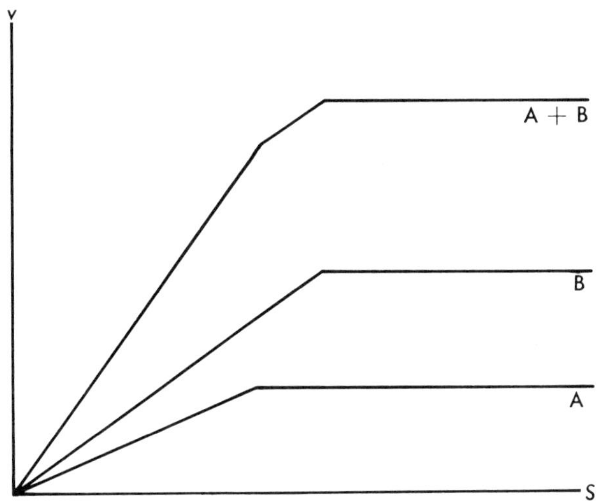

Fig. 6.8. Same case as Fig. 6.7, with Less Difference between Curves A and B, Showing Blending of Curves in sum A + B.

the two fractions by getting a common denominator in the familiar way. Then one inverts the equation. Now, to get the right-hand side of the equation in terms of $1/S$ rather than S, one divides numerator and denominator twice by S. For the sake of simplicity in writing the resulting formula, we will denote $1/S$ by the letter x. Abbreviating $V = V_1 + V_2$ and $p = V_1K_2 + V_2K_1$, we have:

$$1/v = (1 + K_1x)(1 + K_2x)/(px + V) \tag{6.22}$$

This equation may be analyzed in the usual way (binomial approximations). For small values of x, the curve is approximately linear, with a slope $(V_1K_1 + V_2K_2)/V^2$. Letting x become very large, we again approach a linear expression in x; the slope of the asymptote is $K_1K_2/p = K_1K_2/(V_1K_2 + V_2K_1)$. The intercept of the curve on the $1/v$ axis is simply $1/V$, as one might expect, while the intercept of the asymptote is $(V_1K_2^2 + V_2K_1^2)/(V_1K_2 + V_2K_1)^2$.

Theoretically, therefore, a mixture of two enzymes will give a reciprocal plot quite different from that of a single enzyme. The graph will resemble Fig. 6.9. The fact that the initial slope is greater than the final slope can be shown by writing the slope expressions with a common denominator $V^2(V_1K_2 + V_2K_1)$, and comparing numerators, that is $(V_1K_2 + V_2K_1)(V_1K_1 + V_2K_2)$ vs. $(V_1 + V_2)^2K_1K_2$. Ignoring those terms which are the same in both numerators after multiplying out, the remaining terms in the initial slope numerator are $V_1V_2(K_1^2 + K_2^2)$, while in the final slope numerator

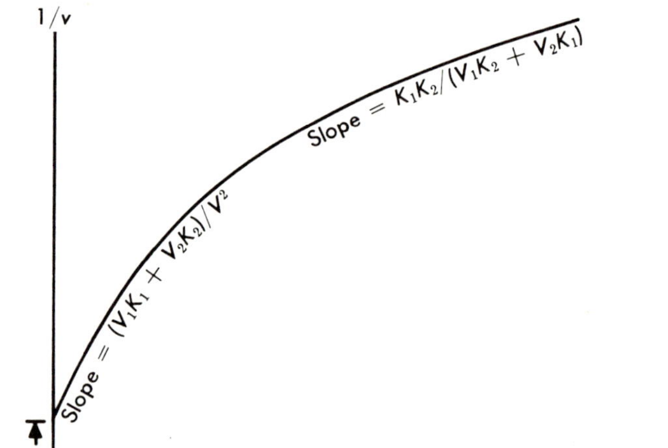

Fig. 6.9. Case of Two Similar Enzymes Acting Simultaneously, Reciprocal Plot.

we have left $V_1V_2(2K_1K_2)$. The first of these expressions is always greater than the second: For $K_1^2 + K_2^2 - 2K_1K_2 = (K_1 - K_2)^2 > 0$ (a squared number is always positive)—hence $K_1^2 + K_2^2 > 2K_1K_2$.

An ambiguous case arises if $K_1 = K_2$. Calling the common value K, we find at once that p reduces to $(V_1 + V_2)K = VK$. Cancelling common factors of numerator and denominator, the right side of 6.22 reduces to $(1/V)(Kx + 1) = (1/V)[K(1/S) + 1]$, which is the expression for a single enzyme already derived in Chapter 2. This is equally obvious from 6.21; for if the K values are the same in both terms, the right side adds up to $VS/(K + S)$.

This being the case, it will be instructive to ask: How much must K_1 and K_2 differ in order that we can distinguish this case from that of a single enzyme? Suppose we write $K_2 = K_1(1 + d)$, so that d is a percentage by which K_2 exceeds K_1, expressed as a decimal fraction. Using this expression for K_2, we can factor K_1 from the numerator of each slope, and factor V from the denominator of each. Thus K_1/V is a common factor of both slopes. The other factor of the initial slope is $1 + V_2\,d/V$, while the other factor of the final slope is $(1 + d)/(1 + V_1\,d/V)$. It is these two factors which must be compared. To illustrate how these functions behave, we have arranged a few calculations in tabular form.

$d =$	0.15	1	4	−0.6
Assume $V_2 = 0.2V_1$				
Initial slope factor		1.167	1.67	
Final slope factor		1.09	1.13	
Assume $V_2 = V_1$				
Initial slope factor	1.075	1.5	3.0	0.7
Final slope factor	1.07	1.33	1.67	0.57
Assume $V_2 = 5V_1$				
Initial slope factor	1.125	1.83	4.33	0.5
Final slope factor	1.12	1.71	3.0	0.44
Assume $V_2 = 10V_1$				
Initial slope factor	1.14	1.91	4.64	0.45
Final slope factor	1.14	1.84	3.7	0.42

Note that the most discriminating results (maximum difference between initial and final slope) are obtained for $V_2 = V_1$. In commonsense terms, if we mix a very weak enzyme with a strong one, or very little of one with a lot of the other, the minor component will escape detection. This conclusion can hardly be very disturbing. We do not need to concern ourselves with a truly minor component as a rule.

In evaluating calculations like those just shown, we need a criterion: How much must two graphically estimated slopes differ for us to tell them apart reliably? Let us recall what we mean by a slope (of a straight line). We pick a point on the line, whose coordinates are, let us say, x_1 and y_1 (x and y being the ordinate and abscissa of our graph). Another point has coordinates x_2 and y_2. The slope of the line then is $(y_2 - y_1)/(x_2 - x_1)$. Suppose the average percent error in the experimental measurement of either x or y is about 2%. Then the error of a difference like $x_2 - x_1$ is about 4% (errors add in forming sums or differences). The percent error of the fraction is then roughly 8% (percent errors add in multiplication or division). That is the percent error of an individual slope. The error in the difference of two slopes is 16%. Conservatively, then, two slopes should differ by about 50% to be reliably distinguished (somewhat less conservatively, by about 35%). If the measurements of x or y or both are better than 2% accurate, we will do better than just estimated; if they are worse, we will come out worse.

OTHER PROBLEMS

Other problems involved in the study of crude enzyme preparations, such as the occurrence of endogenous inhibitors and activators, will be dealt with in the appropriate chapters.

REFERENCES

1. Warburg, O., and Christian, W., Biochem. Z. *314*, 149 (1943).
2. McIlwain, H., Anguainao, G., and Cheshire, J. D., Biochem, J. *50*, 12 (1952)
3. Kratzing, C. C., Biochem. J. *54*, 312 (1953).
4. Reiner, J. M., Arch. Biochem. Biophys. *46*, 53 (1953).
5. Tsuboi, K. K., and Hudson, P. B., Arch. Biochem. Biophys. *61*, 197 (1956).
6. Potter, V. R., in Advances in Enzymology (Nord, F. F., and Werkman, C. H., eds.), vol. IV, Interscience Publishers, New York, 1944.
7. Reiner, J. M., Gest, H., and Kamen, M. D., Arch. Biochem. *20*. 175 (1949).
8. Hanes, C. S., and Isherwood, F. A., Nature *164*, 1107 (1949).
9. Mommaerts, W. F. H. M., and Seraidarian, K., J. Gen. Physiol. *30*, 401 (1947).

7

Bimolecular Reactions

Most enzyme kinetic theory works on the simple assumption that enzyme-catalyzed reactions are unimolecular. This is justified as a first approximation, and its relative mathematical simplicity is helpful. Nevertheless, we must eventually come to grips with the fact that most of the reactions are at least bimolecular. As a typical case, one thinks of hexokinase. The reaction catalyzed by this enzyme is: glucose + ATP → glucose-6-phosphate + ADP. To disregard either of the reactants would make no sense. Even in the case of hydrolytic reactions like those of the esterases and phosphatases, the formally unimolecular character is due to the adventitious circumstance that such reactions generally occur in systems where water, the second reactant, is the solvent, and where its molar concentration is consequently very high and can for all practical purposes be deemed constant.

In this chapter we shall take up the elements of the theory of bimolecular enzyme-catalyzed reactions. Several possible mechanisms will have to be considered.

TERNARY COLLISIONS

The simplest way to conceive a formally bimolecular enzymatic reaction is to suppose that both substrates and the enzyme unite in a single reaction step to form an intermediate complex of the usual type, which then breaks up to release all products in a single step. The stoichiometric scheme is:

$$E + S_1 + S_2 \underset{-1}{\overset{1}{\rightleftharpoons}} C \qquad C \xrightarrow{2} E + \textit{products} \tag{7.A}$$

The conservation equation is simple:

(7.1) $$E + C = E_t$$

Applying the pseudostationary condition to C, we get:

(7.2) $$v = VS_1S_2/(K + S_1S_2)$$

with the customary abbreviations $V = k_2E_t$ and $K = (k_{-1} + k_2)/k_1$.

Scheme 8.A is *a priori* rather improbable, because it demands a ternary collision (usually in somewhat dilute solutions). The likelihood of such a collision, and moreover of one which brings both substrates to precisely the right adjacent active spots on the enzyme molecule at once, is extremely low, and is almost enough to eliminate this model in advance.

However, the model has a simple consequence which suffices to eliminate it. If we let either S_1 or S_2 become very large, the term S_1S_2 in the denominator of 7.2 becomes much larger than K, and v approaches the value $VS_1S_2/S_1S_2 = V$. This means that with large S_1, for instance, and a small value of S_2, the very hugeness of S_1 can somehow compel S_2 to cooperate in saturating the enzyme.

Logically this odd conclusion is not really ridiculous, even though it seems so to our intuition. The probability of a triple collision at the given point of space is (roughly) the probability of finding an enzyme molecule at that point, times the probability of finding a molecule of the first substrate at that point, times the probability of finding a molecule of the second substrate at that point. Each of these probabilities is proportional to the concentration of the corresponding type of molecule. (This is, after all, what we mean when we apply the mass action law to reaction 1 and write its rate as $k_1ES_1S_2$.) Therefore, if S_1 is large enough, it **might** raise the probability of the triple collision high enough for all the enzyme molecules to be kept busy all the time (provided the amount of S_2 is not too small to occupy all the enzyme at least once).

The facts, fortunately, rescue us from this dilemma. A fair number of kinetic studies on bimolecular systems is available, and the findings are in flat contradiction to the conclusions of our model. (See further below.) We can therefore pass on to the next more complicated case.

BIMOLECULAR COLLISIONS—ONE COMPLEX

Suppose the first substrate forms a complex with the enzyme, and subsequently the second substrate collides with this complex, resulting immediately in a reaction which leads to free enzyme and products (**hit-and-run** mechanism). The scheme is:

(7.B) $$E + S_1 \underset{-1}{\overset{1}{\rightleftharpoons}} C \qquad C + S_2 \xrightarrow{2} E + \textit{products}$$

The conservation equation is still 7.1. The differential equation for C is:

$$dC/dt = k_1 E S_1 - k_{-1} C - k_2 S_2 C \tag{7.3}$$

If we set $dC/dt = 0$ as usual, and solve for C, we would get:

$$C = E S_1/(K_1 + g S_2) \tag{7.4}$$

We abbreviate with $K_1 = k_{-1}/k_1$, $g = k_2/k_1$.

If we proceed as usual to eliminate E from 7.4 by means of 7.1, we finally arrive, by the familiar route, at:

$$v = k_2 S_2 C = k_2 E_t S_1 S_2/(S_1 + K_1 + g S_2) \tag{7.5}$$

Once again, let us consider the results of saturation with either substrate. If S_1 becomes very large, the limiting value of v is:

$$V_1 = k_2 E_t S_2 \tag{7.6}$$

If S_2 is allowed to become very large, the saturation value will be:

$$V_2 = k_1 E_t S_1 \tag{7.7}$$

The prediction that the rate will be **always first order** with respect to the limiting substrate is not in general agreement with available experimental data. This model is not as grossly implausible as that involving triple collisions; but it still will not suffice to deal with most experimental situations.

BIMOLECULAR COLLISIONS—TWO DOUBLE COMPLEXES

We get quite different results if we let S_2 hit and run as before, but require an extra step for the release of P_1:

$$E + S_1 \underset{-1}{\overset{1}{\rightleftharpoons}} C_1 \qquad C_1 + S_2 \underset{-2}{\overset{2}{\rightleftharpoons}} C_2 + P_2 \qquad C_2 \overset{3}{\rightarrow} E + P_1 \tag{7.C}$$

Modifying 7.1 to:

$$E_t = E + C_1 + C_2 \tag{7.8}$$

and going through the usual routine, we get the pseudostationary solution:

$$v = k_3 E_t S_1 S_2/(K_p K_m + g S_2 + K_p S_1 + S_1 S_2) \tag{7.9}$$

with

$$K_p = (k_{-2} P_2 + k_3)/k_2 \qquad K_m = k_{-1}/k_1 \qquad g = k_3/k_1 \tag{7.10}$$

The saturation values are:

$$V_1 = k_3 E_t S_2/(K_p + S_2) \tag{7.11}$$

$$V_2 = k_3 E_t S_1/(g + S_1) \tag{7.12}$$

This result is in agreement with experimental data, for example reference 1, in that the rate at saturation with either substrate depends in the conventional way (hyperbola) on the other substrate. Since K_p depends on P_2 (provided $k_{-2} \neq 0$), we may have inhibition by product, as in Chapter 3. Note that the effect would show up only in V_1 but not in V_2. If we had written reaction 2 as a sort of exchange reaction, $C_1 + S_2 \rightleftharpoons C_2 + P_1$, where S_2 replaces S_1 on the enzyme and is released as P_2 in step 3, the equations would be formally identical except for the replacement of P_2 by P_1 in K_p. In either V_1 or V_2, permitting the remaining substrate to grow large results in the same asymptotic value $V = k_3 E_t$.

One is tempted to wonder whether the change in behavior is due to the extra complex in this case. That this is so will become obvious in the next section, where two complexes are formed without the hit-and-run feature.

BIMOLECULAR COLLISIONS—A DOUBLE COMPLEX AND A TRIPLE COMPLEX

We can modify 7.B to eliminate the hit-and-run feature. Instead, we assume the formation of a triple complex of enzyme and both substrates. It is the triple complex which then breaks up to give free enzyme and products:

$$(7.D) \qquad E + S_1 \underset{-1}{\overset{1}{\rightleftharpoons}} C_1 \qquad C_1 + S_2 \underset{-2}{\overset{2}{\rightleftharpoons}} C_2 \qquad C_2 \xrightarrow{3} E + \textit{products}$$

The conservation equation for enzyme now reads:

$$(7.13) \qquad E_t = E + C_1 + C_2$$

We apply the pseudostationary condition first to getting $C_2 = S_2 C_1/K_2$, with the abbreviation $K_2 = (k_{-2} + k_3)/k_2$. Notice that the equation for C_2 involves only C_1 and C_2, while that for C_1 involves C_1, C_2, and E; this may in fact be seen by looking at the scheme 7.C. Taking the C_2 equation first is therefore good tactics from the point of view of carrying out the elimination with the minimum of algebraic manipulation; working the other way round will get the same answer, but with more labor. We now eliminate C_2 from the equation for C_1, and solve for C_1 ($C_1 = ES_1/(k_{-1}/k_1 + k_3 S_2/k_1 K_2)$). The solution is completed by putting C_1 and C_2 (as functions of E) into 7.13 and solving for E. Then we get for the rate expression, $v = dP/dt$ (abbreviating $V = k_3 E_t$):

$$(7.14) \qquad v = k_3 C_2 = VS_1S_2/(K_1K_2 + K_2S_1 + gS_2 + S_1S_2)$$

with the abbreviations $K_1 = k_{-1}/k_1$, $g = k_3/k_1$, $K_2 = (k_{-2} + k_3)/k_2$.

The partial saturation values of the rate can now be calculated as in the preceding section. They are:

(7.15) $$V_1 = VS_2/(K_2 + S_2)$$

(7.16) $$V_2 = VS_1/(S_1 + g)$$

These expressions are formally identical with 7.11 and 7.12, but $k_{-2}P$ of K_p is replaced by k_{-2} in K_2. The explanation is simple. In a one-complex system, saturating with either substrate converts all the enzyme into the rate-determining complex. In a two-complex system, saturation with S_1 maximizes C_1 but not C_2, while saturation with S_2 maximizes C_2 but not C_1.

The rate behaves in a similar way for each substrate even if the other one is not at saturation. To illustrate, suppose we experiment at any **fixed** value of S_2, and measure the rate at various values of S_1. We can rewrite 7.14 as follows:

(7.14′) $$v = (VS_2)S_1/[K_1K_2 + gS_2 + (K_2 + S_2)S_1]$$
$$= [VS_2/(K_2 + S_2)]S_1/\{[K_1K_2 + gS_2)/(K_2 + S_2)] + S_1\}$$

Now the rate expression has the familiar form $V'S_1/(K' + S_1)$, the Briggs-Haldane form. But the apparent maximum rate V' is actually $VS_2/(K_2 + S_2)$, which is the same as V_2 (see 7.15) and is a function of S_2 rather than a constant. Similarly, the apparent K value, K', is actually $(K_1K_2 + gS_2)/(K_2 + S_2)$, and is also a function of S_2 rather than a true constant.

This result has some bearing on the problems of studying enzymes in crude tissue preparations (Chapter 6). It often happens that one finds a crude preparation able to metabolize some added substrate, without necessarily being fully aware of all the properties of the enzyme involved. Suppose that in fact the reaction is bimolecular, and the second substrate is being furnished by the tissue (endogenous substrate). This would mean, as in Chapter 7, that $S_2 = fE_t$. Unlike most of the endogenous cases of Chapter 6, the present situation will give normal-looking v-S_1 curves or reciprocal plots.

But in the present case the apparent constants K' and V' will vary in an unusual way with E_t. Remembering that $V = k_3E_t$, one sees that V' is actually $k_3fE_t^2/(K_2 + fE_t)$, the behavior of which is illustrated in Fig. 7.1. In the case of K', recalling that $g = k_3/k_1$ and $K_1 = k_{-1}/k_1$ (and substituting), we find that the apparent constant K' will become $(k_{-1}K_2 + k_3fE_t)/k_1(K_2 + fE_t)$, which is plotted in Fig. 7.2 for the three possible cases $k_3 > k_{-1}$ (curve A), $k_3 < k_{-1}$ (B), and $k_3 = k_{-1}$ (C). Only in the third case is K' actually independent of E_t.

To see that this is how the figures should go, one need only calculate the initial and final values of K'. When $E_t = 0$, $K' = k_{-1}/k_1$. When E_t becomes very large, $K' = k_3/k_1$. Obviously, if $k_3 > k_{-1}$, the curve will be going from a lower to a higher value, while if $k_3 < k_{-1}$ the opposite will be true. When $k_3 = k_{-1}$, we can write $K' = (k_3K_2 + k_3fE_t)/k_1(K_2 + fE_t) = k_3(K_2 + fE_t) \times k_1(K_2 + fE_t) = k_3/k_1$, which is a constant independent of E_t. The same point

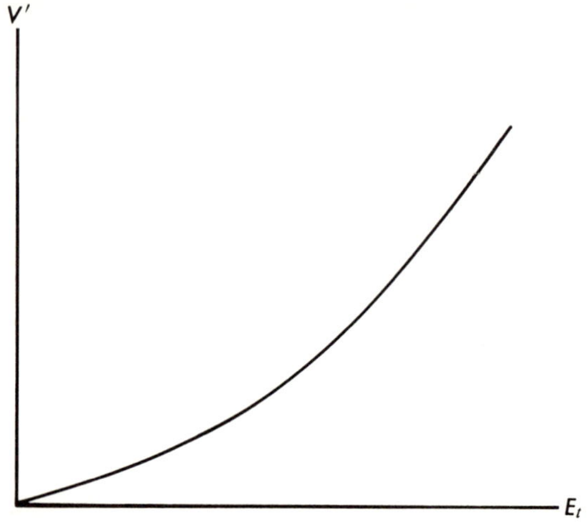

Fig. 7.1. Maximum Rate "Constant" V' as Function of E_t for Bimolecular Reaction, when Rate v is Plotted against One Substrate Concentration S_1 (equation 7.14′, p. 137), and the Other Substrate Is Endogenous (see reciprocal formula, above).

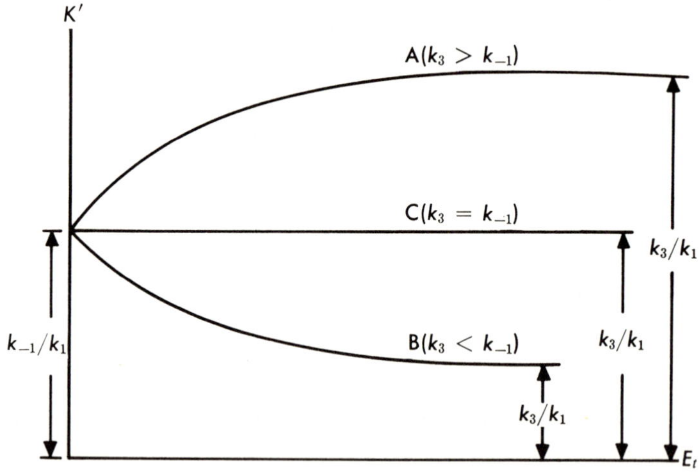

Fig. 7.2. Apparent Haldane "Constant" K' as Function of E_t, for Same Case as Fig. 7.1. Curves A to C are three possible modes of variation which can be found, depending on kinetic constants.

can be demonstrated by calculating the initial slope of the curve, using the binomial expansion of the denominator as shown previously (Chapter 1), and getting for the slope the value $f(k_3 - k_{-1})/k_1K_2$, which will be positive, negative, or zero according as $k_3 > k_{-1}$, $k_3 < k_{-1}$, or $k_3 = k_{-1}$.

We thus have a way of testing, at least in a tentative way, for the possibility that a second substrate may be involved in a reaction which has not been fully identified and explored. We perform the experiment needed to get a reciprocal plot as usual—that is, measure the rate at a series of different S values, with E_t fixed. We then repeat the experiment at other E_t values. From each reciprocal plot we estimate K' and V' graphically as shown in Chapter 2. We plot the set of K' values against the corresponding E_t values, and do likewise for the V' values. Only if V' is directly proportional to E_t and K' is independent of E_t over as wide an E_t range as possible are we at all safe in concluding that we are dealing with a single substrate following Briggs-Haldane kinetics. (But see, for example, Chapter 11 for cases in which, if we do not push the values of S far enough, we will seem to have the simple case, even though in fact both K' and V' are complex **combinations** of more basic constants, which however happen to depend on E_t in a simple way.)

Reactions with cryptic second substrates are by no means likely to be scarce in crude systems. As examples, we may recall the synthesis of glutamine from glutamic acid and ammonia with the aid of ATP, releasing inorganic phosphate, or the activation of amino acids with ATP, releasing primarily pyrophosphate, which is then split to inorganic phosphate by the pyrophosphatase of crude preparations. Either of these reactions, if ATP were the sole **exogenous** substrate in a crude preparation, would at first give the impression of being a simple hydrolytic ATPase reaction. But the test just indicated would show that this is not the case. (It is sometimes feasible to shortcut the problems of endogenous materials by dialysis which will remove all of them. But this is not always practical; destruction of labile enzymes may result, and truly exhaustive removal of endogenous materials within a reasonable time may simply not be possible with certain crude preparations.) This test will then lead to further investigation, which will eventually reveal the true nature of the reaction.

THE RANDOM BIMOLECULAR REACTION

It is conceivable (though not really likely) that the order in which S_1 and S_2 react with the enzyme is not fixed, and may occur at random. That is, not only may S_1 react first and then S_2 react with the resulting complex, but also S_2 may react first and S_1 then combine with this (different) intermediate complex. Scheme 7.C is replaced by:

(7.E)
$$E + S_1 \underset{-1}{\overset{1}{\rightleftharpoons}} C_1 \qquad E + S_2 \underset{-2}{\overset{3}{\rightleftharpoons}} C_2$$
$$C_1 + S_2 \underset{-2}{\overset{2}{\rightleftharpoons}} C_3 \qquad C_2 + S_1 \underset{-4}{\overset{4}{\rightleftharpoons}} C_3$$
$$C_3 \underset{-5}{\overset{5}{\rightleftharpoons}} C_4 \qquad C_4 \overset{6}{\longrightarrow} C_5 + P_1 \qquad C_5 \overset{7}{\longrightarrow} E + P_2$$

The conservation equation naturally becomes:

(7.17)
$$E_t = E + C_1 + C_2 + C_3 + C_4 + C_5$$

Applying the pseudostationary condition to the complexes gives five algebraic equations which, together with 7.17, are sufficient to determine the system. The rate is simply $k_5 C_3$, and is given by:

(7.18)
$$v = k_7 E_t(aS_2F + bS_1G)/K_5K_6[1 + (1 + qaS_2)F + (1 + qbS_1)G]$$

with the abbreviations:

(7.19)
$$K_5 = (k_{-5} + k_6)/k_5 \qquad K_6 = k_7/k_6 \qquad q = 1 + (1 + 1/K_6)/K_5$$
$$a = k_2/(k_{-2} + k_{-4} + k_6/K_5) \qquad b = k_4/(k_{-2} + k_{-4} + k_6/K_5)$$
$$F = [S_1(K_3 + m_2 h_2 S_1) + bk_{-2} S_1 S_2/k_1]/D$$
$$G = [S_2(K_1 + m_1 h_1 S_2) + ak_{-4} S_1 S_2/k_3]/D$$
$$D = (K_1 + h_1 S_2)(K_3 + h_2 S_1) - (k_{-2}\, aS_2/k_1 + k_{-4}\, bS_1/k_3)$$
$$m_1 = (k_{-4} + k_6/K_5)a/k_2 \qquad m_2 = (k_{-2} + k_6/K_5)b/k_4$$
$$h_1 = k_2/k_1 \qquad h_2 = k_4/k_3 \qquad K_1 = k_{-1}/k_1 \qquad K_3 = k_{-3}/k_3$$

This museum specimen is indeed a mess, but not as much of a mess as it seems on first glance. Note that the substrate concentrations occur, other than explicitly, only in F and G. A quick check will show that both numerator and denominator of v are of the third degree jointly in S_1 and S_2; they differ in that the denominator contains all the possibilities, including constant term, first powers, and squares, while the numerator contains only the second- and third-degree cross products. If either substrate is varied, v should begin at the origin and approach a finite asymptote.

PROBLEMS: 1. Analyze the reciprocal graph of 7.18 for either substrate. (Hint: Regroup the denominator to separate $1 + F + G$ first.) Is this case thereby distinguishable from the ordered bimolecular reaction of scheme 7.D? Can the reciprocal plot have an extremum? If so, under what conditions? (Further hint: Use abbreviations for coefficients of various powers of the variable during most of the analysis.)

2. Would your analysis change appreciably if scheme 7.E were modified to give a more chemically plausible case—that the different orders of addition of the substrates give rise to two distinct complexes, which independently yield the products?

3. Would your comparison with 7.D change appreciably if it were modified to unload one product at a time?

REVERSIBLE BIMOLECULAR REACTIONS

Restricting ourselves to the ordered bimolecular case hereafter, we may glance at the problem of reversibility. We shall assume a stepwise complex formation in either the forward or the back reaction. Also, recalling the similar situation of Chapter 3, we postulate that the complex of E, S_1, and S_2 (triple complex for forward reaction) is distinct from the complex of E, P_1, and P_2 (triple complex for back reaction, P_1 and P_2 being the products of the forward reaction and the substrates of the back reaction). The scheme is:

$$E + S_1 \underset{-1}{\overset{1}{\rightleftharpoons}} C_1 \qquad C_1 + S_2 \underset{-2}{\overset{2}{\rightleftharpoons}} C_2 \qquad C_2 \underset{-3}{\overset{3}{\rightleftharpoons}} C_3 \tag{7.F}$$

$$C_3 \underset{-4}{\overset{4}{\rightleftharpoons}} C_4 + P_1 \qquad C_4 \underset{-5}{\overset{5}{\rightleftharpoons}} E + P_2$$

Reaction 3 expresses the fact that the two triple complexes are distinct, and must be converted one into another for the reaction sequence to be completed.

The conservation equation for the enzyme is:

$$E_t = E + C_1 + C_2 + C_3 + C_4. \tag{7.20}$$

If we write the pseudostationary equations for complexes C_1 to C_4, we have four simultaneous linear equations. These can be solved in the usual way. The solutions are all proportional to E. If they are substituted into 7.20, we can now solve for E, and the system is completely specified when the value of E is substituted back into the formulas for C_1 to C_4.

In order that the properties of the solution may be seen as clearly as possible, it is useful to simplify as much as possible by combination or cancellation of terms, and to arrange the terms according to the substrate factors appearing. Thus, we combine terms in S_1S_2, in P_1P_2, and so forth. It is convenient to use the abbreviations $K_2 = (k_{-2} + k_3)/k_2$, $K_3 = (k_3 + k_4)/k_3$, and $r_1 = k_{-1}/k_1$, $r_2 = k_{-2}/k_2$, etc. The last-named ratios are simply the equilibrium constants of the various reactions in scheme 7.F.

We can now express the net rate of the overall reaction in the forward direction (that is, forward rate minus backward rate). We take $v = dP_2/dt = k_5 C_4 - k_{-5} P_2 E$, and obtain:

$$\begin{aligned} v = E_t(k_5 S_1S_2 - k_{-5} r_1r_2r_3r_4 P_1P_2)/\{&[1 + k_5(1 + K_3)/k_4]S_1S_2 \\ &+ r_4 P_1(1 + r_3)S_1S_2 + k_{-5} r_4[r_2r_3/k_1 + r_1r_3/k_2 + r_1r_2/k_3 + r_1/k_2]\, P_1P_2 \\ &+ (k_{-5} r_4/k_1)(1 + r_3)S_2P_1P_2 + r_2r_3r_4 S_1P_1 + (k_{-5}/k_1)\, S_2P_2 \\ &+ k_5(K_2/k_3 + r_2r_3/k_4)S_1 + r_1r_2r_3r_4 P_1 + (k_5/k_1)S_2 \\ &+ (k_{-5}/k_4)(k_4 r_1K_2/k_3 + r_1r_2r_3)P_2 + (k_5 r_1K_2/k_3 + k_5 r_1r_2r_3/k_4)\} \end{aligned} \tag{7.21}$$

This complex expression is not very useful as it stands. However, a considerable amount of information can be squeezed out of it. To begin with, one may calculate the initial rate of the forward reaction by setting $P_1 = 0$ and $P_2 = 0$:

$$v_f = k_5 E_t S_1 S_2 / \{[1 + k_5(1 + K_3)/k_4] S_1 S_2 + k_5(K_2/k_3 + r_2 r_3/k_4) S_1 + (k_5/k_1) S_2 + (k_5 r_1 K_2/k_3 + k_5 r_1 r_2 r_3/k_4)\} \tag{7.22}$$

This equation has the same general form as 7.14, except that the constants are more complex. It is interesting to note, as a check on the reasonableness of the formula, that it does not contain the last two back-reaction rate constants, k_{-4} and k_{-5}, although it does contain k_{-3} (in the form of r_3).

We can also calculate the enzymatic equilibrium constant of the reversible reaction from 7.21, setting the numerator of v equal to zero:

$$K_{ee} = (P_1 P_2 / S_1 S_2)_{equil} = 1/r_1 r_2 r_3 r_4 r_5 \tag{7.23}$$

where we have written, by analogy, $k_{-5}/k_5 = r_5$. It is evident that the constant is, as in 3.15, the product of the equilibrium constants of the successive reactions of the enzyme system ($1/r_1 = k_1/k_{-1}$ is the equilibrium constant of step 1, and so forth). The discussion of K_{ee} in Chapter 3 will apply with appropriate generalization to the present case.

INHIBITION BY PRODUCTS

An interesting feature of 7.21, in which it differs from the analogous 3.11, is that there are terms in the denominator in which substrate and product concentrations appear multiplied together, as $S_2 P_2$. To see how these extra factors operate, let us simplify by saturating the enzyme with one of the substrates. For example, if we let S_1 become very large, we get

$$V_1 = k_5 E_t S_2 / \{[1 + k_5(1 + K_3)/k_4] S_2 + r_4(1 + r_3) P_1 S_2 + r_2 r_3 r_4 P_1 + k_5(K_2/k_3 + r_2 r_3/k_4)\} \tag{7.24}$$

This rate at partial saturation is a function of S_2, but it is also a function of P_1. We can compare this with what happens if we let S_1 become large in 7.22:

$$V_{1f} = k_5 E_t S_2 / \{[1 + k_5(1 + K_3)/k_4] S_2 + k_5(K_2/k_3 + r_2 r_3/k_4)\} \tag{7.25}$$

What is the significance of the two extra terms in P_1 which show up in the denominator of V_1? It is just this: If we are dealing with an enzyme whose rate is quite rapid, it will not be practical to assume that the product concentrations are zero during the time it takes us to measure the forward rate. As the **products** are formed, they will **affect the rate** of the reaction in a way that has **nothing to do with the back reaction.** (Note that the back reaction term, $-k_{-5} r_1 r_2 r_3 r_4 P_1 P_2$ is not present in the numerator of V_1; the product terms are all in the denominator.) The **terms in P_1 in the denominator represent competition** for the enzyme between substrates and products.

If the reader will think back over the way in which the rate formulas in this book have been derived, he will realize that, in the typical cases, the denominator of the rate function consists of terms which are related to the various forms in which the enzyme is found. In particular, the terms of the denominator of 7.21 come from the way in which total enzyme E_t is divided up among free enzyme E and the four complexes C_1 to C_4.

What is happening is simple: The product P_1 has a certain affinity for the enzyme form C_4. To the extent that they tend to combine and make C_3, they leave less enzyme available for step 1. This is one form of **inhibition by products.**

This situation holds even if we saturate the enzyme with both substrates. If in 7.24 we now let S_2 becomes very large, we get:

$$V = k_5 E_t / [1 + k_5(1 + K_3)/k_4 + r_4(1 + r_3)P_1] \tag{7.26}$$

Here the maximum rate of the enzyme is still lowered by the presence of product P_1.

It is of interest to note that the effect of **partial saturation** is not the same for the two substrates. If we go back to 7.21, and let S_2 become large instead of S_1, we get:

$$V_2 = k_5 E_t S_1 / \{[1 + k_5(1 + K_3)/k_4]S_1 + r_4(1 + r_3)P_1 S_1 + (k_{-5} r_4/k_1)(1 + r_3)P_1 P_2 + (k_{-5}/k_1)P_2 + k_5/k_1)\} \tag{7.27}$$

In this case we have inhibition not only by P_1, but also by P_2, and we even find a denominator term proportional to P_1P_2. The reason for this is quite simple. When we saturate with S_1, this reacts with the free enzyme form E (step 1). Thus P_2 cannot find any E to combine with according to step 5; but P_1 can still find some C_4, and so P_1 alone is inhibitory. On the other hand, when we saturate with S_2, this combines with C_1, and does not compete with P_2 for E; therefore both P_1 and P_2 are able to inhibit. (Of course, when we saturate with S_1 in V_2, we now find the limiting formula for V to be the same as in 7.26, as must be the case.)

The **nonsymmetrical effect of partial saturation** is a result, therefore, of the **ordered bimolecular reaction**. This gives a way of testing whether bimolecular reactions are ordered or random. For suppose we generalized 7.E to make the reactions involving the products not only reversible but random as well. Then, if both forms of substrate can react with free enzyme E, and both forms of product can react with the last complexes in the sequence, the substrates and products will all compete. Therefore, in a **random bimolecular** system, **both products** should **be inhibitory**, regardless of which substrate is chosen for partial saturation, and they should also inhibit at complete saturation. In an **ordered bimolecular system**, only P_1 is inhibitory at partial saturation with S_1, and at complete saturation, while both P_1 and P_2 will inhibit at partial saturation with S_2.

An interesting (though negative) example of the way in which data can be analyzed with the help of equations like 7.24 and 7.27 is offered by ATP-creatine phosphotransferase (2). This enzyme catalyzes the reversible bimolecular reaction ATP + creatine $\rightleftharpoons$ ADP + phosphocreatine. The product ADP was found to be a potent inhibitor. However, if it was tested for its effect on reciprocal plots, varying one substrate at a time, it affected only the slope but not the intercept when the variable (unsaturated) substrate was ATP; but it affected both slope and intercept when the variable substrate was creatine.

The reciprocal of V_1 (7.24) is

$$(1/k_5 E_t)[M + r_4(1 + r_3)P_1 + (N + r_2 r_3 r_4 P_1)(1/S_2)],$$

where M and N are abbreviations of complex constants independent of substrate or product concentrations. Similarly, $1/V_2$ (7.27) is

$$(1/k_5 E_t)[M + r_4(1 + r_3)P_1 + (N' + QP_2 + RP_1P_2)(1/S_1)],$$

where N', Q, and R are again constants. The only product appearing in both intercepts is P_1, with the same coefficient, $r_4(1 + r_3)$. There is no way to make this coefficient disappear from one intercept and remain in the other. There is therefore no way to represent the data with 7.24 and 7.27 as they stand.

A careful examination of the experiments shows that creatine was not used at saturation. This suggests going back to the original full-scale equation 7.21. After dropping the back-reaction, term from the numerator $(-k_{-5} r_1 r_2 r_3 r_4 P_1 P_2)$, we can take the reciprocal, divide through by $S_1 S_2$, and pick out the terms which would be slope terms and intercept terms when S_1 is varied, and the slope and intercept terms when S_2 is varied. This gives us:

Varying S_1: Intercept proportional to

$$M + r_4(1 + r_3)P_1 + r_2 r_3 r_4 P_1/S_2 + Q/S_2$$

Varying S_2: Intercept proportional to

$$M + r_4(1 + r_3)P_1 + (k_{-5}/k_1)r_4(1 + r_3)P_1P_2/S_1 + (k_{-5}/k_1)P_2/S_1 + (k_5/k_1)/S_1$$

where M, N, Q, etc., are constants whose value is immaterial just now.

As before, it is evident that we cannot remove P_1 from one intercept expression and not from the other: for the only suitable condition to impose on the constants to remove P_1 is still $r_4 = 0$. But P_2 is already absent from one intercept, present in the other. Let us then identify P_2 with ADP.

The intercept which is affected by P_2 is the one for varying S_2. Comparing with the experimental findings, this means that we should identify S_2 with creatine. By elimination, this identifies S_1 with ATP.

But, in the experiment where creatine was varied, ATP was at saturation level ($0.004M$). This means saturation for S_1. Therefore the intercept terms containing $1/S_1$ should be negligibly small compared with the others. Applying this condition to the intercept for varying S_2, the terms in P_2 drop out, leaving us where we started.

The data, though suggestive, therefore cannot be analyzed in terms of our equations. The fact that the enzyme is magnesium activated, and that magnesium ions form complexes with substances like ATP and ADP, suggests that this system is not completely represented by the equations we have been testing.

PROBLEM: Suppose an enzyme system is tested for product inhibition as follows: One substrate is kept at saturation, and just one product is added to the incubation mixture. The product gives "competitive" inhibition (Chapter 8)—that is, it elevates the slope of the reciprocal plot without affecting the intercept (as compared with the control free of product). If the experiment is repeated with the other substrate kept at saturation and the previously saturated substrate varied, the same product inhibits "non-competitively"—it increases the intercept of the reciprocal plot without affecting the slope.

Is this result consistent with the behavior of the ordered bimolecular system of 7.21?

Try, by looking over 7.F, to correlate the structure of the scheme with the idea of competitive inhibition (product and substrate react with the same form of the enzyme) and uncompetitive inhibition (product and substrate do not compete for the same form of the enzyme). Does this give you any idea of how you might modify a reversible bimolecular scheme so as to account for the experiment just described?

Practical Irreversibility

Enzymes whose action is **practically irreversible** would be particularly interesting to study in connection with 7.21. Cases in point are hexokinase and some of the glycosidases such as invertase (which is of historical interest in that some of the earliest studies of inhibition by products were performed with it). In such enzymatic reactions, the back-reaction term in the numerator of 7.21 will of course always be absent. Which terms of the denominator disappear will depend, however, on just where the irreversibility enters in scheme 7.F. It could conceivably be at any one of the steps, but some are more plausible choices than others.

Whichever assumption we make, the steps can be expressed mathematically in a simple way. If step 3 is irreversible, this means $k_{-3} = 0$, hence $r_3 = 0$. If step 4 is irreversible, then $k_{-4} = 0$ and so $r_4 = 0$. If step 5 is irreversible, then $k_{-5} = 0$. We may in the same way express the assumption that more than one step is irreversible.

If step 3 is irreversible, the terms in P_1 and P_1S_1 drop out. If step 4 is irreversible, these terms and the terms in $P_1S_1S_2$, P_1P_2, and $S_2P_1P_2$ disappear. If step 5 is irreversible, the terms in P_2, S_2P_2, P_1P_2, and $S_2P_1P_2$ vanish.

These effects are most readily detectable in the formulas for partial saturation. Taking V_1 first, if we look at it from the viewpoint of expressing it in terms of "apparent" constants, that is, as $V'S_2/(K' + S_2)$, we divide through top and bottom by $1 + k_5(1 + K_3)/k_4 + r_4(1 + r_3)P_1$. Then

$$V_1 = \{k_5 E_t/[1 + k_5(1 + K_3)/k_4 + r_4(1 + r_3)P_1]\}S_2/$$
$$\{[r_2 r_3 r_4 P_1 + k_5(K_2/k_3 + r_2 r_3/k_4]/$$
$$[1 + k_5(1 + K_3)/k_4 + r_4(1 + r_3)P_1] + S_2\}$$

and so $V' = k_5 E_t/[1 + k_5(1 + K_3)/k_4 + r_4(1 + r_3)P_1]$, while for the apparent constant K' we find

$$[r_2 r_3 r_4 P_1 + k_5(K_2/k_3 + r_2 r_3/k_4)]/[1 + k_5(1 + K_3)/k_4 + r_4(1 + r_3)P_1].$$

Both apparent constants depend on P_1, both being hyperbolas when plotted against P_1 (Figs. 7.3, curve A, and 7.4, curve B). If step 3 is irreversible and $r_3 = 0$, the **functional form** of V' does not change (Fig. 7.3, curve A), though the values of the constants do. But in K' the term in P_1 drops out of the numerator when $r_3 = 0$, so K' is now the same type of hyperbola as V' (Fig. 7.4, curve A): it has P_1 in the denominator only, so that, instead of approaching a finite nonzero limit as P_1 is made large, it sinks always to zero.

If step 4 is irreversible and $r_4 = 0$, the terms in P_1 drop completely out of both V' and K', and they become true constants (Fig. 7.3, curve B, and Fig. 7.4, curve C). If $k_{-5} = 0$, neither of the two apparent constants is affected (Fig. 7.3, curve A, and Fig. 7.4, curve B). The three situations are clearly distinguishable by experiments in which varying amounts of P_1 are added to the incubation mixture and the effect on V' and K' is measured.

Suppose we analyze the matter in terms of V_2 (7.27), the second state of partial saturation. As before, we arrange the formula in the standard form, this time as $V'S_1/(K' + S_1)$. Now V' is just as before. But K' is now a function of both P_1 and P_2. Explicitly

$$K' = [(k_{-5} r_4/k_1)(1 + r_3)P_1P_2 + (k_{-5}/k_1)P_2 + k_5/k_1]/$$
$$[1 + k_5(1 + K_3)/k_4 + r_4(1 + r_5)P_1]$$

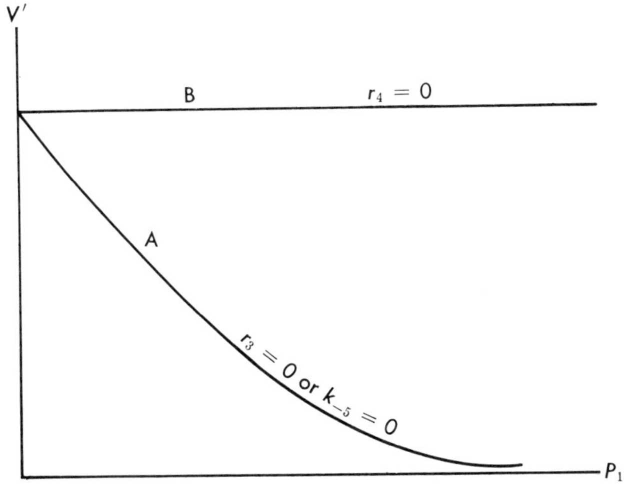

Fig. 7.3. Analysis of **Practically irreversible** Bimolecular Reaction. Apparent maximum rate V' against product P_1, as affected by irreversibility at various reaction steps. From experiments at saturation level of first substrate S_1.

Considered as a function of P_1, K' is again a hyperbola (Fig. 7.5, curve B). If $r_3 = 0$, that form is not affected; it is still a hyperbola of the same kind. If $r_4 = 0$, it becomes a constant, since P_1 drops out (Fig. 7.5, curve C). If $k_{-5} = 0$, P_1 drops out of the numerator, and the hyperbola decreases toward zero as P_1 increases (Fig. 7.5, curve A).

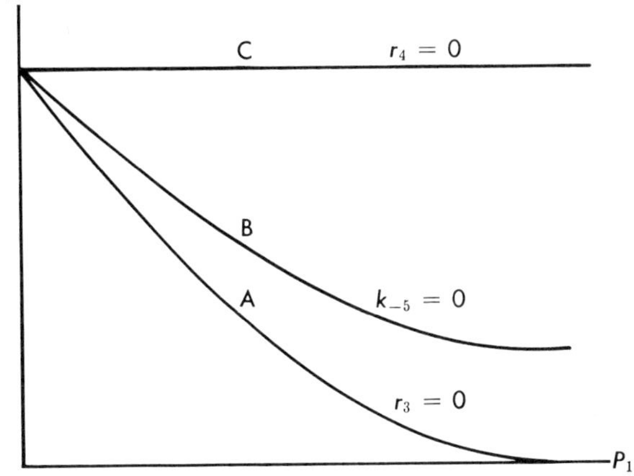

Fig. 7.4. Apparent K' for Same Case as Fig. 7.3.

Considered as a function of P_2, V' of course is always a constant. But K' is a straight line (for P_2 appears only in the numerator (Fig. 7.6, curve A). Neither step 3 nor step 4 irreversibility affects this form, though the constants will be changed as some terms drop out. But if $k_{-5} = 0$, K' becomes a constant (Fig. 7.6, curve B).

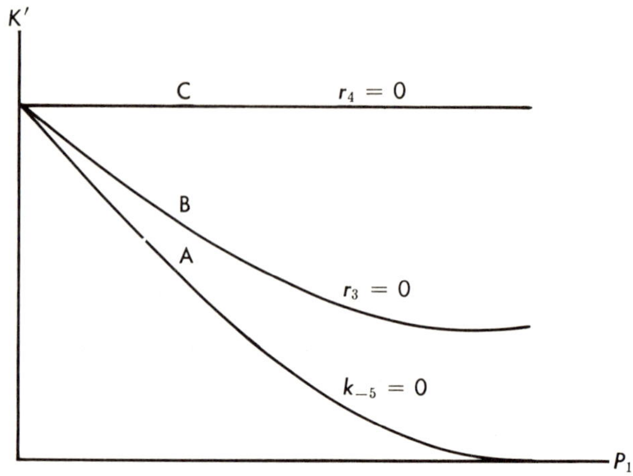

Fig. 7.5. Practically Irreversible Case, Saturated with S_2. Apparent K' as function of product P_1.

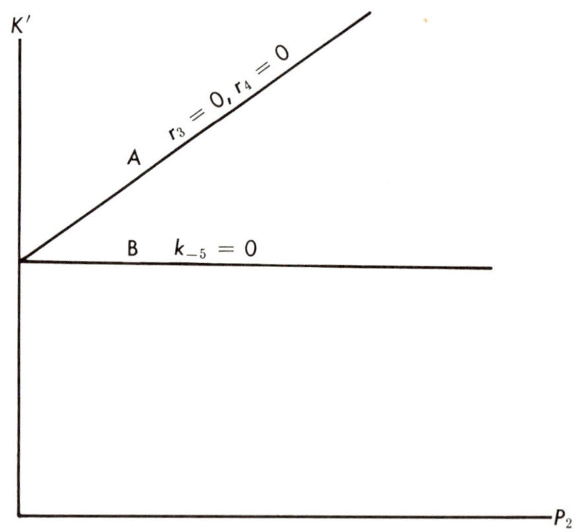

Fig. 7.6. Same Case as Fig. 7.5. K' as function of second product P_2.

The rate at complete saturation (equation 7.26) is independent of P_2, and, as pointed out before, decreases as P_1 increases (Fig. 7.7, curve A). Neither step 3 nor step 5 has any influence on this behavior. But if $r_4 = 0$, V becomes independent of P_1 (Fig. 7.7, curve B).

A natural question to ask at this point is: How does one know, in a concrete situation, which product is P_1 and which is P_2? Or for that matter,

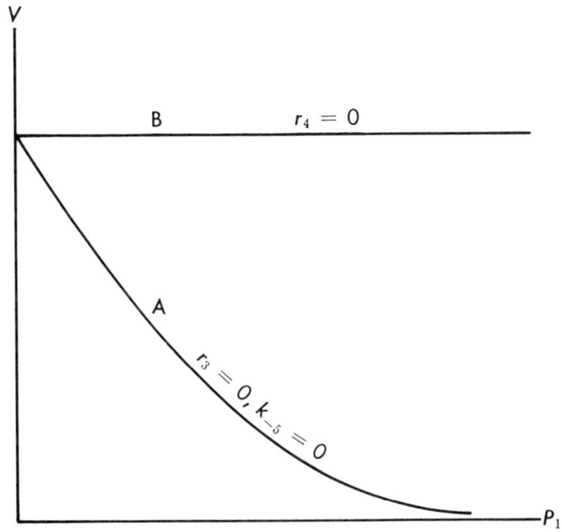

Fig. 7.7. Practically Irreversible Case. Maximum rate V, at saturation with both substrates, as function of product P_1. Effect of irreversibility at various reaction steps.

how does one know which substrate is S_1 and which is S_2? In other words, how does one know which substrate reacts with the enzyme first and which second, which product comes off first and which second?

In the case of a random bimolecular reaction, there is of course no way to tell, and it makes no difference, just because the order **is** random. If the ordered case will fit the data at all, the equations we have just been discussing in this section will answer the questions. From 7.26, the product which is inhibitory at **complete** saturation with both substrates is P_1. Now, from 7.24, if P_1 alone is inhibitory at **partial** saturation with one of the substrates, **that** substrate is S_1.

We can see from the foregoing that careful analysis designed to detect the various effects of irreversibility on product inhibition should enable us to decide at what step, in a scheme like 7.F, the irreversibility occurs. This should eventually contribute to an understanding of the mechanism of enzyme action, and of some of the energy relationships involved in the formation and transformation of complexes.

TIGHTLY BOUND COENZYMES

Although it has been clear from the text of this chapter that coenzymes which are also co-substrates (like the adenine or pyridine nucleotides) fit the assumptions that led to 7.14, we have left out one point. As a rule, the tissue concentration of such coenzymes is low, even when compared with the not very high concentrations of ordinary substrates. Moreover, some of them are found to be rather tightly bound to their enzymes.

This means that we need a conservation equation for the co-substrate, in addition to the one for enzyme. This leaves us with the problem of deciding whether the coenzyme should be S_1 or S_2. If we choose S_1, the conservation relation will be:

$$S_{t1} = S_1 + C_1 + C_2 \tag{7.28}$$

The rest of the system is given by scheme 7.D and equation 7.13. We solve the pseudostationary equation for C_2 as before, getting $C_2 = S_2 C_1/K_2$. We eliminate C_2 from 7.28 with this relation, and then use 7.28 to eliminate S_1 from the pseudostationary equation for C_1. We now have a quadratic equation for C_1. We solve it as usual, taking the negative sign of the radical because C_1 must be zero when $E_t = 0$, and getting

$$C_1 = \tfrac{1}{2}\,[K_2/(S_2 + K_2)]\{(E_t + S_{t1} + M) - \sqrt{(E_t + S_{t1} + M)^2 - 4E_t S_{t1}}\}$$

where

$$M = (K_1 + gS_2/K_2)/(1 + S_2/K_2),\ K_1 = k_{-1}/k_1,\ K_2 = (k_{-2} + k_3)/k_2,\ g = k_3/k_1.$$

From C_1 we get $C_2 = S_2 C_1/K_2$, and hence:

$$\begin{aligned} v = k_3 C_2 &= \tfrac{1}{2}k_3[S_2/(K_2 + S_2)] \\ &\times \{(E_t + S_{t1} + M) - \sqrt{(E_t + S_{t1} + M)^2 - 4E_t S_{t1}}\} \end{aligned} \tag{7.29}$$

If the coenzyme S_1 is tightly bound, this means that k_{-1} is very small, or k_1 is very large, or both. The case where k_{-1} is small, but k_1 is not particularly large, yields nothing very interesting. But if k_1 is very large, then $K_1 = 0$ and $g = k_3/k_1 = 0$; hence $M = 0$. The radicand in 7.29 is now equal to $(E_t - S_{t1})^2$.

Applying the positive square root rule (Chapter 1):

$$\begin{aligned} v &= k_3 S_{t1} S_2/(K_2 + S_2) \qquad \text{if } S_{t1} < E_t \\ v &= k_3 E_t S_2/(K_2 + S_2) \qquad \text{if } S_{t1} > E_t \end{aligned} \tag{7.30}$$

Equation 7.30 is an example of a kind of function well known to mathematicians, and actually quite simple to handle, but rarely discussed in the older texts.

Any rule which relates two sets of numbers defines a function in this general sense. The usual ways to give the rule is by stating a formula, like $y = ax^2$ or $y = \cos x$; and the formula is taken to be valid over the entire possible range of x and y values. But another function could be defined by the formula $y = ax^2$, with the added specification that x shall be confined to rational numbers. One could define a function y by stating that it shall be equal to 1 for all whole number values of x, and equal to 0 for all other values of x; it would be an odd-looking function, but a function nonetheless—and no one would have any trouble in drawing the graph by means of the rule.

Equation 7.20 defines a function of the kind whose slope is **piecewise continuous**. It is defined by one simple formula in one region of x values, by another simple formula in another region of x values, and the two formulas have the same value at the point of juncture between the two x regions. Thus the resulting graph will have a sharp bend at the point of juncture (a sudden change of slope); but there is no actual gap or jump in the graph.

Thus, from the first line of 7.30, the graph of v against S_{t1} is a simple straight line, with slope $k_3 S_2/(K_2 + S_2)$, in the S_{t1} region from 0 to E_t; the line passes through the origin. From the second line of 7.30, which does not contain the symbol S_{t1}, the rate v is a constant, and its graph is therefore a horizontal straight line, for all S_{t1} values from E_t upward. It is easy to see that the two formulas coincide when $S_{t1} = E_t$, and that therefore the two straight lines join, with a sharp bend, at this point. The system behaves as if we were **titrating the enzyme with the coenzyme**; the "equivalence point" or "titre" is at $S_{t1} = E_t$. The appearance of the "titration curve" is given by Fig. 7.8. The curves are drawn for two values of S_2.

On the other hand, if we select S_2 as the coenzyme, we have:

$$S_{t2} = S_2 + C_2 \tag{7.31}$$

The rest of the system is as before.

After we eliminate S_2 from the system by 7.31, we eventually get the solution:

$$\begin{aligned} v = \tfrac{1}{2}k_3\{&[S_1E_t/(S_1+g) + K_2(K_1+S_1)/(S_1+g) + S_{t2}] \\ &- \sqrt{[S_1E_t/(S_1+g) + K_2(K_1+S_1)/(S_1+g) + S_{t2}]^2 - 4S_1S_{t2}E_t/(S_1+g)}\} \end{aligned} \tag{7.32}$$

If we now assume that S_2 is tightly bound to enzyme, we have $K_2 = 0$, and the radicand in 7.32 is simply $[S_1E_t/(S_1+g) - S_{t2}]^2$. The limiting rate formula is therefore:

$$\begin{aligned} v &= k_3 S_{t2} && \text{if } S_1E_t/(S_1+g) > S_{t2} \\ v &= k_3 S_1E_t/(S_1+g) && \text{if } S_{t2} > S_1E_t/(S_1+g) \end{aligned} \tag{7.33}$$

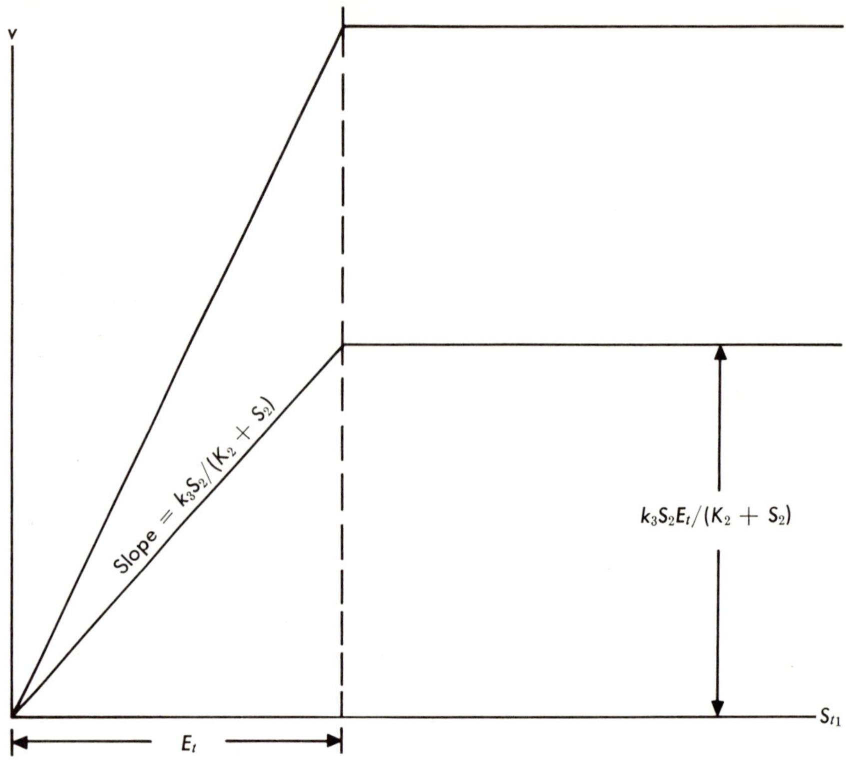

Fig. 7.8. Rate Curve for Tightly Bound Coenzyme, in Case when **Coenzyme Reacts** with Enzyme **Before Substrate** Can react. For two values of substrate concentration, to show effect on slope, maximum level and equivalence point.

The curve (Fig. 7.9) again appears as if we were titrating the enzyme with the coenzyme, with an apparent equivalence point at $S_{t2} = S_1E_t/(S_1 + g)$. The true equivalence point, E_t, is of course obtained if we saturate with S_1. Two curves are drawn, for two values of S_1. (For further discussion of "titration" curves, and for the use of reciprocal plots in this connection, see Chapter 8, and Figs. 8.1 and 8.2.) The beautiful data of Velick *et al.* (3) on the binding of DPN by triose phosphate dehydrogenase follow precisely such a titration curve as we have described. The curves differ only in that our theory predicts a 1 : 1 binding, because we have **assumed** such a relation in the scheme, whereas the data show three equivalents of DPN bound per mole of enzyme. This discrepancy could easily be remedied; but the simpler version will suffice to illustrate the point.

It seems more natural to us that the coenzyme should be bound to the enzyme before the substrate. Nevertheless, the alternative must be considered

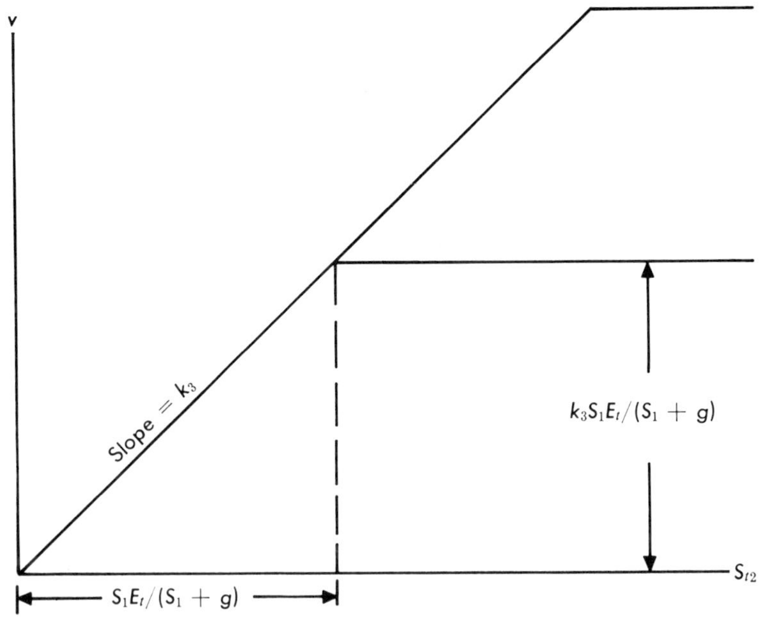

Fig. 7.9. Rate Curve for Tightly Bound Coenzyme, when **Coenzyme Reacts** with Enzyme **After Substrate**. For two substrate levels, to show effect on maximum rate and equivalence point.

equally possible. The distinction, as shown by the comparison of 7.30 and 7.32, is that the equivalence point in the first case is independent of S_2, the concentration of the substrate, while the equivalence point in the second case does depend on the substrate level S_1 but the slope is independent of S_1. Unfortunately, Velick's data do not permit this important point to be tested. The fact that some crystalline preparations of the enzyme contain bound DPN however, argues in favor of the first alternative.

A case in which binding is definitely ordered, and in which it follows our intuitive notion that coenzyme should be bound first, is given by lactic dehydrogenase (4).

An important feature of the titration curves discussed in this section is the following: In each case, at substrate saturation, the slope of the rising portion is k_3, while the equivalence point is E_t. We are therefore able to estimate k_3 and E_t **separately** when such a case arises.

REFERENCES

1. Rowen, J. W., and Kornberg, A., J. Biol. Chem. *193*, 497 (1951).
2. Kuby, S. A., Noda, L., and Lardy, H. A., J. Biol. Chem. *210*, 65 (1954).
3. Velick, S. F., Hayes, J. E., Jr., and Harting, J., J. Biol. Chem. *203*, 527 (1953).
4. Takenaka, Y., and Schwert, G. W., J. Biol. Chem. *223*, 157 (1956).

8

Effects of Nonreactants on Enzymes

In addition to the effects of reactants (substrates, products, coenzymes) on enzymes, we find an enormous variety of compounds capable of modifying enzyme activities. Some of these increase the rate, and are termed **activators**; some decrease it, and are termed **inhibitors**. We shall continue to use these terms; but it will become evident that the categories are not as firm as they sound—there are circumstances under which compounds usually known as inhibitors may activate, or compounds usually known as activators may inhibit. When a neutral word is required, I shall refer to **effectors**. (The hydrogen ion considered as an effector requires special treatment, which is postponed to Chapter 10. What many textbooks refer to as prosthetic groups, such as the haem components of oxidases, may be considered as parts of the enzyme if we wish to ignore their slight dissociability, otherwise as extremely tightly bound co-substrates.)

The action of effectors cannot be explained in terms of ionic strength or buffer effects on proteins, although it is true that many such compounds have historically entered enzyme systems as buffers, or as "neutral salts." The noteworthy aspect of effectors is that they exhibit a great deal of specificity. There are enzyme systems which may be activated only by magnesium ion (or to a very much lesser degree by the closely related manganese); there are enzymes which are activated by potassium ions and inhibited by sodium ions. The specificity is frequently one-many rather than one-one, as a mathematician would put it: To a given enzyme there may correspond just one activator; but to such an activator (for example, an alkali metal or alkaline earth ion) there may correspond many enzymes which it activates.

A THEORY OF EFFECTORS

Let us begin with a minimum of assumptions. We shall suppose that an effector X can combine with either free or substrate bound enzyme, forming E_x and C_x respectively. Let E_x react with S to form an alternative triple complex C'_x. Suppose that C, C_x, and C'_x are all capable of turning over to form product, at specific rates which may in particular instances equal zero. The scheme is:

$$\text{(8.A)}\quad E+S\underset{-1}{\overset{1}{\rightleftharpoons}}C \qquad C\xrightarrow{2}E+P \qquad E+X\underset{-3}{\overset{3}{\rightleftharpoons}}E_x \qquad C+X\underset{-4}{\overset{4}{\rightleftharpoons}}C_x$$

$$E_x+S\underset{-5}{\overset{5}{\rightleftharpoons}}C'_x \qquad C_x\xrightarrow{6}E+P+X \qquad C'_x\xrightarrow{7}E+X+P$$

Note that no violence would be done to the conservation equation:

$$\text{(8.1)}\qquad E_t=E+E_x+C+C_x+C'_x$$

if we let reaction 6 or 7 or both give rise to E_x+P instead.

Following the usual procedure, we get the solution of the pseudostationary equations:

$$\text{(8.2)}\qquad v=[k_2S+k_6SX/K_4+k_7SXu_x/K_5u_s]E_t/[S+u_x+SX/K_4+\sigma Xu_x]$$

with the abbreviations:

$$\text{(8.3)}\qquad u_x=K_1+k_6X/k_1K_4 \qquad \sigma=(1+S/K_5)/u_s \qquad u_s=K_3+k_7S/k_3K_5$$

$$K_1=(k_{-1}+k_2)/k_1 \qquad K_3=k_{-3}/k_3 \qquad K_4=(k_{-4}+k_6)/k_4$$

$$K_5=(k_{-5}+k_7)/k_5$$

(Note: It is easiest to begin solving for C_x and C'_x, then using the results to get E_x and C, finally using 8.1 to get E.)

Consider v as a function of X. In view of the definition of u_x, numerator and denominator are quadratic functions of X, with all terms positive. For $X=0$, v is the conventional $k_2SE_t/(K_1+S)$. When X approaches infinity, v approaches $k_7SE_t/(K_5+S)$, the rate one would expect if the form C'_x alone were forming product. (The terms attributable to C_x are linear in X, and would be expected to be most significant for intermediate X values.)

Now it is elementary to say that the effector is behaving as an activator if $k_7/(K_5+S)>k_2/(K_1+S)$, as an inhibitor if the reverse holds true.

However, a glance at the role of S and the constants indicates that there is more to the matter than that. For instance, the foregoing inequality neither implies nor is implied by $k_7/K_5>k_2/K_1$—that is, the role of the effector at moderate S values is not necessarily the same as at small S values. Moreover, letting S approach infinity in 8.2 gives the limit $(k_2 + k_6X/K_4)/(1 + X/K_4)$, which is no longer quadratic, and whose behavior as a function of X is determined by the relation of k_6 to k_2—that is, by the relation of the turnover rates of C_x and C rather than of C'_x and C.

This prompts an examination of 8.2 to answer the questions: (1) Can $v(X)$, in rising to an activated value or falling to an inhibited value, attain a maximum or a minimum respectively? (2) Can $v(X)$ make a false start, dropping to a minimum on the way to an activated value or rising to a maximum on the way to an inhibited value?

If we collect terms and write the numerator and denominator of v as quadratics in X, and differentiate with respect to X, we get for the numerator of the derivative (the denominator being everywhere positive as usual):

$$-(b\gamma - c\beta)X^2 + 2(c\alpha - a\gamma)X + (b\alpha - a\beta) \tag{8.4}$$

with:

$$a = k_2 S \qquad b = k_6 S/K_4 + K_1 k_7 S/K_5 u_s \qquad c = k_6 k_7 S/K_4 k_1 K_5 u_s \tag{8.5}$$

$$\alpha = K_1 + S \qquad \beta = S/K_4 + k_6/K_4 k_1 + K_1\sigma \qquad \gamma = \sigma k_6/K_4 k_1$$

It is easy to show that the answer to the first question is in the affirmative: Overshoot and undershoot may occur. We can either set 8.4 equal to zero and examine the solution of the quadratic, or we may inspect 8.4 directly. Let us begin with the simpler possibility. Suppose $c\alpha - a\gamma > 0$; this is readily seen to be equivalent to the condition for activation at large X values, and the asymptote of the v-X curve must be above the intercept at $X = 0$. Now suppose that $b\gamma - c\beta$ has the same (positive) sign. We can prove that, in consequence, $b\alpha - a\beta$ is also positive. (Multiply the first inequality by b, the second by a, and add.) Then the initial slope of $v(X)$ is positive, while the final slope is negative. Since the function is continuous, this means that the curve rises to a maximum and drops to the eventual activated value.

If in 8.4 the coefficients of $-X^2$ and of X are both negative, the constant term is likewise negative. We can therefore infer that the curve will decrease to a minimum, from which it will rise to the eventual inhibited value.

These cases, then, represent curves in which an activator or an inhibitor is more effective at intermediate than at maximal concentrations. Now let us ask about the consistency of the crucial inequalities. They may be united into the statement that $k_7/(K_5 + S)$ lies between $k_2/(K_1 + S)$ and $k_6/(S + k_6/k_1)$; the former is the lower bound in the activation case and the upper bound in the inhibition case. Consistency requires that the lower and upper bounds of the $v(X)$ asymptote shall be in the order required. Inverting the fractions (and of course inverting accordingly the direction of the inequality sign), we find for the activation case (noting that $K_1 = k_{-1}/k_1 + k_2/k_1$)

the relation $(k_{-1}/k_1) + (1 - k_2/k_6)S > 0$, while the reverse inequality must hold in the inhibition case.

Clearly, this inequality will always be valid for $k_2 < k_6$; and this is neither implied nor contradicted by the previous relation between k_7 and k_2. If on the contrary $k_2 > k_6$, the consistency condition depends on S, whose coefficient is now negative. For sufficiently great S, the condition will no longer hold. In the inhibition case, where the critical inequality for consistency is in the reverse direction, it will never hold for $k_2 < k_6$, and for $k_2 > k_6$ it will be valid only for S sufficiently high.

To try to see intuitively why S should have the kind of effect it has, note that C_x/C is proportional to X, while C'_x/C is proportional to Xu_x/u_s. Take the activation case. If both C_x and C'_x have relatively high turnover numbers, they will both make a significant contribution to v as X increases to moderate values; at high X and intermediate S values, where C'_x predominates, the depressing effect of S (via u_s) on the ratio to C will lower the activity somewhat, giving a maximum for intermediate X and intermediate S. For infinite S, only C_x is significant, as shown earlier. But if C_x has low turnover, its contribution at intermediate X values will diminish, and the depressing effect of S on C'_x, not yet compensated by the X effect, will iron out the maximum. (I will cheerfully admit that this attempt at a commonsense interpretation sounds complex and fumbling; and I contend that it illustrates the superiority of mathematics to words in expressing the results of a set of complex relationships.)

Maxima in activation curves are well-known phenomena. For examples, see reference 1.

The answer to the second question (Can activators inhibit or inhibitors activate?) is a little less obvious. Suppose once more that the coefficient of X in 8.4 is positive (eventual activation). If the coefficient of $-X^2$ is negative, the asymptote is approached by a rising curve from below. Now, if the constant term is negative, the curve starts downward, only to rise later; by continuity there must be a minimum. The reverse set of conditions will hold for the eventual-inhibition early-activation case. But if the first two coefficients in 8.4 are of opposite sign, the sign of the constant term is no longer determined; and we will have to examine it separately:

$$(8.6) \quad b\alpha - a\beta = [k_7(K_1 + S) - k_2(K_5 + S)]K_1S/K_5u_s$$

$$(k_{-1}/k_1)(k_6S/K_4) + (k_6 - k_2)S^2/K_4$$

The first term is positive in the eventual-activation case; the second is always positive, and the third may be of either sign. For this expression to be negative in the activation case, it is necessary that the last term be negative ($k_6 < k_2$) and fairly large (which it will be for fairly high S values, since

S/u_s has a horizontal asymptote for high S, making the first two terms of 8.6 practically linear while the last remains quadratic. In the eventual-inhibition case, 8.6 must be positive, the first term is negative, and the condition will be satisfied, for sufficiently high S, if $k_6 > k_2$.

We may conclude, then, that activation by inhibitors and the converse are possible, but that they will occur only for sufficiently great values of S.

After taking up another interesting approach to the theory of effectors in the next section, we shall consider, for the remainder of the chapter, a number of special cases.

PROBLEM: 1. If E_x cannot combine with S in reaction 5, or if C'_x cannot turn over, show that the only possibility for large X values is inhibition, but that conditions for activation at lower X values can be found. Is undershoot (maximal inhibition at intermediate X) possible?

2. In the same way, analyze the case where C_x cannot turn over.

3. Develop the special cases of **compulsory activation** ($k_2 = 0$) and **optional activation** ($k_2 \neq 0$) in which either (a) complex E_x is formed, but C_x is not (**pre-activation**) or (b) C_x is formed but E_x is not (**post-activation**). In case (b), is it still possible to let C_x yield E_x plus products instead of E plus X plus products?

4. Develop the general theory of this section on the assumption that C'_x yields $E_x + P$ rather than $E + X + P$ in reaction 7. How does the result differ from 8.2?

5. Derive the pre-activation cases (compulsory and optional) that arise from the theory of problem 4.

POLYMERIC ENZYMES AND "ALLOSTERIC" EFFECTS

In recent years a dramatic set of effector phenomena came, not from conventional enzymology, but from the study of genetic and metabolic control systems. The experiments and hypotheses have been considered in more detail elsewhere (2). Suffice it for our purposes to say that small molecules were inferred to act as regulators or as essential parts of regulators for crucial metabolic phenomena, in particular for biosynthetic reactions. Contributions to this analysis of intracellular regulation came from many laboratories; one of the major sources, in any case, was a group at the Pasteur Institute (Jacob, Monod, and their associates), who approached the problem through studies in microbial genetics. A consequence of this activity was an interest in the mechanism of such effects and in the question of how widespread they might be. Examination of a number of cases of so-called "end-product inhibition" and closely related examples (3) brought out an intriguing situation: The effector action could be abolished by treatments of the enzyme which did not abolish enzyme activity, and which by their nature suggested that they were producing changes in conformation, and in particular changes in the degree of polymerization of proteins made up of protein subunits. To this must be added the fact that the effectors are usually not conceivable analogues or homologues of the substrates, so that their effects are very unlikely to come from direct competition between them and the substrates for places on the enzyme.

On the basis of these considerations, Monod and his associates (4) proposed the following model: The enzyme is a polymer of identical protein subunits. Each subunit contains a group capable of interacting with and binding substrate; the various groups in this class are identical in their properties, and independent of one another. Each subunit also contains other groups, capable of interacting with effectors; and again the various groups of each effector-binding class are identical, and independent of one another—and of the substrate-binding groups. Finally, the enzyme polymer is capable of existing in two configurations, whose properties with respect to substrates and effectors are not necessarily the same. The free enzyme can change from one configuration to the other; and each configuration, as described above, can give rise to a series of complexes containing various numbers of substrate and/or effector molecules.

The polymeric character of the enzyme is introduced in order to account for the effects of agents which dissociate such polymers on the effector phenomena. The authors make further assumptions about the precise nature of the polymers, which are perhaps relevant to a discussion of the relation between enzyme activity and protein structure, but which need not be considered further at this point. Formally, each protein may be viewed as a single molecule (polymeric or not, as one pleases) with a multiplicity of substrate-binding and effector-binding groups, all independent of one another. The important feature of the model for our present discussion is this: The model proposes to account for the properties of effectors by the fact that they combine to different extents with the two forms of the enzyme, hence bias the amounts of the two forms available to the substrate. If the effector produces a bias toward the form with higher turnover rate, it will behave like an activator, if the reverse, like an inhibitor. The classical models of effector action, as usually expressed, amount to this: the effector combines with the enzyme, thereby modifying its capacity to combine with substrate and/or to convert the substrate to product. If these modifications increase the reaction rate, the effector is an activator, otherwise an inhibitor.

Leaving aside for the moment an analysis of the differences between these models, let us proceed with that of Monod. To begin with, let us consider interaction with just one ligand (to use Monod's terminology); let it be the substrate. Since the model is formally identical to the multivalent enzyme model of Chapter 5, we may adopt the notation and procedures of that model. To this we must add the postulate that there are two forms of free enzyme, and correspondingly two series of complexes. Following Monod, we shall call them the R and T forms; and use R_i or T_i where in Chapter 5 we used C_i. We have two parallel schemata like 5.B, one for the R and one for the T series. In addition, we have:

(8.B) $$R_0 \rightleftharpoons T_0$$

with rate constants k and k', and equilibrium constant $L = k/k'$.

As pointed out in Chapter 2, it is neither obligatory nor correct to deal with such systems by assuming that each reaction is in equilibrium, since the pseudostationary equations are readily solved, as in Chapter 5. If one follows out the two sets of equations backwards, it turns out that the two new equations (for dR_0/dt and dT_0/dt) boil down to:

(8.7) $$T_0 = LR_0$$

which is the same as the equilibrium relation.

The conservation equation is:

(8.8) $$E_t = \sum_{i=0} R_i + \sum_{i=0} T_i$$

With the assumption that the "microscopic constants" (Monod's terminology—what I have called "intrinsic constants") are all the same for the various groups, we get relations like 5.25 and 5.26′ for both the R and the T series. It follows that:

(8.9) $$v = nE_t[m_r(S/M_r)(1 + S/M_r)^{n-1} + Lm_t(S/M_t)(1 + S/M_t)^{n-1}]/[(1 + S/M_r)^n + L(1 + S/M_t)^n]$$

where we distinguish the constants by subscripts r and t, and m_r is the same as the m_a of Chapter 5 relative to the R series, etc.

Except for the constants m_r and m_t and the factor n, this expression is formally identical with Monod's equation 2, which he calls the "saturation function." (In the notation of that paper, $S/M_r = \alpha$, $M_r/M_t = c$.) Our M_r and M_t, however, are Haldane constants as defined in Chapter 5, while Monod's K_r and K_t are equilibrium constants.

The difference between the two expressions is significant. Monod's equation puts the onus for all differential effects on the constants M_r and M_t, while 8.9 also disposes of m_r and m_t. (Monod alludes to these distinct mechanisms as K and V effects, but does not treat them further.) Thus the extreme case of differential action as between the R and T series would be expressed in Monod's case by letting his K_t (formally analogous to our M_t)

become infinitely great, meaning that S does not combine with T. Then S/K_t vanishes wherever it appears; the effect on 8.9, if we treated M_t the same way, would be to remove the second term in the numerator and to reduce the second term in the denominator to L (footnote on p. 92 of Monod's paper). We have another possibility: to set $m_t = 0$, removing the second term from the numerator of 8.9 but leaving the second term of the denominator intact.

It is easy to see how the results would compare. Consider the expression in Monod's footnote: say $x(1+x)^{n-1}/[L+(1+x)^n]$, where $x = S/M_r$. To get a rough idea of how this will behave, rewrite it by dividing through by $(1+x)^{n-1}$, getting $x/[L/(1+x)^{n-1}+1+x]$. Now if we had $x/(L+1+x)$, we know what happens: The numerator and denominator both increase until they become practically equal ($y = ax/(b+x)$ of Chapter 1); the increase in the numerator is faster for small x because the term $L+1$ in the denominator is independent of x, and gives the denominator a certain inertia, as it were. If, as above, we have $L/(1+x)^{n-1}$, however, the rate advantage of the numerator over the denominator for small x is pulled down by the rapid increase of $(1+x)^{n-1}$. Thus the fraction does not rise as fast, and tends to rise with a positive second derivative (concave up), especially if L is large so that the effect on the L term is important. Eventually the L term in the denominator becomes negligibly small, and the function behaves practically like $x/(1+x)$.

Compare this result with the case that comes from setting $m_t = 0$ in 8.9. Treating it as above, the L term in the denominator becomes

$$L(1+cx)^n/(1+x)^{n-1},$$

where $c = M_r/M_t$. Unless c is quite small, the L term in the denominator will not only not decrease, but will increase with x, since it involves an nth degree expression divided by one of degree $n-1$. Thus, if the previous curve had an accelerated phase and an inflection point, it will now rise more rapidly, and may lose the inflection point entirely. (Thus if $c = 1$, the denominator becomes $(L+1)(1+x)$.)

Now the curves of, for example, 8.9 are quite interesting just because, for some values of the parameters, they are deeply scalloped at the beginning, with a long lag followed by a rapid rise and an inflection point. This is characteristic of what Monod calls "the cooperative homotropic effect of the ligand." In plain English, this means that the enzyme form most favorable to activity is stabilized against the less favorable form by combination with the substrate itself. Thus, as R combines with substrate, more R forms become available than in the absence of substrate. At low substrate concentrations, there is a minimum of R forms, and the curve rises slowly. As substrate increases, more R forms are available, and the formation of enzymatically effective complexes avalanches. The effect is "cooperative," because every time a substrate molecule ties up an R form, it makes it available for other substrate molecules and keeps it from being converted to the T form. The effect is called "homotropic" because the molecules that cooperate are of the same molecular species; if an effector and a substrate cooperate, the effect is "heterotropic."

Such positive and negative heterotropic effects are readily derived. We may adopt Monod's approximation: We assume that S combines only with R forms, that an activating effector A combines only with R forms, and that an inhibitory effector I combines only with T forms.

Let the equilibrium constants for A and I combinations be called P and Q respectively; and let us for convenience introduce $u = A/P$ and $w = I/Q$ Then the concentration of complexes in which i molecules of A are combined with an R_j, X_{ij}, is:

(8.10) $$X_{ij} = C_{in}\, u^i R_j$$

where R_j is the same as in the derivation of 8.9. For the inhibitor case we have:

(8.11) $$Y_{ij} = C_{in}\, w^i T_0$$

The C_{in} are combinatorial coefficients as in Chapter 5 $(= n!/i!(n-i)!)$. The factor is T_0 rather than T_j because substrate does not combine with T forms.

To 8.8 we must add the sum of all X_{ij} and Y_{ij}. Now, repeating the procedures already used in Chapter 5 and in the "homotropic" case:

(8.12) $$v = nE_t m_r x(1+x)^{n-1}/[L' + (1+x)^n]$$

with $x = S/M_r$ and:

(8.13) $$L' = L(1+w)^n/(1+u)^n$$

Recalling our analysis of the "homotropic" case, we can see that the effect of inhibitor (w) will be to increase the so-called cooperative effect by increasing the importance of the L term in the denominator, while the effect of activator (u) will be to decrease the effect and remove the inflection from the $v - x$ curve. To put the matter in simple terms: The effect of inhibitor is to favor T forms, requiring relatively more substrate to get a respectable concentration of active R forms, while the activator favors R forms and thus enables active (substrate-combined) R forms to accumulate more rapidly as substrate is increased.

There is ample evidence (for which we refer the reader to the paper of Monod *et al.* (4)) for the occurrence of numerous enzyme systems which can be analyzed successfully in terms of the formalism of this section. The immediate question is whether—and how—the underlying model is distinguishable from the classical model.

The essence of the model is to be found in the "homotropic" case; once given that, the "heterotropic ligands"—that is, effectors—either exaggerate or diminish the effects, as we have just seen. The apparently unique feature of the model is the "cooperative effect"—that is, the possibility of a v-S or v-x curve with an initial upward curvature and an inflection point.

Comparison with the multivalent enzyme case of Chapter 5 shows that there is no "homotropic cooperative" effect—**if** we accept the identity of the intrinsic constants for all sites. But if we do not make this assumption, then an inflected curve may well be obtained, as is illustrated by curve B of Fig. 5.1. However, if we proceed to extend the cases of Chapter 5 (either the n-place case or the special two-place case) by including effectors, we soon find clear distinctions.

Suppose, for example, that we elaborate on the two-place case, with $k_{2a} > k_{1a}$, and distinct intrinsic constants, by adding a complete non-competitive inhibitor—that is, an inhibitor that can add to either of the complexes as well as to free enzyme, and whose effect is complete inhibition of activity, whether one molecule of inhibitor is added or two molecules are added. Assuming for simplicity that the intrinsic constants of binding by inhibitor are the same regardless of the molecular species bound, we find:

$$(8.14) \qquad v = (k_{1a}S/K_1 + k_{2a}S^2/K_1K_2)/(1+S/K_1+S^2/K_1K_2)(1+w)^2$$

where w, as before, stands for I/Q. The effect of inhibitor binding is to multiply the denominator of v by a quantity greater than one. This will indeed emphasize the "cooperative effect" of the substrate, inasmuch as it will decrease all values of v for any value of S. But it will not shift the position of the inflection point, whereas in the comparable case of Monod's theory (8.12, $n=2$, $u=0$) there will be an **additive** term $L(1+w)^2$ in the denominator, which is easily shown to shift the inflection point to the right. Another easy way to see the difference is to note that the ratio of v without inhibitor to v with inhibitor in 8.14 is a function of w only, while in the comparable version of 8.12 it is a function of x and w both.

It is therefore evident that the cases of applicability of the "allosteric" effector theory are likely to be distinguishable from those where the classical effector theory holds. Examples of the necessary theoretical and experimental analysis, leading to the conclusion that quite a number of "regulatory" enzymes represent variants of the classical mechanism rather than of the Monod-Changeux model, are to be found in the work of Sanwal and Atkinson and their respective co-workers (5).

Koshland *et al.* (6) have published an elaborate extension of the polymeric model, which takes account of possible geometric relations of the polymerized subunits, and which stresses the use of graphical methods of the diagnostic type. The theory, although restricted to the tetramer case, is more general than the original polymer theory in that it takes account of sub-unit interactions and conformational changes in specific sub-units of a polymer. The spirit of this treatment is therefore closer to that of the classical mechanisms (e.g., in the form adumbrated in Chapter 5).

PROBLEMS: 1. In the foregoing discussion, will the conclusion be affected if we do not assume that all intrinsic binding constants of the inhibitor are the same?

2. In the foregoing discussion, the assertion that inhibitor displaces the inflection point to the right is made without proof (as also in Monod). Prove this assertion.

3. In the two-place system with $k_{2a} > k_{1a}$ and distinct intrinsic constants, introduce (under the classical assumptions) a competitive inhibitor which combines only with free enzyme. Demonstrate that addition of inhibitor displaces the inflection point to the **left**.

4. For the general multivalent enzyme, under classical assumptions, develop the case of a **microscopically competitive** inhibitor—one which can combine with any C_j, but only with places not already occupied by substrate. For simplicity, assume that all intrinsic substrate and inhibitor constants are identical, and that the inhibition is complete, all enzyme molecules with one or more inhibitor molecules bound to them being inactive.

5. In problem 4, suppose inhibition affects only the groups bound competitively by inhibitor, the groups bound by substrate being normally active. Derive rate function.

6. Carry out derivations similar to those of problems 4 and 5 for the allosteric case. Assume that the competitive inhibitor combines only with free groups on R_j molecules, and does not combine with molecules of the T series. Does either of these formulations support Monod's assertion that low concentrations of competitive inhibitor will activate at low substrate concentrations?

THE EFFECT OF ACTIVATOR LIMITATION

As in Chapter 4, we may inquire what happens if the amount of bound activator is an appreciable fraction of total added activator. In any of our models, we add a conservation equation for activator, and use it to eliminate A.

For example, take the model in which activator combines with free enzyme, and substrate combines with this complex afterwards (**pre-activation**). The stoichiometric scheme is

$$(8.C) \qquad A + E \underset{-1}{\overset{1}{\rightleftharpoons}} E_a \qquad E_a + S \underset{-2}{\overset{2}{\rightleftharpoons}} C_a \qquad C_a \overset{3}{\rightleftharpoons} E_a + P$$

We have two conservation equations:

$$E_t = E + E_a + C_a \tag{8.15}$$

$$A_t = A + E_a + C_a \tag{8.16}$$

where A_t stands for total activator added to the reaction mixture. This amount of activator now exists in three forms—free activator, activator bound to enzyme, and activator bound to enzyme with substrate also bound.

The solution (treating E_a as being in equilibrium with E and A for simplicity) is:

$$v = \tfrac{1}{2}k_3[S/(K_2 + S)]\{A_t + E_t + K_1K_2/(K_2 + S) - \sqrt{[A_t + E_t + K_1K_2/(K_2 + S)]^2 - 4A_tE_t}\} \tag{8.17}$$

It is easy to see that this expression will be zero whenever either S, A_t, or E_t is zero. The formulas for partial saturation with A_t and E_t are obtained, as in Chapter 4, by rearranging the radicand and using the binomial expansion. Then:

$$V_a = k_3 E_t S/(K_2 + S) \qquad V_e = k_3 A_t S/(K_2 + S) \tag{7.18}$$

As in Chapter 4, we have the theoretical possibility of "saturation with enzyme"—achievement of a rate that is independent of enzyme concentration. This rate is first-order with respect to activator A_t. Note that 8.17 is completely symmetrical with respect to A_t and E_t, because we now have 8.16 as well as 8.15.

Saturation with substrate gives something new. If we let S become very large, the factor $S/(K_2 + S)$ in 8.17 approaches 1, while the term $K_1K_2/(K_2 + S)$ approaches zero. The radicand approaches the expression $(A_t + E_t)^2 - 4A_tE_t = (A_t - E_t)^2$. Using the positive square root rule of Chapter 1

$$\begin{aligned} V_s &= k_3 A_t \qquad \text{when } A_t \leqslant E_t \\ V_s &= k_3 E_t \qquad \text{when } A_t > E_t \end{aligned} \tag{8.19}$$

Like Figs. 7.8 and 7.9, the graph of 8.19 (Fig. 8.1) is a titration curve: We are **titrating the enzyme with the activator**. The break in the curve is the "equivalence point." The value of A_t at this point is the "titre"—the value of E_t.

If this kind of curve is found experimentally, it will obviously be quite useful. For it permits us to **calculate k_3 and E_t independently**—as the **slope** of the initial portion of the curve and as the **titre** respectively. It is therefore worth while to consider one practical problem. The reader will notice that the broken curve of Fig. 8.1 has a general resemblance to an ordinary hyperbola of the $ax/(b + x)$ type. Both curves rise from the origin, and then flatten out. With experimental error blurring the sharp outlines of the theoretical curves, will we be able to tell when we have the curve of Fig. 8.1?

The answer is found by using the Lineweaver-Burk reciprocal plot. As the reader will see by writing the reciprocal formulas of 8.19, or directly by reasoning from Fig. 8.1, the reciprocal plot will look like Fig. 8.2, curve A. It will have a horizontal portion of height $1/k_3 E_t$, followed by a straight line with slope $1/k_3$; the two will join at the equivalence point when $1/A_t = 1/E_t$. Notice that the straight-line portion, if extrapolated back, should pass through the origin, as indicated by the dot-and-dash line. This will readily be distinguished, even in the presence of experimental error, from the reciprocal plot of a conventional hyperbolic formula, which will be a simple straight line, also having intercept at $1/k_3 E_t$, and resembling curve B of Fig. 8.2.

Let us now turn to the second (**post-activation**) model, in which enzyme first combines with substrate and then picks up activator. The stoichiometric scheme and the conservation equation are

(8.D) $$E + S \underset{-1}{\overset{1}{\rightleftharpoons}} C \qquad C + A \underset{-2}{\overset{2}{\rightleftharpoons}} C_a \qquad C_a \overset{3}{\rightarrow} E + A + P$$

(8.20) $$E_t = E + C + C_a$$

We must add the conservation equation for A_t, which is divided among free activator and activator bound to enzyme-substrate complex:

(8.21) $$A_t = A + C_a$$

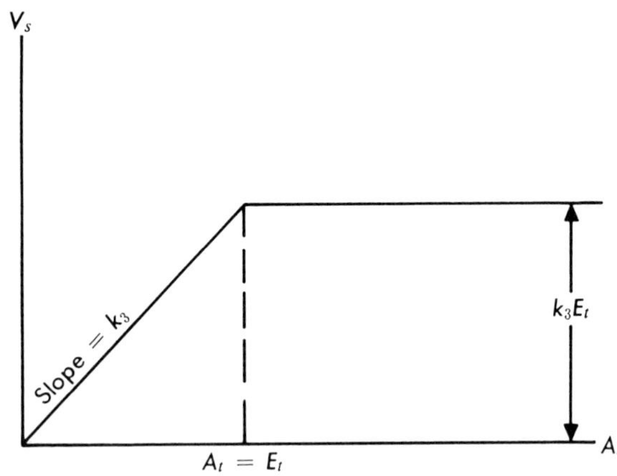

Fig. 8.1. Rate at Saturation with Substrate S, as Function of Activator Concentration A_t. Shows titration form of curve obtained when appreciable amount of total activator is bound by enzyme.

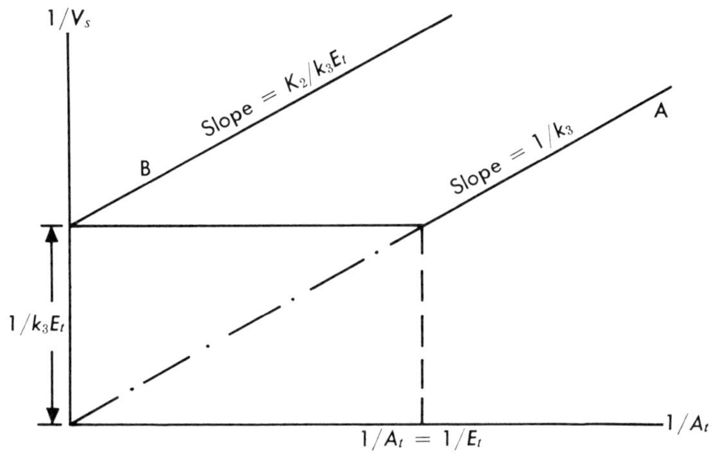

Fig. 8.2. Reciprocal Plot of Fig. 8.1, Showing How Titration Case (Curve A) Can be Distinguished from Rectangular Hyperbola Type of Relation, Plotted as Curve B. Dot-dashed line is prolongation of final part of curve A, showing that it passes through origin of graph.

The solution is:

$$v = \tfrac{1}{2}k_3\{SE_t/(g+S) + A_t + K_2(K_1+S)/(g+S) - \sqrt{[SE_t/(g+S) + A_t + K_2(K_1+S)/(g+S)]^2 - 4SE_t A_t/(g+S)}\} \tag{8.22}$$

The partial saturation formulas for A_t and E_t are:

$$V_a = VS/(g+S) \qquad V_e = k_3 A_t \tag{8.23}$$

Letting S become very large, we get:

$$V_s = \tfrac{1}{2}k_3\{E_t + A_t + K_2 - \sqrt{(E_t + A_t + K_2)^2 - 4E_t A_t}\} \tag{8.24}$$

In this instance V_s is not a titration type of curve, but a conventional saturation type. However, a special case arises if the binding of activator is rather tight. This would mean (check scheme 8.D) that k_2 is large compared with k_{-2} or k_3. In that event, $K_2 = (k_{-2} + k_3)/k_2$ approaches zero. The formula 8.24 would then have precisely the same form as 8.19, and would assume the titration form.

Note that these two models differ in another respect. The V_e of 8.18 is proportional to $S/(K_2 + S)$, while the V_e of 8.23 is independent of S. If it is practical to add enough enzyme to the incubation mixture to get saturation with enzyme, it should therefore be possible to distinguish between the models by varying S.

Can we infallibly tell the two models apart? When activator is not tightly bound, we certainly can, since V_s of the first model is a titration curve and V_s of the second is not. If activator is tightly bound (K_2 in the second model approximately zero) then V_s of the second model is also a titration curve, and the distinction disappears.

But let us go back to the complete rate formulas 8.17 and 8.22, when the substrate is not at its saturation level. In the case of tight activator binding, $K_1 \cong 0$ in the first model and $K_2 \cong 0$ in the second. Both rate formulas assume the titration form. They are nevertheless different.

The limiting form (when $K_1 \cong 0$) of 8.17 is:

(8.25) $$v = k_3 SA_t/(K_2 + S) \qquad \text{if } A_t < E_t$$
$$v = k_3 SE_t/(K_2 + S) \qquad \text{if } A_t > V_t$$

The limiting form (when $K_2 \cong 0$) of 8.22 is:

(8.26) $$v = k_3 A_t \qquad \text{if } A_t < SE_t/(g + S)$$
$$v = k_3 SE_t/(g + S) \qquad \text{if } A_t > SE_t/(g + S)$$

The curves to be expected are illustrated in Fig. 8.3 and Fig. 8.4. Three curves, for three different S values, are included in each figure. The distinction is simple: Although the slope and the final level of the graphs increase

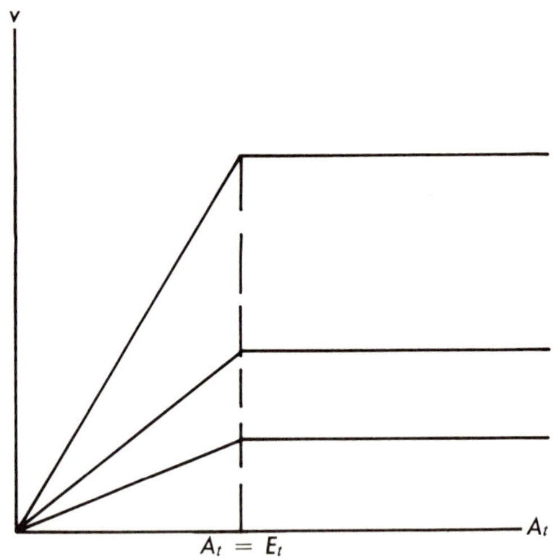

Fig. 8.3. Titration Curves as in Fig. 8.1, Drawn for Three Different Non-saturating Values of S, Showing How S Affects Final Level and Equivalence Point. (Activation model *a*.)

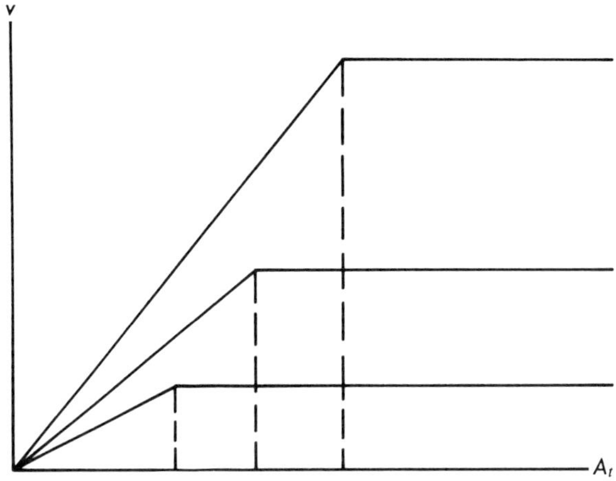

Fig. 8.4. Titration Curves at Different S Values for Activation Model b, Showing Effects of S on Final Level and Equivalence Points.

as S increases in both cases, in the first model the equivalence point is independent of S, while in the second the apparent equivalence point also increases with S. In the second model, the true equivalence point, $A_t = E_t$, is found only when the system is saturated with substrate.

PROBLEM: Develop the model of 8.C, treating E_a as pseudostationary rather than in equilibrium. How are the conclusions affected?

ACTIVATION OF SUBSTRATE

So far we have considered the role of an activator to be a combination with some form of the enzyme (either free enzyme or enzyme-substrate complex). An alternative possibility is that the activator combines with the substrate, and that it is the **compound substrate** which then successfully combines with enzyme and is transformed. For instance, one might suppose, in the case of some magnesium-activated enzymes acting on phosphate esters, that it is the magnesium salt of the ester which combines with and is acted on by the enzyme. A demonstration of this possibility was carried out with ATP-creative phosphototransferase (7).

Let us begin by considering a simple system, in which S and A combine to form an activated form S_a. This form then reacts with enzyme in the classical way. We have a scheme for activation of substrate:

(8.E) $$S + A \underset{-1}{\overset{1}{\rightleftharpoons}} S_a$$

and for enzyme action:

(8.E) $$E + S_a \underset{-2}{\overset{2}{\rightleftharpoons}} C_a \qquad C_a \xrightarrow{3} E + A + P$$

We also have a conservation equation for enzyme:

(8.27) $$E_t = E + C_a$$

We can solve the steady-state equations for C_a (scheme 8.E) by inspection ($C_a = ES_a/K_2$), and use equation 8.27 to get

(8.28) $$C_a = E_t S_a/(K_2 + S_a)$$

with the usual abbreviation $K_2 = (k_{-2} + k_3)/k_2$.

We now introduce the conservation equations for activator and substrate:

(8.29) $$A_t = A + S_a + C_a$$

(8.30) $$S_t = S + S_a + C_a$$

We now use the equilibrium equation from scheme 8.E1:

(8.31) $$SA = K_1 S_a$$

to eliminate A from 8.29. Solving 8.28 for S_a, we eliminate that variable from 8.29 and 8.30. We now use 8.30 to eliminate S from 8.29, obtaining:

(8.32) $$A_t = C_a + K_2 C_a/(E_t - C_a) + [K_1 K_2 C_a/(E_t - C_a)]/[S_t - C_a - K_2 C_a/(E_t - C_a)]$$

It is pointless to try to solve this for C_a; the equation is of degree 4 in that variable. We shall therefore use the device of studying the inverse function, introduced in Chapter 1. We first simplify by converting to a relative velocity, $r = k_3 C_a/k_3 E_t = C_a/E_t$. (We shall show subsequently that $k_3 E_t$ is indeed the maximal rate of the enzyme.) We have:

(8.32′) $$A_t = rE_t + K_2 r/(1 - r) + K_1 K_2 r/[(1 - r)(S_t - rE_t) - K_2 r]$$

Consider A_t as a function of r. When $r = 0$, $A_t = 0$. As r increases, each of the terms increases because of the r in the numerator. Also, the second term has $1 - r$ in the denominator; this becomes smaller as r increases, and so the fraction increases still faster. The denominator of the third term starts with the value S_t (when $r = 0$); it also gets smaller, so the term becomes still larger.

When one of the denominators (terms two and three) becomes very small, the term itself will become very large, and so A_t will become very large corresponding to the case of partial saturation with activator. The only problem is which value of r to assign. Setting the denominator of the second term equal to zero, we have $r = 1$ or $v = V_a = V$. Setting the denominator of the third term equal to zero gives a quadratic, and the solution is:

$$(8.33) \qquad r_a = \tfrac{1}{2}\{[1 + (S_t + K_2)/E_t] \pm \sqrt{[1 + (S_t + K_2)/E_t]^2 - 4S_t/E_t}\}$$

The limiting value of r, for which A_t becomes very large, will be either $r_a = 1$, or r_a as given by 8.33, whichever is smaller. To see what will happen, let us refer to the two roots of the quadratic expression (given by 8.33) as a and b (a is the smaller root, with minus sign of radical). We can then write 8.32 as:

$$(8.32'') \qquad A_t = rE_t + K_2 r/(1 - r) + K_1K_2 r/(a - r)(b - r)E_t$$

Suppose $a < 1$ and $b > 1$. Then A_t will be positive, and getting larger, till r reaches the value a. at which point A_t increases without bound. This behavior is shown by branch A in Fig. 8.5.

As soon as r gets a little larger than a, the third term of 8.32″ is negative, but very large. As r increases further, and approaches 1, the second term of 8.32″ is becoming very large, and it is positive. Thus it soon outweighs the negative third term, and the whole expression for A_t reaches the value 0 and then becomes positive. It increases without bound as r arrives at the value 1 (branch B on Fig. 8.5). Again, as r passes 1, the second term is large and negative.

In the meantime r is approaching b, and the third term is getting very large. This term is already negative, since r is greater than a; so the function stays negative, and becomes extremely large, but still negative, as r reaches b. This part of the function is shown in branch C of Fig. 8.5.

As soon as r passes b, the third term of A_t is positive (product of two negative factors) and very large. It begins by decreasing, since the third term is now approaching $K_1K_2 r/r^2E_t$ as r significantly exceeds a and b. But meanwhile the first term of A_t is increasing linearly, and this linear term eventually dominates (branch D of Fig. 8.5).

Of the branches of the curve, only A has any significance, since it has $r = 0$ when $A_t = 0$. Branch B gives r as finite when $A_t = 0$, which cannot be true, since we have set up our stoichiometric scheme so that the active complex C_a could not possibly be formed unless activator were present. Branch C exists only for negative, and hence physically meaningless, values of A_t, and so on. This is another illustration of the fact that an equation, besides giving a meaningful solution, may give other solutions which are mathematically valid, but have no relation to reality.

It only remains to see where the roots a and b actually are with relation to the value $r = 1$. Take a to be the root 8.33 with the minus sign in front of

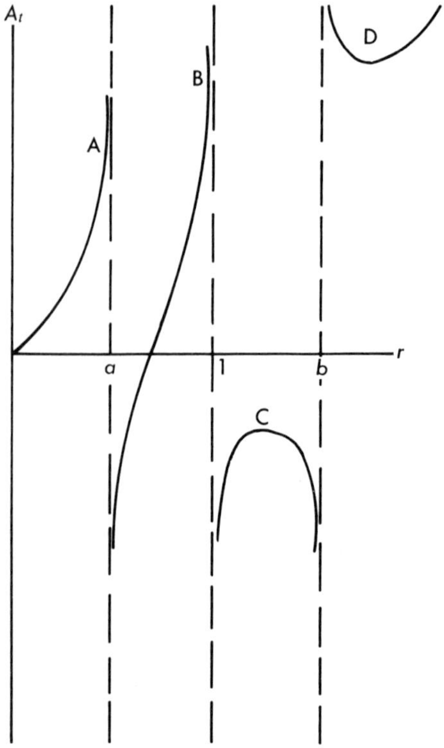

Fig. 8.5. Relation between Activator Concentration A_t and Relative Rate r. Activation via activator-substrate combination. Amounts of activator and substrate bound by enzyme are significant compared to total amounts present.

the radical (the smaller root). If we write the condition that this root shall be < 1, an elementary rearrangement gives

$$(S_t + K_2)/E_t - 1 < \sqrt{[1 + (S_t + K_2)/E_t]^2 - 4S_t/E_t}\,.$$

We do not know whether the left side of the inequality is negative or positive. But if the left side is negative, the inequality obviously holds—any negative number is algebraically less than any positive number. On the other hand, if the left side is positive, we then have the right to square[1] both sides in

[1] For example, suppose we are comparing a number whose actual value is -10 with a number whose actual value is 2. Obviously $-10 < 2$. But if we square, $(-10)^2 = (10)^2 > 2^2$. Squaring a negative number to eliminate a radical sign for purposes of comparison may therefore lead to completely false results.

order to get rid of the radical sign and cancel some terms. Doing this, and cancelling as many terms as possible, we get finally $-(K_2/E_t) < 0$, which is obviously true (any negative number is algebraically less than zero by definition). We have therefore shown that the inequality holds always—that root $a < 1$.

If we apply the same procedure to root b, with the plus sign in front of the square root, we get the condition

$$(S_t + K_2)/E_t - 1 < -\sqrt{[1 + (S_t + K_2)/E_t]^2 - 4S_t/E_t}\,.$$

If the left side is negative, we can make both sides positive by multiplying through by -1 and changing the $<$ sign to a $>$ sign. Then we square both sides. Cancelling terms where possible, we now come out with $-K_2/E_t > 0$, which is certainly untrue. If the left side of our original inequality is positive, then again the inequality is untrue, since a positive number is certainly not algebraically less than a negative number. Thus root b is greater than 1. It follows that the roots are in the order in which we showed them in Fig. 8.5. Branch A of that figure is the correct picture of the activation curve of our system, and the limiting value of r for very large values of A_t is given by the smaller of the roots 8.33 (minus sign before radical).

If we solve 8.32′ for S_t, we get the curve for substrate concentration against rate:

$$(8.34) \qquad S_t = rE_t + K_2 r/(1-r) + K_1K_2 r/[(1-r)(A_t - rE_t) - K_2 r]$$

which has exactly the same form as 8.32′, except that the symbols A_t and S_t have been interchanged. The curve of S_t against r will start at the origin, and rise to large (saturation) values of S_t when r approaches a limiting value which we can get from 8.33 simply by interchanging the symbols S_t and A_t:

$$(8.33') \qquad r_s = \tfrac{1}{2}\{[1 + (A_t + K_2)/E_t] \pm \sqrt{[1 + (A_t + K_2)/E_t]^2 - 4A_t/E_t}\}$$

The physically significant root is again the one with the minus sign.

We have shown that our present model leads to reasonable forms for the relation between rate and A_t or S_t. Since $r = v/k_3 E_t$, then $r_a = V_a/k_3 E_t$, $r_s = V_s/k_3 E_t$. The partial saturation rates are therefore obtained from 8.33 and 8.33′ by multiplying by $k_3 E_t$:

$$(8.33'') \qquad V_a = \tfrac{1}{2}k_3\{(E_t + S_t + K_2) - \sqrt{(E_t + S_t + K_2)^2 - 4S_t E_t}\}$$

$$(8.33''') \qquad V_s = \tfrac{1}{2}k_3\{(E_t + A_t + K_2) - \sqrt{(E_t + A_t + K_2)^2 - 4A_t E_t}\}$$

These rates have the correct kind of dependence on the remaining variables: V_a goes from 0 to V as S_t increases, and V_s goes from 0 to V as A_t increases. Both curves will be of the titration type if K_2 is almost zero (tight binding of S_a to enzyme). The fact that V_a and V_s will both be of titration type if either one is will serve to distinguish this model of activation from others in this chapter.

The fact that S_t and A_t enter symmetrically into 8.32′ is of course due to the fact that they enter symmetrically into reaction 1, scheme 8.E1, and that they are both taken up simultaneously (as the double molecule S_a) by enzyme in reaction 2. This symmetry can be expressed by saying that anything which is true of S_t is true of A_t, and any curve with S_t as a variable can be superimposed on the corresponding curve with A_t as a variable. This is a property of this model which none of the models for activator-enzyme combination has, and is a way of telling which activation mechanism holds in any given experimental case. For an exception due to an interesting inhibition mechanism, see reference 8.

PROBLEM: In the system of this section, a number of simplifications could have been made. For example, the conservation equation for substrate might have been omitted, that for activator might also have been omitted, or the activator conservation equation might have been simplified by dropping the term C_a (on the ground that activator bound by substrate is quantitatively more important than activator bound indirectly by enzyme). Develop these cases, and decide whether any of them is as satisfactory as the one actually derived. For example, do the partial saturation values seem always as faithful to fact?

COMPETITIVE ACTIVATORS

Frequently more than one compound will activate a given enzyme. We have already mentioned the partial interchangeability of magnesium and manganous ions for some enzymes; another likely pair is Mg^{++} and Ca^{++}.

The presence of two activators requires an extended stoichiometric scheme. For the pre-activation model, this is:

$$(8.F) \qquad E + A_1 \underset{-1}{\overset{1}{\rightleftharpoons}} E_{a1} \qquad E_{a1} + S \underset{-2}{\overset{2}{\rightleftharpoons}} C_{a1} \qquad C_{a1} \xrightarrow{3} E_{a1} + P$$

$$E + A_2 \underset{-4}{\overset{4}{\rightleftharpoons}} E_{a2} \qquad E_{a2} + S \underset{-5}{\overset{5}{\rightleftharpoons}} C_{a2} \qquad C_{a2} \xrightarrow{6} E_{a2} + P$$

The conservation equation for enzyme reads:

$$(8.35) \qquad E_t = E + E_{a1} + E_{a2} + C_{a1} + C_{a2}$$

The pseudostationary rate formula is:

$$(8.36) \quad v = k_3 C_{a1} + k_6 C_{a2} = (k_3 A_1/K_1 K_2 + k_6 A_2/K_4 K_5) S E_t / [1 + (1 + S/K_2) A_1/K_1 + (1 + S/K_5) A_2/K_4].$$

We use the abbreviations $K_1 = K_{-1}/k_1$, $K_4 = k_{-4}/k_4$, $K_2 = (k_{-2} + k_3)/k_2$, and $K_5 = (k_{-5} + k_6)/k_5$.

This rate is a normal function of S, and approaches the partial saturation value:

$$V_s = (k_3 A_1/K_1K_2 + k_6 A_2/K_4 K_5)E_t/(A_1/K_1K_2 + A_2/K_4 K_5) \tag{8.37}$$

Suppose that the second activator were absent—that is, $A_2 = 0$. In this special case, V_s would become:

$$V_{s1} = k_3 E_t . \tag{8.38}$$

Is it possible that $V_s < V_{s1}$—that the rate with both activators present should be less than with only one? If we use 8.37 and 8.38 to express this inequality, we find after simplifying:

$$V_s < V_{s1} \qquad \text{if } k_6/k_3 < 1 \tag{8.39}$$

In other words, the presence of a **second activator of lower efficiency** will **lower the rate.**

If we went to work with the post-activation model, we must appropriately rewrite scheme 8.B:

$$(8.G) \qquad E + S \underset{-1}{\overset{1}{\rightleftharpoons}} C \qquad C + A_1 \underset{-2}{\overset{2}{\rightleftharpoons}} C_{a1} \qquad C_{a1} \xrightarrow{3} E + A_1 + P$$

$$C + A_2 \underset{-4}{\overset{4}{\rightleftharpoons}} C_{a2} \qquad C_{a2} \xrightarrow{5} E + A_2 + P$$

The enzyme conservation equation is of course:

$$E_t = E + C + C_{a1} + C_{a2} \tag{8.40}$$

$$\begin{aligned} v &= k_3 C_{a1} + k_5 C_{a2} \\ &= (k_3 A_1/K_2 + k_5 A_2/K_4)SE_t/ \\ &\quad [K_1 + S + (S+g)A_1/K_2 + (S+h)A_2/K_4] \end{aligned} \tag{8.41}$$

with the abbreviations $K_1 = k_{-1}/k_1$, $K_2 = (k_{-2} + k_3)/k_2$, $K_4 = (k_{-4} + k_5)/k_4$, $g = k_3/k_1$, and $h = k_5/k_1$. The rate for partial saturation with respect to substrate is:

$$V_s = (k_3 A_1/K_2 + k_5 A_2/K_4)E_t/(1 + A_1/K_2 + A_2/K_4) \tag{8.42}$$

If the second activator is omitted, so that $A_2 = 0$, we have:

$$V_{s1} = k_3 A_1 E_t/(K_2 + A_1) \tag{8.43}$$

The condition $V_s < V_{s1}$, after simplification, reduces to:

$$V_s < V_{s1} \qquad \text{if } k_5/k_3 < A_1/(K_2 + A_1) \tag{8.47}$$

The fraction $A_1/(K_2 + A_1)$ is itself less than 1, and k_5/k_3 must be still smaller. If we raise A_1 to the saturation level, the fraction becomes equal to 1, and the inequality becomes $k_5/k_3 < 1$, similar to 8.39.

More complicated versions of the two models, in the presence of two activators, give relations too messy to repay investigation. For the time being, we may satisfy ourselves by drawing our conclusions from 8.39 and 8.47: If an enzyme system contains two activators, one of which gives active enzyme with a lower turnover number than the other, the net activity of the enzymes will be lower than in the presence of the more effective activator alone.

An excellent example of this situation is the behavior of the trans-glutamylase or glutamine exchange enzyme first described by Stumpf (9). Both the arsenate and the phosphate ion will serve as activators for this enzyme. Phosphate is the poorer activator of the two—that is, the rate with phosphate, while higher than the blank rate, is lower than that with arsenate. It is readily demonstrated that adding phosphate to the arsenate-activated system progressively lowers the net rate.

The experimental literature is full of examples of inhibition of ion-activated enzymes by somewhat similar ions (10), and closer investigation may show that some of these cases are illustrations of activator competition of the kind studied in this section.

STRUCTURAL RELATIONS IN ACTIVATOR BINDING

Many, perhaps most, of the more common ionic activators are polyvalent: Mg^{++}, Ca^{++}, Mn^{++}, Fe^{++}, HPO_4^{--}. It is conceivable, therefore, that some of them may have more than one point of attachment to the enzyme. It is true that one can visualize them as grasping enzyme with one hand and substrate with the other, so linking the two together. But this is assuredly not always the case; nor does the function of linking enzyme to substrate demand the use of one valency or charge on each of the two partners. If a divalent cation, for example, is bound to enzyme in two places, in chelate fashion, it can nevertheless serve as an electropositive region which will permit the binding of an electronegative group such as the phosphate group of a phosphate ester. It is reasonable, then, to ask what may be the kinetic consequences of twofold binding of activator. (It should be borne in mind that such twofold binding is also possible for monovalent ions: a chelate type of binding does not necessarily imply neutralization.)

One set of possibilities is illustrated in Fig. 8.6. The first stage is binding of one valence of activator by one valence or binding site of enzyme (we have pictured a bivalent cation interacting with two anionic groups of enzyme, but this should be regarded as illustrative only)—this forms complex C_1. Now the other end of the activator may become anchored, forming complex C_{12}, which will then react with substrate. On the other hand, a second activator may occupy the second binding group of the enzyme site, resulting in the inactive complex C_{11}. The complete stoichiometric scheme is written as follows:

(8.H) $$E + A \underset{-1}{\overset{1}{\rightleftharpoons}} C_1 \quad C_1 \underset{-2}{\overset{2}{\rightleftharpoons}} C_{12} \quad C_1 + A \underset{-3}{\overset{3}{\rightleftharpoons}} C_{11}$$
$$C_{12} + S \underset{-4}{\overset{4}{\rightleftharpoons}} C_2 \quad C_2 \overset{5}{\rightarrow} C_{12} + P$$

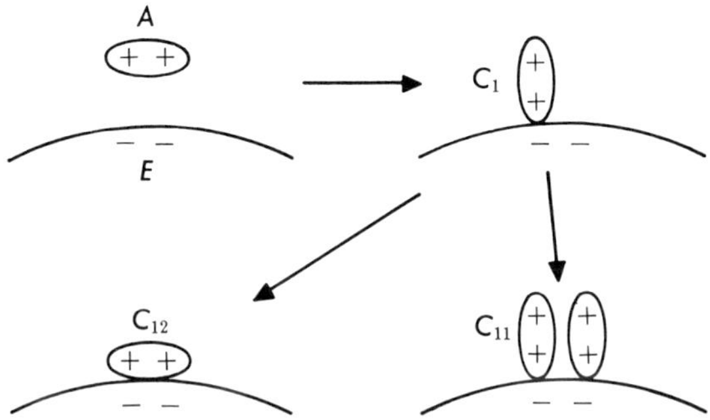

Fig. 8.6. Diagram Illustrating Steps in Activation when Activator Makes Two-point Contact with Enzyme. Includes formation of inactive form with two molecules of activator on enzyme.

The conservation equation for enzyme is:

(8.45) $$E_t = E + C_1 + C_{12} + C_{11} + C_2$$

The familiar route brings us to:

(8.46) $$v = k_5 C_2 = k_5 AxE_t/[K_1\sigma + \rho_1 x + A(1+x) + \sigma A(1 + A/K_3)]$$

with the abbreviations:

(8.47) $$K_1 = k_{-1}/k_1 \quad K_2 = k_{-2}/k_2 \quad K_3 = k_{-3}/k_3 \quad K_4 = (k_{-4} + k_5)/k_4;$$
$$\rho_1 = k_5/k_1 \quad \rho_2 = k_5/k_2 \quad \sigma = K_2 + \rho_2 x \quad x = S/K_4$$

Inasmuch as $v = 0$ when $A = 0$ and v approaches 0 as A approaches infinity, a maximum is indicated. Setting $dv/dA = 0$, the value of A at the maximum is obtained:

(8.48) $$A_m = \{[K_1(K_2 + \rho_2 x) + \rho_1 x]K_3/(K_2 + \rho_2 x)\}^{\frac{1}{2}}$$

This value depends on x (hence on S), increasing from $(K_1K_3)^{\frac{1}{2}}$ for $S = 0$ to $[K_3(K_1 + \rho_1/\rho_2)]^{\frac{1}{2}}$ as S approaches infinity. This is in agreement with common sense. The occurrence of a maximum (fall after rise) is due to the conversion of C_1 to the useless C_{11}. The substrate competes with A by combining with C_{12}, the stabilized (doubly anchored) form of C_1. Hence the eventual fall due to self-inhibition is delayed as S increases; as S rises, we get a family of curves with peaks shifting to the right.

Scheme 8.H follows a pre-activation model, modified to take into account the existence of two points of contact for the activator. If we were to follow a post-activation mechanism (formation of enzyme-substrate complex **before** combination with activator), we would get:

(8.I)
$$E + S \underset{-1}{\overset{1}{\rightleftharpoons}} C \qquad C + A \underset{-2}{\overset{2}{\rightleftharpoons}} C_1 \qquad C_1 + A \underset{-3}{\overset{3}{\rightleftharpoons}} C_{11}$$
$$C_1 \underset{-4}{\overset{4}{\rightleftharpoons}} C_{12} \qquad C_{12} \overset{5}{\rightarrow} E + A + P$$

with the conservation equation:

(8.49)
$$E_t = E + C + C_1 + C_{11} + C_{12}$$

The solution is:

(8.50)
$$v = k_5 C_{12}$$
$$= k_5 S E_t A / [(K_1 + S)(K_2 K_4 + h) + gA + (1 + K_4)SA + K_4 SA^2/K_3]$$

where $g = k_5/k_1$, $h = k_5/k_2$, and the other constants are as before.

If we set dv/dA equal to zero and solve for A, we get:

(8.51)
$$A_m = \sqrt{K_3(K_1 + S)(K_2 K_4 + h)/K_4 S}$$

The S saturation limits of 8.50 and 8.51 are:

(8.52)
$$V_s = k_5 E_t A / [(K_2 K_4 + h) + (1 + K_4)A + K_4 A^2/K_3]$$

(8.53)
$$A_{ms} = \sqrt{K_3(K_2 K_4 + h)/K_4}$$

It should be noted that 8.51 predicts a peak value A_m moving towards **lower** values as S is made larger. We are not aware of any cases which behave in this way.

MULTIPLE ACTIVATION

We may briefly consider the possibility that more than one molecule (or ion) of activator is bound (and required) per molecule of enzyme, but only one molecule of substrate. Let the scheme be:

(8.J)
$$E + A \underset{-1}{\overset{1}{\rightleftharpoons}} E_1 \qquad E_1 + A \underset{-2}{\overset{2}{\rightleftharpoons}} E_2 \quad \ldots \quad E_{n-1} + A \underset{-n}{\overset{n}{\rightleftharpoons}} E_n;$$
$$E_n + S \underset{-(n+1)}{\overset{n+1}{\rightleftharpoons}} C_a \qquad C_a \overset{n+2}{\longrightarrow} E_n + P$$

The conservation equation for enzyme is of course:

(8.54)
$$E_t = E + E_1 + E_2 + \cdots + E_n + C_a$$

It will be seen that the pseudostationary relations for the set $E_1, \ldots, E_n$ are recursive, as were the relations for $C_1, \ldots, C_n$ in Chapter 5. Using the notation of Chapter 5, we get by successive elimination $E_j = EA^j/p_j$, while $C_a = SE_n/K_{n+1}$, where we abbreviate with $K_{n+1} = (k_{-(n+1)} + k_{n+2})/k_{n+1}$. We use 8.54 to solve for E, and get finally for the rate formula:

$$(8.55) \qquad v = (k_{n+2}E_tS/K_{n+1})\,(A^n/p_n)\Big/\left[1 + \sum_1^n A^j/p_j + (S/K_{n+1})\,(A^n/p_n)\right]$$

We can approximate this formula by a simpler one, as in Chapter 5, if we assume that the binding sites for A do not influence one another. The successive constants for binding and release of A would then be the same except for the usual statistical factors:

$$(8.56) \qquad k_j = (n - j + 1)m \qquad k_{-j} = jm_-$$

Consequently, $1/p_j = C_{jn}/M^j$, with $M = m_-/m$, very much as in Chapter 5. Hence, expressing the summation in terms of the binomial

$$(8.57) \qquad v = (k_{n+2}E_tS/K_{n+1})(A/M)^n/[(1 + A/M)^n + (S/K_{n+1})(A/M)^n]$$

This function is zero when A or S is zero, and approaches the usual partial saturation values $k_{n+2}E_tS/(K_{n+1} + S)$ and $k_{n+2}E_t$ as A or S becomes very large. The interesting aspect of the function is its behavior for large n. Consider the expression $(A/M)^n$ for large n—which for practical purposes may mean a number between 10 and 20. When $A/M < 1$, the power is very small; when $A/M > 1$, it is very large.

This means that the numerator of 8.57 will be very small when $A/M < 1$, and in consequence so will the whole expression. When $A/M > 1$, on the other hand, the nth power will be so large that the saturation value of the denominator, and hence of the whole fraction, is practically realized. We have, therefore, a function which hugs the A axis when $A < M$, rises very rapidly when $A \cong M$, and is practically at its maximum value for all $A > M$. The form is illustrated in Fig. 8.7, curve A. This shape is the more pronounced as n is larger, and approaches (as n approaches infinity) a **step function** which is exactly equal to zero for $A < M$, and exactly equal to V_a for $A > M$ (curve B of the figure).

A system like this behaves as if it had a **critical** or **threshold value**. No activity is seen below $A = M$, and full activity above that value. Not only is there a threshold, therefore, but we have an example of **all-or-none** behavior—in one state, the system does not work, in another it works to the fullest degree. Moreover, if the normal value of A in a tissue were to lie slightly below M, any slight change in environment or internal behavior which would change the value of A to be slightly above M would cause the hitherto inactive system to be fully active—an example of a **trigger mechanism**.

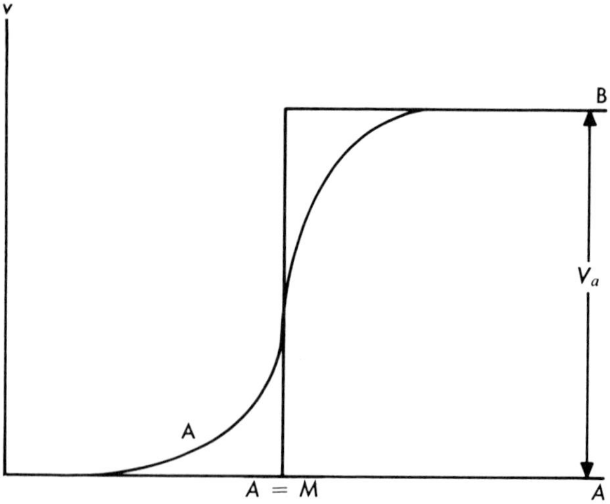

Fig. 8.7. Effects of Multiple Activation. Curve A for moderate number of activator molecules per enzyme, curve B for large number, showing activation threshold and all-or-none behavior.

The foregoing was an extension of the pre-activation model. The same kind of result comes from an extension of the post-activation model:

$$E + S \underset{-(n+1)}{\overset{n+1}{\rightleftharpoons}} C \qquad C + A \underset{-1}{\overset{1}{\rightleftharpoons}} C_1 \quad \ldots \quad C_{n-1} \underset{-n}{\overset{n}{\rightleftharpoons}} C_n \tag{8.K}$$

$$C_n \xrightarrow{n+2} C_{n-1} + P$$

The pseudostationary treatment of this scheme leads to the rate formula:

$$v = (k_{n+2} S E_t / K_{n+1})(A^n/p_n) \Big/ \left[1 + (S/K_{n+1}) \left(1 + \sum_1^n A^j/p_j \right) \right] \tag{8.58}$$

which by the assumptions leading to 8.56 simplifies to:

$$v = (k_{n+2} S E_t / K_{n+1})(A/M)^n / [1 + (S/K_{n+1})(1 + A/M)^n] \tag{8.59}$$

This function will have different partial saturation values from 8.57, in accordance with the difference of the fundamental model. But the effect of large n values on the behavior with respect to A will be as before.

ENDOGENOUS ACTIVATORS

In the study of crude or partially purified enzymes, the occurrence of endogenous activators is likely. We have already treated endogenous substrates in Chapters 6 and 7; the same lines are followed here. The amount of endogenous activator is proportional to the amount of enzyme added to the incubation mixture:

(8.60) $$A_e = fE_t$$

Taking any of the rate equations developed in this chapter, we put A_e for A or A_t (whichever the case may be) if no exogenous activator is added, or put $A_t + A_e$ for A_t, or $A + A_e$ for A, if exogenous activator is added.

An interesting example is the equation 8.17. Substituting:

(8.17) $$v = \tfrac{1}{2}k_3[S/(K_2+S)]\{A_t + (1+f)E_t + K_1K_2/(K_2+S) - \sqrt{[A_t + (1+f)E_t + K_1K_2/(K_2+S)]^2 - 4(A_t + fE_t)E_t}\}$$

Using the usual binomial approximation, we find for small E_t the form $pA_tE_t/(A_t+q)$, with $p = k_3S/(K_2+S)$ and $q = K_1K_2/(K_2+S)$. Observing the positive square root law (Chapter 1), we find that, as E_t approaches infinity, the rate v approaches $p[fE_t + A_t - fq/(1-f)]$ if $f < 1$ and $p[E_t - q/(f-1)]$ if $f > 1$.

If $f > 1$, clearly the final slope is greater than the initial slope, since $A_t/(A_t+q)$ is always less than 1. The intercept of the asymptote on the v axis is negative. The form of the curve is shown in Fig. 8.8.

If $f < 1$, the final slope is greater than the initial slope if $f > A_t/(A_t+q)$. Since q is a function of S, which approaches 0 as S approaches infinity, there

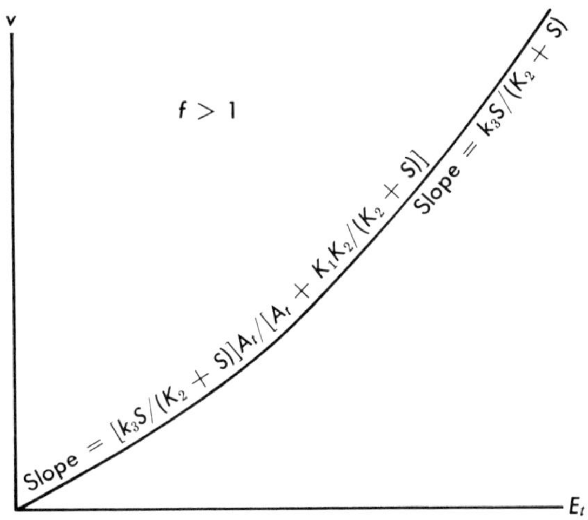

Fig. 8.8. Effect of Endogenous Activator in a Model *a* System where Activator Bound by Enzyme is Significant Fraction of Total Activator Present. Ratio of endogenous activator to enzyme (f) is greater than 1.

must be a value of S for which the final slope equals the initial slope, and for greater values of S the ratio of the slopes will be reversed. It is easy to see that the intercept reverses sign under the same conditions. This situation—a change in the form of the v-E_t curve as S is varied—is sketched in Fig. 8.9.

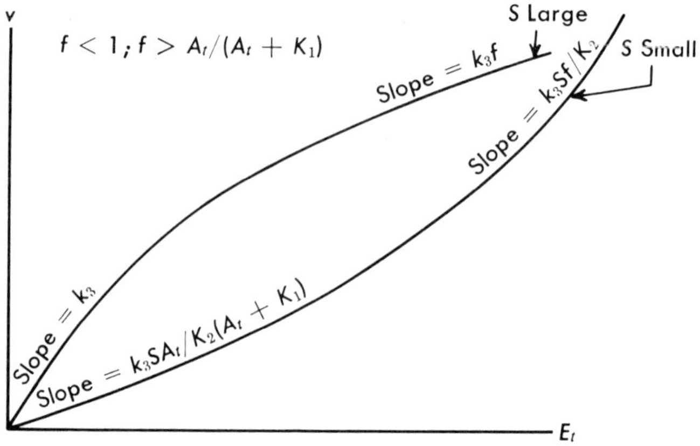

Fig. 8.9. Same System as Fig. 8.8 ; Ratio of Endogenous Activator to Enzyme Is Less Than 1. Illustrating how shape of curve can be changed by varying substrate concentration S from small to large values.

Note the physical significance of f as the ratio of endogenous activator to enzyme; the case $f < 1$ implies that the endogenous activator alone is insufficient to activate all enzyme even if it were 100% bound. The condition on A_t implies that the added A_t together with A_e is not enough to saturate the enzyme when S is small. We then get a v-E_t curve showing the characteristic dilution effect. Making S large enough compensates for the low value of $A_t + A_e$ by raising the **rate of utilization** of activated enzyme E_a; hence the shape of the curve changes. If A_t is added at saturation level, naturally the dilution effect will never be seen.

When S is at saturation level, we again have to consider the sign of the radicand in 8.17. The radicand is now the square of $A_t - (1-f)E_t$. When $f > 1$, this will always be positive, and $v = k_3 E_t$. If $f < 1$, the two possibilities are open (that is, $A_t - (1-f)E_t > 0$ and $(1-f)E_t - A_t > 0$), and we find

$$v = k_3(A_t + fE_t) \qquad \text{if } A_t < (1-f)E_t \tag{8.61}$$

$$v = k_3 E_t \qquad \text{if } A_t > (1-f)E_t$$

The rising part of the titration curve has an intercept on the v axis—which corresponds to the simple fact that the enzyme is partially activated when

$A_t = 0$. The equivalence point is $(1 - f)E_t$ (Fig. 8.10). When A_t is below this point, it means (putting in the definition of f) that $A_t + A_e < E_t$. To find the true equivalence point E_t from the experimental data is simple: (a) For small A_t get the intercept $k_3 f E_t$; for large A_t, get the maximum rate $k_3 E_t$; (c) divide one by the other, and so get f, and hence $1 - f$; (d) knowing $1 - f$, calculate E_t from the equivalence point $(1 - f)E_t$.

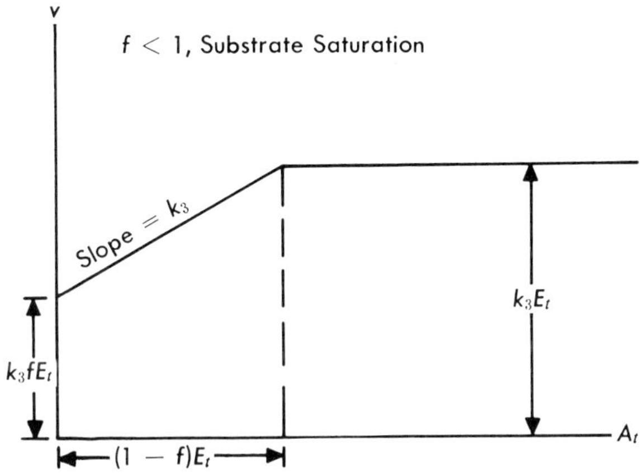

Fig. 8.10. Same System as Fig. 8.9, at Saturation Level of S, Showing Titration Type of Curve with Intercept.

PROBLEM: Develop endogenous activator theory of the simplest preactivation model (problem 5 of first section).

Tightly Bound Endogenous Activator

In connection with the occurrence of endogenous activators, we may consider one practical problem that often arises. One finds that an enzyme has a certain activator (that is, the activity of the enzyme increases when this substance is added). However, when one attempts to free the enzyme completely from activator which may have been present in the tissue that served as the enzyme source material (for example, by prolonged dialysis), there is residual activity which apparently cannot be removed without such drastic measures that the enzyme protein would be fatally damaged. The question which then arises is this: Is there a small amount of activator tightly bound to the enzyme, or is the enzyme capable of some activity in the complete absence of activator? A direct answer is not always feasible because of technical limitations; moreover the direct approach demands an absolutely pure enzyme preparation, whereas we would sometimes like to decide the issue without exhaustive purification.

The reader will readily satisfy himself that the cases are indistinguishable in the absence of exogenous activator. The first alternative can be treated by means of 8.A, with the added conservation equation $fE_t = E_x + C'_x$, with $K_3 \sim 0$, with reactions 4 and 6 omitted, and with $k_2 = 0$, while the second is obtained by putting $X = 0$ and leaving k_2 nonzero. The results are formally identical. If we add activator, and put $X + fE_t$ for X, we can use the same relations without the added conservation equation. In one case $k_2 = 0$, and $K_3 = 0$; in the other, both are non-zero. The equations are somewhat complex; but it is easily shown that the residual-activity case approaches the usual $k_2 E_t$ as S approaches infinity, while the tight-binding case approaches zero.

PROBLEM: The demonstration of the assertions of the last paragraph is left as an exercise.

INHIBITOR LIMITATION: COMPLETE INHIBITION

If we assume complete inhibition (inhibitor complexes do not function enzymatically), we drop reactions 6 and 7 of 8.A. Let us also drop reaction 5, and add a conservation equation for inhibitor:

$$I_t = I + E_i + C_i \tag{8.62}$$

The rate equation becomes:

$$v = k_2\{-(k - mD) + [(k - mD)^2 + 4kmE_t]^{\frac{1}{2}}\}/2km \tag{8.63}$$

where $k = 1 + K_1/S$, $m = 1/K_4 + K_1/K_3 S$, $D = E_t - I_t$, and the other abbreviations are as in the first section. The rate goes from $k_2 E_t/k$ when $I_t = 0$ to zero as I_t approaches infinity.

A more elegant form is obtained if we introduce the notion of **fractional residual activity** (ratio of inhibited rate to uninhibited rate) and of **fractional inhibition** (1 − fractional residual activity), which we call r' and i respectively. In the present case it is obvious that:

$$r' = kC/E_t \qquad i = 1 - r' \tag{8.64}$$

where C is determined by the pseudostationary equations in the presence of inhibitor.

In the present case E_i and C_i are dead-end compounds, since they do not turn over. Thus we have simply:

$$E_i = E_3/K_3 \qquad C_i = C_3/K_4 \tag{8.65}$$

Using these relations, 8.1 with $C'_x = 0$, and 8.62, as well as the definitions 8.64, and noting that $E + C = kC$ by definition of k, we can eliminate all the unknowns from 8.62 in favor of i, and write:

$$I_t = iE_t + (k/m)i/(1 - i) \tag{8.66}$$

This reminds us of 2.21, which relates S and r in a way similar to the relation between I_t and i (except for the additional term iE_t).

Introducing $u = i/(1-i)$ as a new variable, so that $i = u/(1+u)$, we may analyze 8.66 in the customary fashion (Fig. 8.11).

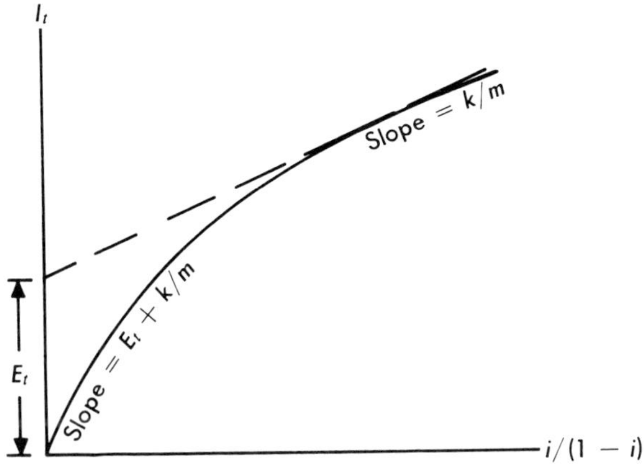

Fig. 8.11. Inhibition Curve of Simple Inhibitor, when Significant Fraction of Total Inhibitor is bound to Enzyme.

Equation 8.66 gives the amount of inhibitor I_t which must be added to the incubation mixture in order to produce a given degree of inhibition i. Suppose i is fixed at some arbitrary value, say $i = 0.5$ (50% inhibition). Then:

$$I_t = 0.5E_t + k/m,$$

which shows how the amount of inhibitor required to get 50% inhibition depends on the other factors of the system.

I_t for 50% inhibition increases directly with E_t, and of course is usually greater than $0.5E_t$. The extra term k/m is simply the amount of inhibitor which is still unbound when $i = 0.5$, while $0.5E_t$ is the amount bound.

The term k/m is a function of S. In fact, from the definitions of k and m:

$$k/m = (S + K_1)/(S/K_4 + K_1/K_3) \tag{8.67}$$

This function has the value K_3 when $S = 0$, and approaches the value K_4 as S becomes very large. It will therefore increase, remain constant, or decrease with the change in S according as $K_3 < K_4$, $K_3 = K_4$, or $K_3 > K_4$ (Fig. 8.12).

Now K_3 and K_4 are the dissociation constants of the complexes E_i and C_i. We may therefore interpret the graph of k/m against S—which except for a constant is also the graph of I_t against S—as follows: **If E binds I more tightly than does C, increasing S requires more inhibitor** to get a given degree of inhibition (for example, 50%); **if C binds I more tightly than does E, increasing S requires less inhibitor** to get a given degree of inhibition; and if **E and C bind I equally well, substrate has no influence** on the inhibitor requirement for a given degree of inhibition. We will refer to these three situations as **substrate-antagonized**, **substrate-promoted**, and **substrate-indifferent inhibition.** The last type (substrate-indifferent) is known in the older terminology as "non-competitive".

Complete Inhibition: Exclusive

Exclusive inhibition is readily seen to be a special case of inclusive inhibition. Suppose inhibitor reacts only with free enzyme E, and not with C. This can be expressed by saying that the compound of C and I, namely C_i, is dissociated to an infinite extent. Its dissociation constant is K_4. Hence in this case we say that K_4 is infinitely large. The term S/K_4 in the denominator of 8.67 is therefore vanishingly small, and we have:

$$k/m = K_3(1 + S/K_1) \tag{8.67'}$$

This increases with S regardless of the values of the constants, and so does I_t. Hence this case—*complete exclusive E-type* inhibition—is **always substrate-antagonized.**

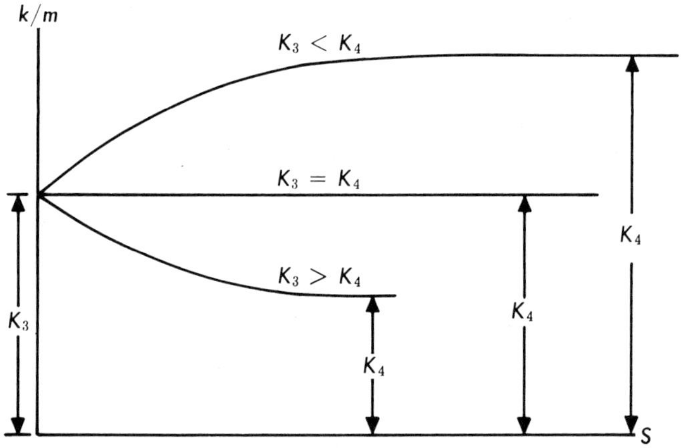

Fig. 8.12. Variation of k/m, the Slope of Fig. 8.11, with Substrate Concentration S. The three curves, reading from the top, represent cases in which inhibitor attacks free enzyme more than enzyme-substrate complex, attacks both equally, attacks free enzyme less.

Suppose on the other hand that I combines only with C and not with E. Then, by reasoning similar to that of the preceding paragraph K_3 becomes infinitely large, K_1/K_3 in 8.67 becomes vanishingly small, and we get:

$$k/m = K_4(1 + K_1/S) \qquad (8.67'')$$

This decreases with increasing S, and so does I_t. Hence this case—**complete exclusive C type inhibition**—is **always substrate-promoted.**

The fact that exclusive E inhibition is always substrate-antagonized (to put it another way, the fact that inhibitor can combine with enzyme in this case only if the enzyme is free) tempts one to refer to this sort of inhibition as "competitive," and to argue that the inhibitor and substrate are competing for the same site on the enzyme molecule. This interpretation is far too narrow, and the suggestive term "competitive" suggests more than it has a right to. Any influence exerted by substrate which will prevent binding of inhibitor, or by inhibitor which will prevent binding of substrate, would result in exclusive E inhibition. The substrate might deform the enzyme in combining with it, or change its electrical properties, or offer steric hindrance, without occupying the sites usually occupied by inhibitor. The inhibitor might act similarly without occupying the sites usually occupied by substrate.

The useful term "competitive" should be reserved for those cases where there is some definite presumption that substrate and inhibitor do in fact combine with enzyme through the same groups, and hence do compete. There are at least two lines of evidence which would support such a presumption: (a) direct evidence bearing on the nature of the binding sites, such as similar or identical pH curves for substrate and inhibitor binding; (b) evidence of a suitable similarity between substrate and inhibitor in electronic properties which are relevant to binding by enzyme. Most cases in which some presumption of competition holds are supported by evidence of type (b)—the terms "substrate analogues" or "metabolite analogues" are often used in this connection. In evaluating evidence of type (b), it would be useful to have **negative controls**—cases in which some of the supposedly essential binding groups of substrate or inhibitor are blocked or altered, and in which binding or inhibition does in fact fail to occur.

Complete Inhibition: Tight Binding

Suppose that inhibitor is bound very tightly to either E or C (or both), so that binding is *practically irreversible*. (This does not necessarily mean that binding is truly irreversible: Thus, binding may be so tight that the amount of free I in a closed system is practically nil, yet it may be possible to free the enzyme from inhibitor by dialysis.) In that case, either $K_3 \cong 0$ or $K_4 \cong 0$ (or both). In either case, the corresponding term in the denominator of 8.10 will be extremely large, and hence k/m will be practically equal to zero.

Looking at 8.66, we see that two possibilities exist if k/m approaches zero. Either $1 - i$ approaches zero (i approaches 1, meaning 100% inhibition) and hence the ratio $(k/m)i/(1 - i)$ remains finite and non-zero; or this is not true, the second term of 8.66 is zero, and $I_t = iE_t$. In other words, $i = I_t/E_t$ (provided $I_t < E_t$), or $i = 1$ (provided $I_t \geqslant E_t$).

We appear to be dealing with something similar to the titration cases. This is confirmed if we go to 8.63, and divide through by the factor m in the denominator. As k/m becomes zero, the radicand becomes simply $D^2 = (E_t - I_t)^2$, and the square root is $E_t - I_t$ or $I_t - E_t$ according as one or the other expression is positive. Hence:

$$v = (k_2/k)(E_t - I_t) \qquad \text{if } I_t < E_t \tag{8.68}$$

$$v = 0 \qquad \text{if } I_t \geqslant E_t$$

This behavior is illustrated in Fig. 8.13, where v is plotted against I_t. The equivalence point in the titration of enzyme with inhibitor is a true equivalence point, and furnishes an independent way of measuring E_t.

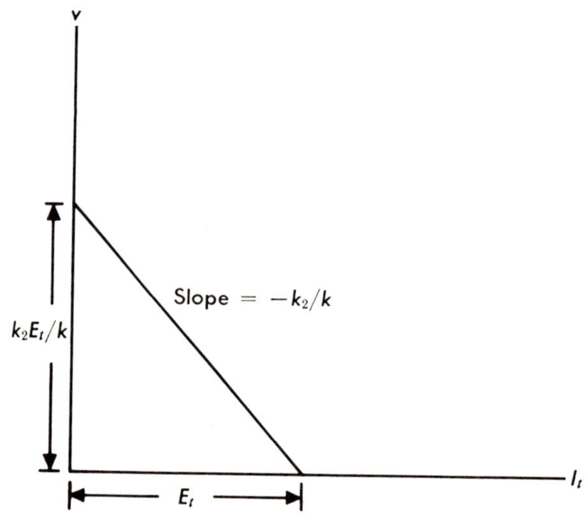

Fig. 8.13. Titration Type of Inhibition Curve Obtained when Inhibitor Is Very Tightly Bound (practically irreversible inhibition).

Complete Inhibition: Inhibitor Binding Neglected

If we can assume that the amount of bound inhibitor is negligible compared with the amount of free inhibitor, and hence drop the conservation equation 8.62, all formulas become appreciably simpler. Carrying out the calculation of v under the new condition:

$$v = k_2 E_t S/[K_1(1 + I/K_3) + S(1 + I/K_4)] \tag{8.69}$$

which can be rewritten with the help of 8.64 as

$$I = (k/m)i/(1 - i) \tag{8.70}$$

giving an interpretation to the missing term iE_t as bound inhibitor, which we have now agreed to regard as negligible.

The form of 8.69 lends itself to a reciprocal plot, and such plots have been widely employed in the analysis of inhibition, since most theories of inhibitor action have been at least as restricted as 8.69, and usually even more so. The reciprocal version has the form:

$$1/v = (1/k_2 E_t)[K_1(1 + I/K_3)(1/S) + (1 + I/K_4)] \tag{8.71}$$

It is linear in $1/S$, as in Chapter 2. The effect of the inhibitor is to change both the intercept and the slope of the straight line (Fig. 8.14). In the

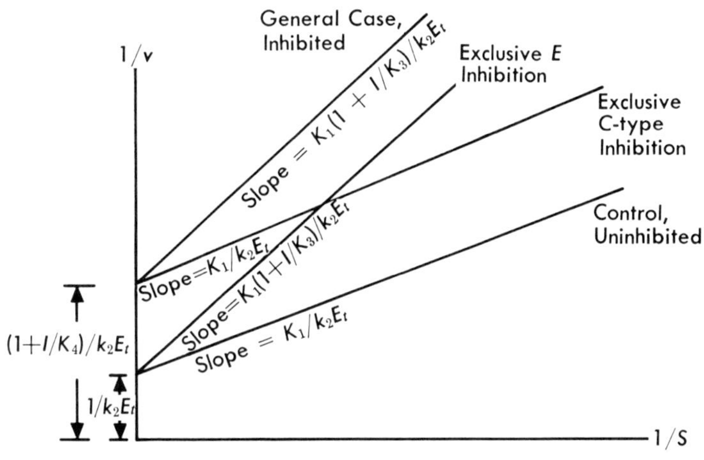

Fig. 8.14. The Effect of Inhibitor on the Reciprocal Plot of a Simple Enzyme System, when the Amount of Inhibitor Bound by Enzyme Is a Small Fraction of Total Inhibitor.

exclusive E type of inhibition ($1/K_4 \simeq 0$), the term I/K_4 drops out, so that the intercept is $1/k_2 E_t$ as in the case of the uninhibited enzyme. In exclusive C type of inhibition, the term I/K_3 drops out, and the slope is $K_1/k_2 E_t$ as in uninhibited case. In substrate-indifferent inhibition, where $K_3 = K_4$, the intercept and slope are both changed by the **same factor**.

DISPLACEMENT OF INHIBITOR BY SUBSTRATE

Implicit in the theory of the first section of this chapter was a mechanism of *partial inhibition*, resulting from the possibility that the complexes with inhibitor might exhibit enzymatic activity. A different type of partial inhibition results if we assume that the substrate itself may replace the inhibitor. For the moment, let us assume that this displacement is reversible.

(8.L) $E + S \underset{-1}{\overset{1}{\rightleftharpoons}} C \qquad C \overset{2}{\longrightarrow} E + P \qquad E + I \underset{-3}{\overset{3}{\rightleftharpoons}} E_i \qquad E_i + S \underset{-4}{\overset{4}{\rightleftharpoons}} C + I$

Reaction 4 is the new feature of this scheme. The conservation equation is $E_t = E + C + E_i$.

If we assume that reaction 3 is always in equilibrium ($E_i = EI/K_3$) and neglect the conservation equation for inhibitor, we get the simple result:

(8.72) $$v = k_2 C = k_2 E_t MS/(1 + MS + I/K_3)$$

with:

(8.73) $$M = (k_1 + k_4 I/K_3)/(k_{-1} + k_2 + k_{-4} I)$$

This rate formula has some novel features. As I approaches infinity, the rate approaches zero, since M approaches the constant value $k_4/k_{-4}K_3$, so that the denominator of 8.72 is dominated by the large term I/K_3. In this respect, the behavior is like that of complete exclusive inhibition, and unlike that of the usual mechanism of partial inhibition. On the other hand, if S is allowed to approach infinity, the rate approaches $k_2 E_t MS/MS = k_2 E_t$, which is the same as the maximum rate of the uninhibited enzyme. (If we were to take E_i in the steady state instead of equilibrium—governed by reactions 3 and 4 jointly—the rate formula would be more complex, but would have the same characteristics.)

The reciprocal form of 8.72, $1/v = (1/k_2 E_t)[(1 + I/K_3)(1/S)/M + 1]$, is linear in $1/S$, just as in 8.71. The intercept is simply $1/k_2 E_t$, independent of I (Fig. 8.15, curve B). The slope y is interesting: it is $y = (1/k_2 E_t)(1 + I/K_3)/M$. Consider y as a function of I. The initial slope of **its** graph against I (applying the binomial theorem to the denominator and discarding higher powers of I as usual) is $[k_{-4}K_3 + K_1(k_1 - k_4)]/k_1 K_3 k_2 E_t$, with $K_1 = (k_{-1} + k_2)/k_1$. The asymptote of the curve (letting I approach infinity) is also linear, with slope $k_{-4}/k_4 k_2 E_t$.

The final slope less the initial slope is $(K_3 K_4 - K_1)(k_1 - k_4)/k_1 K_3 k_2 E_t$ where $K_4 = k_{-4}/k_4$. This is positive provided $K_3 K_4 - K_1$ has the same sign as $k_1 - k_4$; it is negative if they have opposite signs, and it will be zero if either factor vanishes. Thus (Fig. 8.16, A or B) the curve of y plotted against I may be concave upward or downward.

If $k_1 = k_4$, the initial slope of the $y - I$ curve,

$$[k_{-4}K_3 + K_1(k_1 - k_4)]/k_1 K_3 k_2 E_t,$$

becomes equal to the final slope $k_{-4}/k_4 k_2 E_t$. This should make the $y - I$ curve a straight line. This is verified by seeing that the numerator of M (8.73) becomes $k_1(1 + I/K_3)$ when $k_1 = k_4$; hence $(1 + I/K_3)/M$ becomes simply $(k_{-1} + k_2 + k_{-4}I)/k_1$, and therefore y is a linear function of I (Fig. 8.16, C).

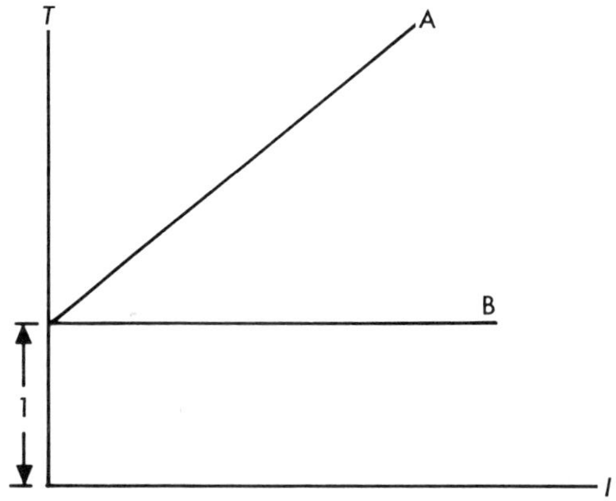

Fig. 8.15. The Intercept b of Reciprocal Plots, as a Function of Inhibitor Concentration I. Curve A is for the case of 8.71 (reciprocal plot of Fig. 8.14). Curve B is for the reciprocal form of 8.72. (Substrate displacement).

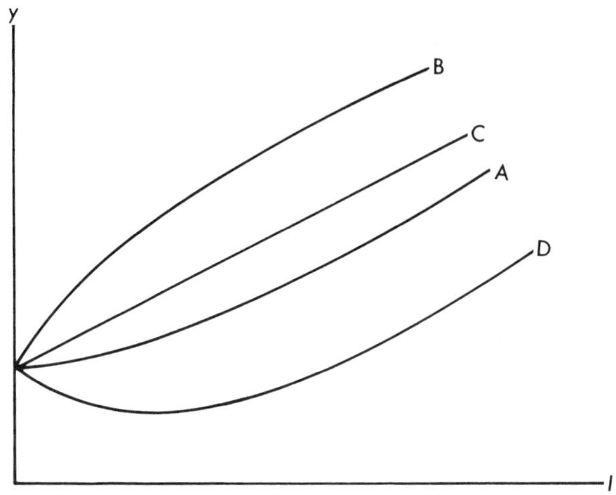

Fig. 8.16. The Slope y of Reciprocal Plots. Curves A to D are various cases of the reciprocal of 8.72.

Again, if $K_3 K_4 = K_1$, we can eliminate K_1 from the initial slope expression, cancel K_3 from numerator and denominator, and see at once (since $K_4 = k_{-4}/k_4$) that the initial slope is equal to the final slope. Similarly, going back to the formula for M with the condition $K_3 K_4 = K_1$, the denominator of M reduces to $K_3 K_4(k_1 + k_4 I/K_3)$, keeping in mind the definitions of K_4 and K_1. Hence a factor in M cancels, and M reduces to $1/K_3 K_4$. Therefore y is linear in I (Fig. 8.16, C) under this condition also.

If $k_1 < k_4$, it is possible for the initial slope of the y-I plot to be negative: The condition for this is of course $k_{-4} K_3 + K_1(k_1 - k_4) < 0$. This would give a curve (Fig. 8.16, curve D) with a minimum. The location of the minimum is at

$$I = K_3\{-k_1/k_4 + \sqrt{(k_1/k_4)^2 - [k_1/k_4 + (k_{-1} + k_2)(k_1 - k_4)/k_{-4} k_4 K_3]}\}$$

In this case, 8.72 would have a maximum with respect to I.

INHIBITOR-SUBSTRATE COMPLEXES

An inhibitor, like an activator, may affect enzyme action through the substrate, and this may occur in at least two ways: (a) The inhibitor combines with substrate, and decreases its effective concentration, or (b) the inhibitor combines with substrate, and the complex is itself an inhibitor for the enzyme.

Decrease of Effective Substrate

Consider case (a). We have the scheme:

$$\text{(8.M)} \qquad E + S \underset{-1}{\overset{1}{\rightleftharpoons}} C \qquad C \xrightarrow{2} E + P \qquad S + I \underset{-3}{\overset{3}{\rightleftharpoons}} S_i$$

together with the conservation equations:

$$\text{(8.74)} \qquad E_t = E + C$$

$$\text{(8.75)} \qquad S_t = S + S_i$$

$$\text{(8.76)} \qquad I_t = I + S_i$$

We are neglecting the amount of S bound in C, and considering only how S is distributed between free substrate and inhibitor-complexed substrate.

We find:

$$\text{(8.77)} \qquad v = k_2 E_t S/(K_1 + S)$$

with

$$\text{(8.78)} \qquad S = \tfrac{1}{2}\{-(I_t - S_t + K_3) + \sqrt{(I_t - S_t + K_3) \;\; + 4K_3 S_t}\}$$

with the abbreviation $K_3 = k_{-3}/k_3$.

When $I_t = 0$, $S = S_t$, as should be the case. As I_t increases, S decreases, and approaches zero for very large values of I_t. Finally, S approaches S_t when S_t becomes very large.

It is easy to write down the more elegant relation between I_t and i. We use the usual definition of r' and i, and, remembering that S in 8.77 is S_t in the absence of inhibitor, we find $r' = [S/(K_1 + S)]/[S_t/(K_1 + S_t)]$, and so:

$$(8.79) \qquad i/(1 - i) = K_1(S_t - S)/(K_1 + S_t)S$$

At the same time, we observe that 8.76 can be written, in view of reaction 3 ($S_i = SI/K_3$, hence $I = K_3 S_i/S$) and of 8.75 ($S_i = S_t - S$), as:[2]

$$(8.76') \qquad I_t = K_3(S_t - S)/S + (S_t - S)$$

Solving 8.79 for S (writing $i/(1 - i)$ as u to facilitate the manipulation), and substituting in 8.76′, we have:

$$(8.80) \qquad I_t = (1 + S_t/K_1)\{K_3\, i/(1 - i) + S_t\, i/(1 - i)/[1 + (1 + S_t/K_1)i/(1 - i]\}$$

We can readily see what the graph of I_t against $i(1 - i)$ will be like. When the variable $i/(1 - i)$ is small, the second term in the braces is approximately equal to $S_t\, i/(1 - i)$. Thus the initial portion of the curve is linear, with slope $y_1 = (K_3 + S_t)(1 + S_t/K_1)$. As $i/(1 - i)$ becomes infinite (as will be the case when i approaches 1), the entire second term in the braces approaches a constant. The first term is still linear. Hence the asymptote of 8.80 is again linear, with slope $y_2 = K_3(1 + S_t/K_1)$, which is smaller than the initial slope. If we prolong the asymptote back to the axis of I_t, it intercepts it at the value S_t. The curve is sketched in Fig. 8.17.

We can easily distinguish this case from any case of complete inhibition by the intercept S_t, that of the complete inhibition case being E_t, independent of S_t (and usually very much smaller). The only case with which confusion might arise is that of partial inhibition by substrate displacement, which has the general characteristics that the rate drops to zero for large I, and approaches the uninhibited maximum rate for large S. These two general statements are also true for the present case (see sentences following 8.78). To make a sharper distinction than these qualitative remarks permit, let us go back to 8.72 and work out the relation between I and $i/(1 - i)$, so as to compare the relation to 8.80.

[2] Note that 8.76′ and 8.80 (which depends on 8.76′) are usable only when S_t is finite. Since, as we have seen, S approaches S_t as S_t approaches infinity (8.78), this would make v become equal to $V = k_2 E_t$ by 8.77, and so $r' = 1$, $i = 0$. The reason why 8.76′ and 8.80 fail for infinite S_t is that they depend on the use of the relation $S_i = S_t - S$, which becomes indeterminate when both S_t and S are infinitely large.

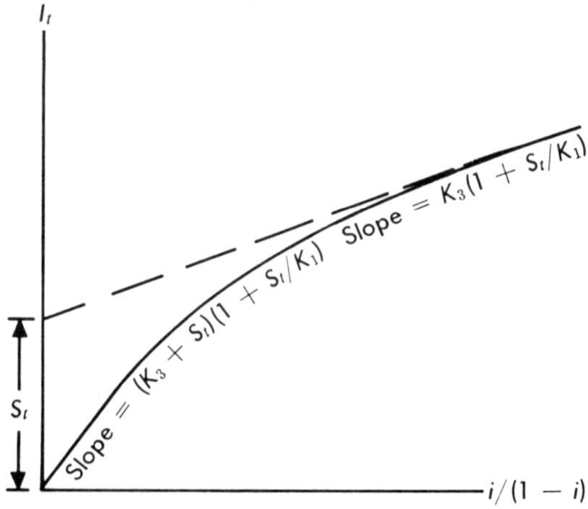

Fig. 8.17. Inhibition via Substrate, Type *a* (inhibitor decreases effective substrate concentration), Equation 8.80.

Defining r' and i as usual and using 8.72 and 8.73, we are able to get $i/(1-i)$ as a function of I (the reverse relation in this case is too complex)

$$(8.81) \quad i/(1-i) = [(1 + I/K_3)(k_{-1} + k_2 + k_{-4}I)/ (k_1 + k_4 I/K_3) - K_1]/(K_1 + S)$$

Using the binomial approximation, we find that the initial portion of the curve of $i/(1-i)$ against I is linear, with slope

$$[k_{-4}K_3 + K_1(k_1 - k_4)]/(K_1 + S)k_1K_3 .$$

The curve passes through the origin. The final portion of the curve is linear in I, with slope $K_4/(K_1 + S)$, where, as previously, $K_4 + k_{-4}/k_4$.

Looking at the situation backwards, we may say that the plot of I against $i/(1-i)$ will start at the origin, have an initial linear portion with a slope of $k_1K_3(K_1 + S)/[k_{-4}K_3 + K_1(k_1 - k_4)]$, and have a linear asymptote with a slope of $(K_1 + S)/K_4$ (Fig. 8.18).

In the curve of $i/(1-i)$ against I, the asymptote will intercept the $i/(1-i)$ axis at $(K_1k_4 - K_3k_{-4})(k_1 - k_4)/k_4(K_1 + S)$, and it will intercept the I axis at $(K_1k_4 - K_3k_{-4})(k_1 - k_4)/k_{-4}$.

We can now distinguish between the present case of substrate inhibition 8.80 and the case of partial inhibition by displacement. The 8.80 graph (Fig. 8.17) has an intercept of S_t on the I_t axis for the asymptote. The partial inhibition case for which we have just worked out the details has an I intercept of $-(K_1k_4 - K_3k_{-4})(k_1 - k_4)/k_{-4}$, independent of S.

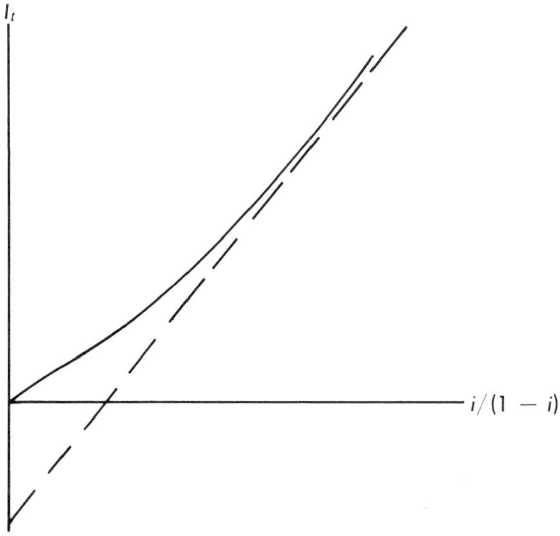

Fig. 8.18. Inhibition Curve of Substrate Displacement Type, Equation 8.81.

Inhibition by *S-I* Complex

The second way in which substrate can interact with inhibitor is shown in the stoichiometric scheme:

$$(8.M) \quad E+S \underset{-1}{\overset{1}{\rightleftarrows}} C \qquad C \xrightarrow{2} E+P \qquad S+I \underset{-3}{\overset{3}{\rightleftarrows}} S_i \qquad E+S_i \underset{-4}{\overset{4}{\rightleftarrows}} C_i$$

with the conservation equation 8.76

$$(8.76) \qquad I_t = I + S_i$$

and

$$(8.82) \qquad E_t = E + C + C_i$$

We omit the conservation equation for substrate (8.75) in the interests of mathematical simplicity, since the tying up of substrate by inhibitor is not the essence of this mechanism, but the inhibition of enzyme by the modified substrate S_i.

$$(8.83) \qquad v = k_2 E_t S/[K_1 + S + K_1 S I_t/K_4(K_3 + S)]$$

where K_3 and K_1 are as before, and $K_4 = k_{-4}/k_4$.

This rate goes from the normal value $k_2 E_t S/(K_1 + S)$ towards zero as I_t varies from zero to very large values. If S approaches infinity while I_t

remains finite, the term $K_1SI_t/K_4(K_3+S)$ approaches I_tK_1/K_4, and v approaches the uninhibited maximum value k_2E_t.

When we calculate the relation of I_t to $i/(1-i)$ as usual, we get:

$$I_t = [(K_1+S)(K_3+S)K_4/K_1S]i/(1-i) \tag{8.84}$$

Unlike most of the other curves of this sort which we have discussed, this is a straight line passing through the origin (Fig. 8.19). The slope y

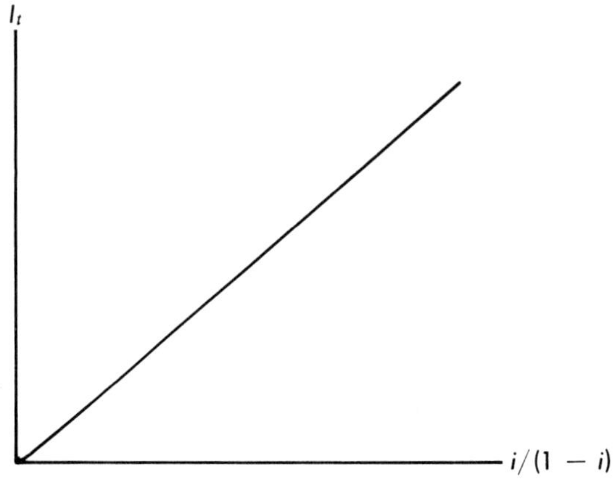

Fig. 8.19. Inhibition via Substrate (Type *b*) if Substrate-Inhibitor Complex Is the Actual Inhibitor (equation 8.84).

approaches infinity as S approaches zero (Fig. 8.20). This is what should be expected; since the inhibitor S_i cannot be formed without S, we would expect that, for small S, I_t must be very large to give any degree of inhibition. As S increases, the slope of 8.84 reaches a minimum when $S=\sqrt{K_1K_3}$, and then approaches a linear asymptote. Equation 8.70 is a straight line passing through the origin also; but the slope k/m is quite different from the slope of 8.84.

Combination of Mechanisms

If in the foregoing case we restore the conservation equation 8.75, we get a somewhat different result. The easiest way to express this result is to go directly to the I_t against $i/(1-i)$ relation.

$$I_t = K_4Ru\{K_3/(S_t+K_4)+S_t/[S_t+K_4(1+Ru)]\} \tag{8.85}$$

where we have abbreviated with $R = 1+S_t/K_1$.

Considering I_t as a function of $u=i/(1-i)$, we see that it is, as so often before, a curve with an initial linear portion and an asymptote. The initial

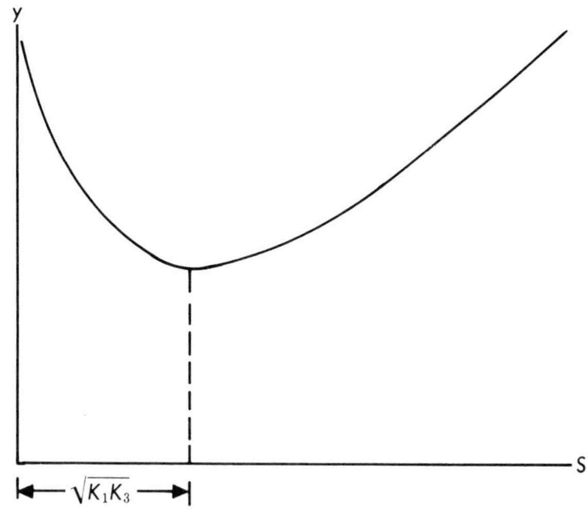

Fig. 8.20. Slope y of Fig. 8.19 as a Function of Substrate Concentration S.

slope is $y_1 = K_4 R(K_3 + S_t)/(K_4 + S_t)$. This expression, considered as a function of S_t, has some resemblance to the slope of 8.84 (bearing in mind that R stands for $1 + S_t/K_1$). But it does not become infinite for $S_t = 0$ because of the extra term K_4 in the denominator. It may have a minimum under certain conditions[3], and it has a linear asymptote. Since

$$dy_1/dS_t = (K_4/K_1)[S_t^2 + 2K_4 S_t + K_4(K_1 + K_3) - K_1K_3]/(K_4 + S_t)^2,$$

the initial value of dy_1/dS_t is $(K_4/K_1)[K_4(K_1 + K_3) - K_1K_3]/K_4^2$, and the final value is K_4/K_1. Hence the y_1-S_t curve (Fig. 8.21) will be concave up or down according as $K_4(K_1 + K_3) - K_1K_3$ is less or greater than K_4^2.

The asymptotic slope of 8.85 is simply $y_2 = K_4 RK_3/(S_t + K_4)$. It is less than the initial slope, since they are identical except for having the factor K_3 in place of $K_3 + S_t$. The intercept is S_t. The appearance of the curve of 8.85 is therefore the same as that of 8.80, Fig. 8.17. The variation of the slopes with S_t distinguishes this case from 8.80: Figs. 8.21 and 8.22 compare the initial slopes y_1, and Fig. 8.23 compares the final slopes y_2.

[3] If we set the derivative dy_1/dS_t equal to zero and solve for S_t, we get $S_{t,\,min} = -K_4 + \sqrt{(K_4 - K_3)(K_4 - K_1)}$. This is real if $K_4 - K_3$ and $K_4 - K_1$ have the same sign. It is positive only if the radicand is greater than K_4^2, which condition can be written as $K_4K_1 + K_3(K_4 - K_1) < 0$. This can surely be true only if at least $K_4 - K_1 < 0$, whence $K_4 - K_3$ must also be negative. But this does not guarantee the critical inequality, which depends on the specific values of the constants.

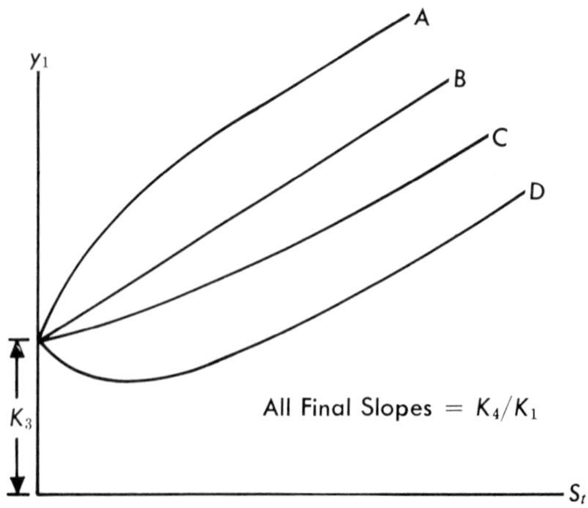

Fig. 8.21. Initial Slope y_1 of Equation 8.85 (inhibition via substrate, inhibitor lowers effective S and also inhibitor-substrate complex is enzyme inhibitor), as Function of Total Substrate Concentration S_t. Terminal portions of all curves are linear.
Curve A, $K_4(K_1 + K_3) - K_1K_3 > K_4^2$;
Curve B, $K_4(K_1 + K_3) - K_1K_3 = K_4^2$;
Curve C, $K_4(K_1 + K_3) - K_1K_3 < K_4^2$;
Curve D, $K_4(K_1 + K_3) - K_1K_3 < 0$.

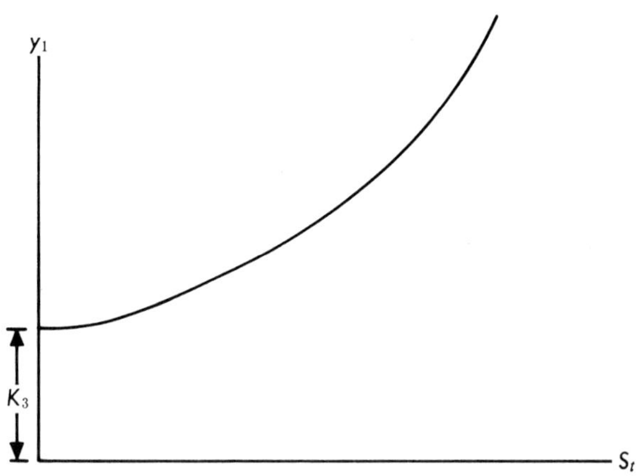

Fig. 8.22. Initial Slope y_1 of Equation 8.80 (inhibition via substrate type a, inhibition curve in Fig. 8.17) as Function of Substrate Concentration S_t. Final portion of curve varies proportional to S_t^2.

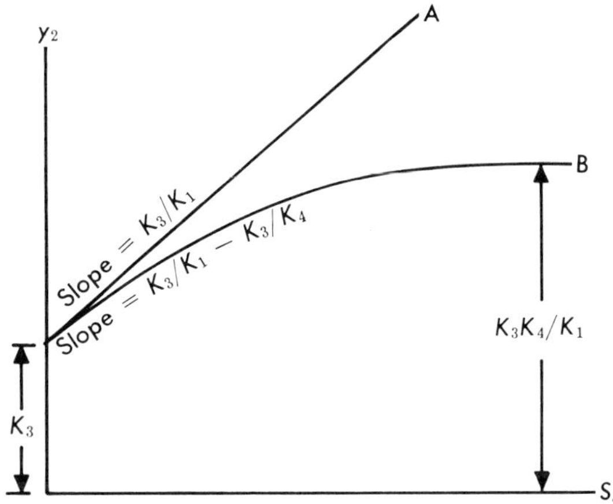

Fig. 8.23. Final Slopes y_2 of Inhibition Curves 8.80 and 8.85 (inhibition type *a* via substrate, and inhibition of types *a* and *b* combined) as Functions of Substrate Concentration S_t. Curve A (straight line) applies to equation 8.80 and Fig. 8.17; curve B applies to equation 8.85.

INHIBITOR-ACTIVATOR REACTIONS

Inhibition by interaction of inhibitor and an enzyme activator may again involve one of two mechanisms: (a) The activator is tied up, and thus kept from activating the enzyme, or (b) the activator-inhibitor complex can combine with enzyme and so inhibit it directly. A combination of the two mechanisms is also possible.

Decrease of Effective Activator

Consider first mechanism (a):

$$E + A \underset{-1}{\overset{1}{\rightleftharpoons}} E_a \qquad E_a + S \underset{-2}{\overset{2}{\rightleftharpoons}} C_a \qquad C_a \overset{3}{\longrightarrow} E_a + P \qquad A + I \underset{-4}{\overset{4}{\rightleftharpoons}} A_i \tag{8.N}$$

The conservation equations are:

$$E_t = E + E_a + C_a \tag{8.86}$$

$$A_t = A + A_i \tag{8.87}$$

$$I_t = I + A_i \tag{8.88}$$

We find:

$$I_t = u[K_4/w + A_t/(u + w)] \tag{8.89}$$

with the abbreviation $w = 1/(1 + RA_t/K_1) = K_1K_2/(K_1K_2 + A_tK_2 + A_tS)$; now R is defined as $1 + S/K_2$.

The equation 8.89 is of the familiar type of 8.80 and 8.85. The initial slope is $(K_4 + A_t)/w$, therefore a linear function of S. It increases more strongly with A_t, since it contains a term in A_t^2(the denominator of w contains A_t).

The final slope is K_4/w, which is linear both in S and in A_t. The intercept of the asymptote is A_t.

Inhibition by S-A Complex

If mechanism (b) holds, we add to scheme 8.N the reaction:

$$E + A_i \underset{-5}{\overset{5}{\rightleftharpoons}} E_{ai} \tag{8.N$'$}$$

E_{ai} being enzymatically inactive. We drop 8.87, but revise 8.86 to read:

$$E_t = E + E_a + C_a + E_{ai} \tag{8.90}$$

We treat this just as we did the previous case of mechanism (a), and find

$$I_t = (1 + K_4/A)K_5\,u/w \tag{8.91}$$

where $K_5 = k_{-5}/k_5$, and w is as before, but using the notation A instead of A_t, since we are treating them as synonymous.

This curve, just like that of mechanism (b) for substrate-mediated inhibition, is a straight line passing through the origin. The slope has the same kind of behavior with respect to A as the slope of 8.84 had with respect to S. It is $(K_4 + A)(K_1 + RA)K_5/K_1A$. As A varies, the slope expression will start with very large values, descend to a minimum when $A = \sqrt{K_1K_4/R}$ and then approach the linear expression K_5RA/K_1. For all A values, the slope is linear in S because of the term in R.

Combination of Mechanisms

If we assume a mixture of the two kinds of mechanism for activator-mediated inhibition, we still include 8.N′ in scheme 8.N and we still use 8.90, but we also use 8.87 once again. We get:

$$I_t = A_t\,u\,[K_4/A_t\,w' + 1/(u + w')] \tag{8.92}$$

where $w' = w(1 + A_t/K_5)$.

Just as 8.80 was a special case of 8.85, so we can see that 8.89 is a special case of 8.92, which would be obtained by assuming K_5 infinitely large (which amounts to saying that the compound E_{ai} is not formed). The initial slope of 8.92 is $(K_4 + A_t)/w'$ and the asymptotic slope is K_4/w', while the intercept of the asymptote is again A_t.

On the one hand, the initial slope of 8.89 is

$$(K_4 + A_t)[K_1K_2 + (K_2 + S)A_t]/K_1K_2,$$

which is proportional to A_t^2 for large values of A_t. On the other hand, the initial slope of 8.92 is $(K_4 + A_t)[K_1K_2 + (K_2 + S)A_t]/K_1K_2(1 + A_t/K_5)$, which approaches $RK_5A_t/K_1 + K_5[1 + (K_4 - K_5)R/K_1]$ as A_t approaches infinity. Even easier to detect is the behavior of the final slope

$$K_4[K_1K_2 + (K_2 + S)A_t]/K_1K_2(1 + A_t/K_5)$$

As A_t increases, this approaches the constant value RK_4K_5/K_1, which now depends only on S via R and is independent of A_t, whereas the final slope of 8.89 is $K_4[K_1K_2 + (K_2 + S)A_t]/K_1K_2$, which increases steadily with A_t.

The three cases of inhibition via activator are formally similar to those of inhibition via substrate; their graphs will resemble Figs. 8.17 and 8.19 in appearance. They can be distinguished by the dependence of slopes and intercepts on activator concentration, as discussed in the text.

ENDOGENOUS INHIBITION

Crude preparations will frequently contain substances that are inhibitory for enzymes which are present in them. Such internal or endogenous inhibition has not been studied very systematically—every enzymologist knows of a lot of cases, but has momentarily mislaid the references.

One would expect that such inhibitors might act by any of the pathways discussed in this chapter—that is, they may act on enzyme, on substrate, or on an activator. Not only such compounds as heavy metal contaminants and stray metabolic intermediates may be considered; in view of the well-known habit of complex formation among proteins, one might expect that some of the proteins in crude preparations are capable of inhibiting others, just as they are in fact capable of protecting each other against denaturation (see Chapter 11).

There is no difficulty in converting any of the formulas of this chapter into the endogenous form: It is simply a matter of writing $I = fE_t$ or $I_t = fE_t$, depending on the formula. As the simplest example, let us take 8.69 and make the substitution, getting

$$v = k_2E_tS/[K_1(1 + fE_t/K_3) + S(1 + fE_t/K_4)] \tag{8.93}$$

Considered as a function of E_t, the rate, instead of increasing steadily, approaches a saturation value (Fig. 8.24). A more illuminating way to see the effect is to plot the **specific activity** of the enzyme, defined as **rate per unit of enzyme**, or v/E_t. Since we have already introduced the symbol $r = v/k_2E_t$, the relative rate, we may as well use this, since it differs only by

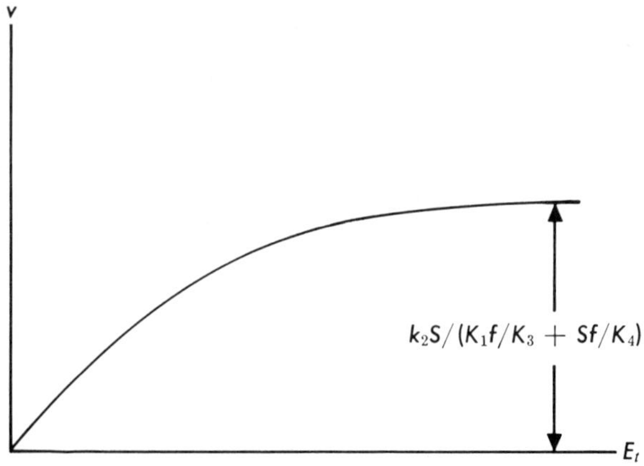

Fig. 8.24. Inhibition by Endogenous Inhibitor (simple system of equation 8.69, as modified in 8.93): Rate v as Function of Enzyme Concentration E_t.

the factor $1/k_2$, but has the same properties with respect to E_t. This function, as shown in Fig. 8.25, begins at $S/(K_1 + S)$ and decreases steadily to zero as E_t increases. Concentrating the enzyme in the incubation mixture therefore results in a decrease of the **specific activity**, while the uninhibited enzyme has the value $r = S/(K_1 + S)$ irrespective of the value of E_t. We may turn this statement around and say that the **specific activity is increased by**

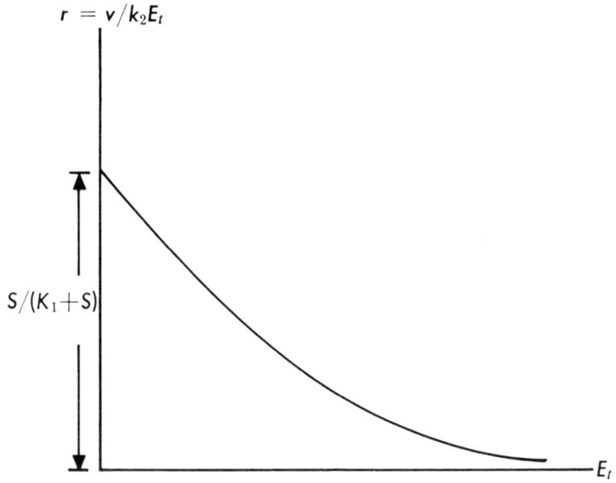

Fig. 8.25. The Curve of Fig. 8.24 Expressed in Terms of Relative Rate r Instead of Rate v, to Exhibit Second Dilution Effect.

dilution of enzyme when there is an endogenous inhibitor present—an effect which we will call the *second dilution effect.* (We remind the reader that the **first dilution effect** was found in connection with endogenous substrates or activators).

The popular reciprocal plot technique is less direct in this case, but can also be used, provided that each of its "constants" is tested for its dependence on E_t. Thus

$$1/v = (1/k_2 E_t)[K_1(1 + fE_t/K_3)(1/S) + (1 + fE_t/K_4)] \tag{8.94}$$

But the intercepts and slope depend on E_t in an unconventional way. For an uninhibited enzyme: $1/v = (1/k_2 E_t)[K_1(1/S) + 1]$. The intercept on the $1/v$ axis in an uninhibited enzyme is $1/V = 1/k_2 E_t$. The intercept on the $1/S$ axis is $1/K_1$, and is independent of E_t. In the inhibited case, the intercept on $1/v$ is $1/V'$, where the apparent constant $1/V' = (1+fE_t/K_4)/k_2 E_t$. The intercept on the $1/S$ axis in the inhibited case is $1/K' = (1+fE_t/K_4)/K_1(1+fE_t/K_3)$. The way the two intercepts depend on E_t is sketched in Figs. 8.26 and 8.27, and compared with the behavior of the same intercepts when no endogenous inhibitor is present.

In adapting to the endogenous case the various relations where I_t is plotted against $i/(1-i)$, the substitution of fE_t for I_t is correct, but another

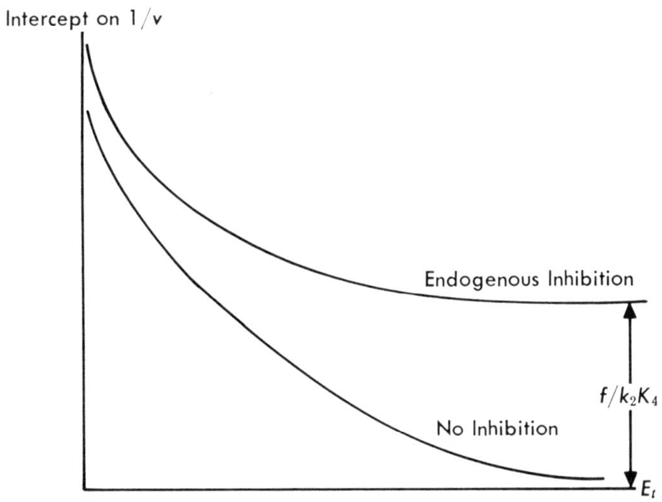

Fig. 8.26. Intercept on $1/v$ Axis from Reciprocal Plot of Endogenous Inhibition, Plotted as Functions of Enzyme Concentration E_t. Compared with intercept behavior with no endogenous Inhibitor.

step should also be taken. For as a rule one would not discover an **endogenous** inhibitor in a study of **inhibition**, but in a study of **activity.** It would therefore be useful also to convert i into a symbol more directly related to activity, as in 8.93 and 8.94. The easiest way to do this is to introduce the relative rate $r = v/V$.

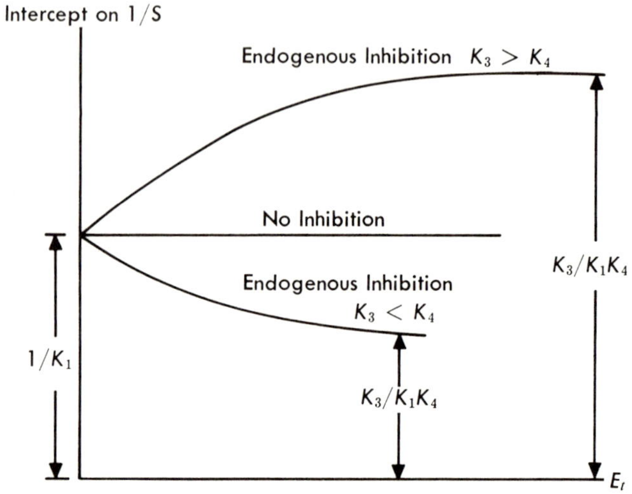

Fig. 8.27. Intercept on $1/S$ Axis from Reciprocal Plot of Endogenous Inhibition (same situation as Fig. 8.16), Plotted as Function of Enzyme Concentration E_t. Compared with intercept in absence of endogenous inhibitor.

As an example, consider the case of 8.66. Recall that we defined the residual fraction of activity as $r' = v_{inhibited}/v_{uninhibited}$. Since $v_{un} = k_2 E_t/k$ in this instance, we have $r' = kv/k_2 E_t$. But $r = v/V = v/k_2 E_t$. Therefore $r' = kr$. Then $i = 1 - kr$. With this substitution, and $I_t = fE_t$, we have (collecting terms in E_t):

(8.95) $$E_t = (k/m)(1 - kr)/kr(f - 1 + kr)$$

The graph of E_t against r, illustrated in Fig. 8.28, again shows the **second dilution effect** Looking at it as a plot of r against E_t, it begins at $r = 1/k = S/(K_1 + S)$, and decreases to 0 if $f \geqslant 1$, and decreases to $(1 - f)S/(K_1 + S)$ if $f < 1$.[4]

[4] For $f < 1$, there is a negative branch to the left of the dashed line, which has been omitted from the figure as meaningless.

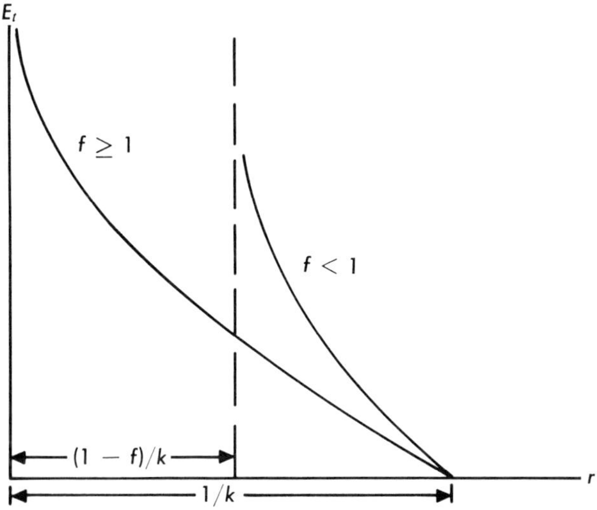

Fig. 8.28. Endogenous Inhibition, Bound Inhibitor Is Significant Fraction of Total Inhibitor (case of equation 8.66 as modified in 8.95). Relation of E_t to r, exhibiting second dilution effect. Two cases: endogenous inhibitor concentration less than ($f < 1$) or greater than or equal to ($f \geq 1$) enzyme concentration.

SOME EXPERIMENTAL COMPARISONS

Inhibitor relations of the form of 8.80 and 8.85, or of 8.89 and 8.92, seems to be frequent occurrences. Recalculations of the data of Greenberg on adenylic acid (11), of Singer *et al.* on succinoxidase and lysozyme (12), of Reif on antimycin A (13), of Town *et al.* on suramin (14), and of Altman on sulfonamides (15) give curves of this form resembling Fig. 8.17. The intercepts seem too large to be attributed to E_t, and so the curves are probably not fitted by 8.66 (Fig. 8.11).

Recalculation of Mackworth's data on ethyl iodoacetate (16), of Kachmar and Boyer's data on Ca^{++} inhibition (17), and of Reiner's data on azide and iodoacetate (18) gives curves of the form of 8.81, Fig. 8.18, with negative I intercepts. Singer's data on uranium inhibition of urease (12) may also fall in this category.

Unfortunately, the critical tests which would permit positive identification with precisely one equation (by the methods of this chapter) are not applicable, since the experiments were not designed with such tests in mind.

A somewhat disguised form of the titration equation (tight inhibitor binding) 8.68 was derived by Berry (19), applied to inhibition of cholinesterase by dicyclohexylfluorophosphate (DCFP), and used to calculate turnover constants.

COMPETITIVE SUBSTRATES AND ANTIMETABOLITES

Many inhibitors in current use are analogues of the substrates normally acted on by the enzyme in question. In some instances the analogues are what might be called "electronic analogues," and are structurally so different from the substrate that the enzyme will not act on them. An example might be the use of methionine sulfoxide as an analogue for glutamine (20). These analogues behave as simple inhibitors of the exclusive E type.

Some analogues, however, are what might be called "modified substrates," and the enzyme may act on them to some degree. It is in this sense that one may view, for instance, methyl-β-D-galactoside as an analogue of lactose, or thymine as an analogue of uracil.

When such **competitive substrates** are used as inhibitors, care must be taken in interpreting the kinetic constants obtained from experimental data, as we shall see.

Consider substrate 2 in equation 6.12 as inhibitory analogue of substrate 1. The conventional way to study a "competitive" system is to use the reciprocal plot of the rate equation. If we treat 6.12 in this way, by plotting $1/v$ against $1/S_1$, we have:

$$(8.96) \quad 1/v = (1/V)[1 + (1 + S_2/K_2)K_1(1/S_1)]/[1 + g(S_2/K_2)K_1(1/S_1)]$$

with $V = k_{1a}E_t$, $K_1 = (k_{-1} + k_{1a})/k_1$, $K_2 = (k_{-2} + k_{2a})/k_2$, and $g = k_{2a}/k_{1a}$.

The graph of this relation is not linear (Fig. 8.29). It has an initial linear portion, with slope $(K_1/V)[1 + (1 - g)(S_2/K_2)]$, and eventually it approaches the limiting value $(1/V)(K_2 + S_2)/gS_2$.

If g is very small, the curve may approximate its initial linear course over the **practical** domain. We can readily calculate what the order of magnitude of g must be if the binomial approximation is to hold to within 1%. Suppose for instance $S_2 \sim 0.01M$, $K_2 \sim 0.0001$, $K_1 \sim 0.001$. Then if $S_1 \sim 0.01M$, g must be 0.01 in order that the neglected quadratic term have a magnitude of about 0.01. If we push S_1 down to $\sim 0.0001M$, g would have to drop to 0.0001.

We can get the initial slope $(K_1/V)[1 + (1 - g)(S_2/K_2)]$ from experimental data, whether the plot is approximately linear or not. Suppose we try to compare such results with the kind of primitive formula which is in common use for analysis of so-called competitive inhibitors (21). This formula would be a special case of 8.71, obtained by setting $1/K_4 = 0$ (no inhibitor combination with C), and would be of the form $1/v = (1/V)[1 + K_1(1 + I/K_i)(1/S_1)]$, with a slope $(K_1/V))(1 + I/K_i)$, where I is the inhibitor concentration, and K_i is the dissociation constant of the E-I complex (K_3 in 8.71). Knowing the slope of the uninhibited enzyme curve, namely K_1/V, and measuring the slope for the inhibited enzyme, one can calculate $1 + I/K_i$ and hence calculate K_i—this is the reasoning generally

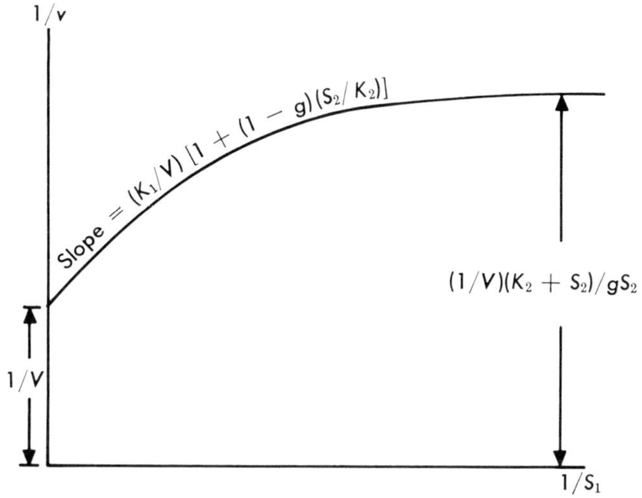

Fig. 8.29. Reciprocal Plot for System with a Competitive Inhibitor Which Is Also a Substrate. The main substrate is S_1, the competitive substrate is S_2. This type of curve is obtained if a non-specific method of analysis is used to follow the activity of the enzyme

applied. If we apply it here, remembering that the second substrate is the inhibitor, hence I is S_2, we can see, by comparison with the initial slope of 8.96, $(K_1/V)[1+(1-g)S_2/K_2]$, that K_i corresponds to $K_2/(1-g)$.

In other words, the apparent dissociation constant of the enzyme-inhibitor complex is not a true dissociation constant, and among other things depends on the ratio of turnover numbers $k_{2a}/k_{1a}=g$.

The foregoing discussion has assumed that the analytical method used will estimate the products formed from both S_1 and S_2, as for example if we were to study the effect of the methyl galactoside on the enzymatic hydrolysis of lactose by measuring the increase in reducing sugar as the glucosidic bond is broken. We may on the other hand use a method which is specific for the substrate, as would be the case if our study of the β-galactosidase were to employ the chromogenic compound *o*-nitrophenyl-β-D-galactoside and we would measure colorimetrically the amount of *o*-nitrophenol released. This would not measure the splitting of the methyl galactoside, and so the **measured** rate would only include the term $k_{1a}C_1$ but not the term $k_{2a}C_2$. In such a case k_{2a} will not appear in 8.96, and so g will not appear. The $1/v$ against $1/S_1$ graph will be a true straight line, and $K_i=K_2$.

However, K_2 has been defined as $(k_{-2}+k_{2a})/k_2$. This definition for K_2 is not changed if we use an analytical method that fails to measure the reaction via S_2 (because the balance that determines the value of $C_2=ES_2/K_2$

depends on whether the reaction via k_{2a} is going on, regardless of whether that reaction is being measured or not). Therefore K_2 is **not** in general equal to k_{-2}/k_2, and is **not** a dissociation constant, and its reciprocal does **not** measure the affinity of the enzyme for the analogue S_2. The discussion of Chapter 2, on the relation between K and K_m, applies here *in toto*.

In short, there is no general justification for using experiments with competitive substrates to estimate the effect of chemical structure on affinity for enzyme. Such calculations make sense only if the analogue is actually not a substrate at all, and one can thus safely conclude that $k_{2a}=0$ and $K_2=k_{-2}/k_2$. Conclusions based on the apparent enzyme affinities of various potentially competitive substrates (22) might well be true, but have no logical support.

STIMULATION BY INHIBITORS: SPECIAL CASES

In this section we consider some special instances of stimulation by inhibitors, besides those considered in the first section of this chapter.

Protective Partial Inhibition

Anticipating some of the results of Chapter 11, we may remark that any protein is probably at all times undergoing some denaturation, depending on temperature, pH, and other environmental factors as well as on the intrinsic structural properties of the protein. Furthermore, denaturation of proteins is known to be reversible under some conditions (23). Finally, it is known that substrates and activators protect enzymes against various inactivating influences, including proteolysis and denaturation (24). That this is also true for inhibitors (indeed probably for all compounds combining with a protein in an appropriate way) has been demonstrated (25).

We may therefore enlarge the model of the first section by adding the scheme:

(8.A′) $$E \underset{-8}{\overset{8}{\rightleftharpoons}} E_d$$

where E_d is the denatured and inactive form. For simplicity, we shall deal only with the exclusive E reaction of the inhibitor—that is, we drop reactions 4 and 6 of 8.A.

The result is to add E_d to the conservation equation. It is a dead-end compound, and does not affect the numerator of 8.2; to the denominator it adds simply $u_x/K_8(K_8=k_{-8}/k_8)$, which in the present instance is K_1/K_8. The condition for activation by effector then becomes:

(8.97) $$k_7/(K_5+S)>k_2/(K_1+K_1/K_8+S)$$

Consequently, it is possible that the condition is not satisfied in the absence of reaction 8, so that the effector is an inhibitor by our usual definition, but that 8.97 is satisfied by virtue of a sufficiently great value of $1/K_8$.

Since denaturation is known to have a very high temperature coefficient (26), it follows that a temperature increase of a few degrees might serve to convert a purely inhibitory effector into a purely stimulatory effector. It is obvious that this mechanism offers the possibility for a powerful regulatory mechanism, possibly relevant to the behavior of poikilotherms or to hibernation.

The same line of reasoning is applicable to a situation in which enzyme is diverted, not by denaturation but by still another (complete exclusive E) inhibitor. The added reaction would be:

$$E + Z \underset{-8}{\overset{8}{\rightleftharpoons}} E \tag{8.A''}$$

and the denominator of 8.2 is now enlarged by a term $K_1 Z/K_8$. The reasoning leading to 8.97 and the conclusions flowing therefrom are immediately applicable.

Observe that these mechanisms do not raise the left side of 8.97 but rather lower the right side. The kind of situation one has to think about, therefore, is one in which an enzyme is already low in activity because of denaturation or an inhibitor Z, and has its activity partially restored by the addition of the otherwise inhibitory effector X.

Without wishing to push the analogy too hard, I would like to point out the resemblance of these situations to some of the inductive phenomena observed during embryonic development. To develop this theme further would of course demand identification of the enzymatic target and a demonstration of inhibitory action by the inducer under suitable circumstances.

Another class of possible exemplifications of this mechanism is found in mammalian physiology—for instance, the effects of adrenaline on circulatory tone in the skeletal muscle on the one hand and in the gastrointestinal tract on the other (27).

PROBLEM: Apply the mechanisms of this subsection to the allosteric model (second section of chapter). Does the inhibitory allosteric effector protect?

Inhibitor as Anti-Inhibitor

Consider the case of two complete non-exclusive inhibitors, along the lines of the first section, with the turnover constants (k_6 and k_7 and their analogues for the new inhibitor) set equal to zero. The numerator of 8.2 reduces to $k_2 S$, while the denominator gains two terms in the new inhibitor concentration Z similar to those in X: $SZ/K_4' + \sigma' Z K_1$ (note that u_x reduces to K_1).

Now let X combine with Z to form a harmless compound Z_x (say one that cannot combine with enzyme); and assume that Z is sufficiently limited in concentration to require a conservation equation:

$$Z_t = Z + Z_x \tag{8.98}$$

Combining this relation with the $Z - X$ equilibrium, we get:

$$Z = Z_t/(1 + X/K_{10}) \tag{8.99}$$

where K_{10} is the dissociation constant of Z_x.

The numerator of v is independent of X; the denominator, after substitution of Z by 8.99, may be differentiated to see whether there is an extremum. We get:

$$X_e = [K_{10} Z_t(\sigma' K_1 + S/K_4')/(\sigma K_1 + S/K_4)]^{\frac{1}{2}} - K_{10} \tag{8.100}$$

It may be left to the reader to prove that: (a) X_e may indeed have positive values; (b) the condition for such values may be switched in and out by varying S; and (c) the extremum of the denominator is a minimum, making the fraction a maximum. We have here, therefore, a new example of stimulation—in a certain range of concentrations—by an inhibitor.

Rescue by Substrate Displacement

An interesting situation arises if we introduce into the substrate displacement mechanism (8.73) the additional complete exclusive E inhibitor Z. Repeating the procedure used initially, we find that a term Z/K is added to the denominator of 8.73, K being the dissociation constant of the E-Z complex; there are no other changes.

It was already shown in the original discussion of substrate displacement that the rate might have a maximum with respect to I. With the additional term in Z, we may write for the optimal value of I, after some simplifications:

$$I_m = \{-1 + [p(1 + Z/K - k_1/k_4)]^{\frac{1}{2}}\}/(k_4/k_1K_3) \tag{8.101}$$

with $p = (K_1k_4 - K_3k_{-4})/k_1$.

The conditions for I_m to be real and positive are: p and the other factor of the radicand must have the same sign, and the radicand must be greater than 1. By getting d^2v/dI^2 and introducing into it the condition $dv/dI = 0$, it is easy to see that the sign of the second derivative is that of $-p$, so that it is required that $p > 0$ if the extremum is to be a maximum. The crucial condition for I_m to be real is therefore $1 + Z/K > k_1/k_2$. The condition that the radicand in I_m be greater than 1 is $p(1 + Z/K - k_1/k_4) > 1$. The realization of any maximum, therefore, depends on Z in an obvious way. Evidently the conditions might fail for $Z = 0$ but be fulfilled for $Z \neq 0$. The theory has been applied (28) to account for the effects of citrate and oxalate on inhibition of prostatic acid phosphatase by fluoride. That citrate—not an inhibitor, with an affinity constant for the enzyme of about 10^6 compared to the fluoride affinity constant of about 10^{16}—could totally abolish fluoride inhibition is a nice demonstration of the superiority of good applied mathematics to unaided intuition. The result is perfectly reasonable, even if not obvious without theory: The citrate-enzyme complex can react with substrate and turn over—unlike the fluoride-enzyme complex, it is not a metabolic dead end. Hence citrate "saves" the enzyme in the presence of fluoride. Oxalate, a relatively poor inhibitor, is not quite as effective as citrate but behaves in an essentially similar fashion.

OPTIMAL INHIBITOR CONCENTRATIONS: THE POLYMERIC INHIBITOR

A large part of this chapter was devoted to cases in which activity is a maximum, and so inhibition is a minimum, at some inhibitor concentration. A phenomenon which is almost the mirror image of this is the case where inhibition is a maximum at some inhibitor concentration, and then decreases.

We postulate that the inhibitor I itself acts in conventional fashion of complete exclusive E type; but it also forms a polymer I_n which is a partial inhibitor by displacement. The stoichiometric equations are:

(8.O) $$E + S \underset{-1}{\overset{1}{\rightleftharpoons}} C \qquad C \overset{2}{\rightarrow} E + P \qquad E + I \underset{-3}{\overset{3}{\rightleftharpoons}} E_i \qquad nI \underset{-4}{\overset{4}{\rightleftharpoons}} I_n$$

$$E + I_n \underset{-5}{\overset{5}{\rightleftharpoons}} E_i' \qquad E_i' + S \overset{6}{\rightarrow} C + I_n$$

where we have neglected the back reaction of step 6 for the sake of simplicity.

The conservation equation is:

(8.102) $$E_t = E + C + E_i + E_i'$$

We find

(8.103) $$v = k_2 C = k_2 E_t S(1 + hI^n/K_4 K_5)/[K_1 + S(1 + hI^n/K_4 K_5) + K_1 I/K_3 + K_1 I^n/K_4 K_5]$$

where $h = k_6/k_1$, and the other constants have the usual obvious meanings.

The curve of v against I (Fig. 8.30) begins with the conventional $k_2 E_t S/(K_1 + S)$ when $I = 0$, and asymptotically approaches $k_2 E_t S/(K_1/h + S)$, which differs only by the presence of K_1/h instead of K_1 in the denominator.

The rate must decrease as I increases when I is small: for the initial slope is $-K_1k_2E_tS/K_3(K_1+S)^2$, which is negative.

It follows that v may have a minimum, which would mean that the degree of inhibition would have a maximum. Finding dv/dI and setting it equal to zero, we get the equation for the critical value of I:

$$h(n-1)I^n-(1-h)nK_3I^{n-1}-K_4K_5=0 \tag{8.104}$$

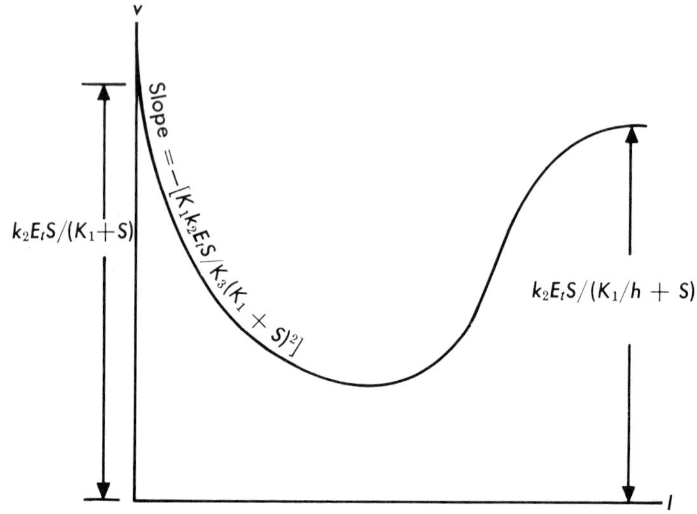

Fig. 8.30. Relation of v to I for an Inhibitor Which Is Less Inhibitory at High Concentrations than at Medium Concentrations (fluoride is an example).

This type of equation always has exactly one positive solution (by Descartes' Rule of Signs). To find the solution, we will have to estimate the numerical values of the constants, since there is no general form for the solution of an nth degree equation that would be very practical.

From the initial slope of the v-I curve, which we found to be $-K_1k_2E_tS/K_3(K_1+S)^2$, we can of course get K_3, the other constants being known or obtainable from the rate of the uninhibited enzyme. The asymptotic value of v, $k_2E_tS/(K_1/h+S)$, gives h, the other constants being known (S is held below saturation level).

To estimate n, a simple graphical trick is available. When I is large, we can neglect the constant (I-independent) terms in the denominator of v in equation 8.103, and keep only the terms in I and in I^n. Thus

$$v \simeq k_2E_tShI^n/[(K_1+hS)I^n+K_1K_4K_5I/K_3].$$

Now if we form $1/v$, we get:

(8.105) $$1/v = (1/k_2 E_t)[(K_1 + hS)/hS + (K_1 K_4 K_5/K_3 h)(1/I^{n-1})]$$

If $1/v$ is plotted against $1/I^{n-1}$, the result should be a straight line. One would thus try various values of n, plot $1/v$ agains $1/I^{n-1}$, and see which gives the best straight line. Alternatively, one may rewrite 8.105 by transferring the constant (I-independent) term $(1/k_2 E_t)(K_1 + hS)/hS$ to the left side and taking logarithms. This gives:

(8.105′) $$\log\,[1/v - (K_1 + hS)/k_2 E_t hS] = \log\,(K_1 K_4 K_5/k_2 E_t K_3 h) - (n-1)\log I$$

It follows that a plot of $\log\,[1/v - (K_1 + hS)/k_2 E_t hS]$ against $\log I$ will be a straight line whose slope will be $-(n-1)$. We already know K_1, h, and the other constants, so that this plot is perfectly feasible without assuming any trial values. Finally, the intercept of the straight line on the ordinate axis is the logarithm of the composite quantity $K_1 K_4 K_5/k_2 E_t K_3 h$. If one gets the value of the intercept from the graph, takes antilogarithms to get the composite constant, then one can calculate $K_4 K_5$, the other constants being known. All the constants of the system are now known, and a numerical solution is feasible.

It should be noted that, if we took E_i' to be determined by the steady state of reactions 5 and 6 together, we would still get $E_i' = EI_n/K_5$. But K_5 would now be defined as $(k_{-5} + k_6 S)/k_5$ instead of k_{-5}/k_5. Using the procedure for estimating $K_4 K_5$ just outlined, we could test which assumption fits best by looking for variation with S.

There are not many known inhibitors which have the polymerization properties that would fit our scheme. The theory has been successfully applied to inhibition of prostatic acid phosphatase by fluoride (28). This inhibitor is of course the one which one would have selected on theoretical grounds as most likely to behave as postulated in this section (29), and the inhibition curves all showed well-marked maxima. Lohmann (30) seems to have found a similar fluoride effect many years ago; but his published data do not suffice for a check on the theory.

The inhibition of the acid phosphatase showed an interesting secondary aspect of the tendency for fluorides to form hydrogen-bonded polymers. The inhibitor itself turned out to be, not F^-, but the ionized dimer HF_2^-. This was shown by a plot in the region of small inhibitor concentrations. Consulting 8.103, we see that, if one discards terms in I^n, and denotes by v_u the uninhibited rate $k_2 E_t S/(K_1 + S)$, the fractional inhibition i (defined as usual by $i = (v_u - v)/v_u$ is related to I by a familiar type of equation

(8.106) $$I = K_3(1 + S/K_1)i/(1 - i)$$

If I is the same as F^-, the plot of F^- (fluoride added as NaF) against $i/(1-i)$ should be a straight line. However, this was not the case, but instead a plot of $(F^-)^2$ against $i/(1-i)$ gave a straight line. Thus the inhibitor I must be a dimer of F^-. The fact that it is a dimer, and not simply two fluoride ions on the same enzyme molecule, is indicated by the pH dependence of the inhibition (31). It could be either HF_2^- or H_2F_2; but the latter is rendered less probable by the fact that at pH 5.5 its concentration is about 0.5% of that of the former, as well as by the reasonableness of a charged inhibitor as competitor for a charged phosphate ester (the other major inhibitor of the enzyme being tartrate ion).

When the techniques for finding n were applied to the fluoride case, the best fit was given by $n = 2$. The partial inhibitory I_n is therefore a tetramer of F^-.

INHIBITOR-ACTIVATOR PAIRS

Up to the present time we have treated activation of an enzyme as a simple formation of a complex between enzyme and activator. There are cases, however, which suggest that activation may sometimes modify some group of the enzyme, while the activator is converted to a compound which would tend to produce the opposite effect. For instance, some enzymes require the presence of a reducing agent for activation. This suggests that the enzyme may have a sulfydryl group which must be in the reduced state for full enzyme activity. Obviously, in the activating reaction, while the —SH group is produced on the enzyme (for example, from an —SS— group), the reducing agent is in turn converted into an oxidizing agent. (Some cases of activation by reducing agents are actually cases of binding of heavy metal inhibitory contaminants, and fall under the earlier sections of this chapter.)

The stoichiometric relations which we shall consider are therefore:

$$E + S \underset{-1}{\overset{1}{\rightleftharpoons}} C \qquad C \xrightarrow{2} E + P \qquad E' + A \underset{-3}{\overset{3}{\rightleftharpoons}} E + A' \tag{8.P}$$

where A' is the "oxidized" form of the activator, and E' is the "oxidized" and inactive form of the enzyme.

The conservation equations are:

$$E_t = E + C + E' \tag{8.107}$$

$$A_t = A + A' \tag{8.108}$$

As we set up the usual equations in order to solve for the various quantities, we get a sense of going around in circles. A count of variables shows five to be determined. A count of equations shows only four: an equation for C, the equilibrium equation for A and A' (or E and E'), and two conservation equations.

A simple consideration gives the missing equation. If the enzyme preparation with which we started, before adding activator, had an amount of free "reduced" enzyme E_0, after activator and substrate are added there will be a larger amount of reduced enzyme, which will be distributed between the two forms E and C, in a ratio determined by reactions 1 and 2. The increase in the amount of active enzyme will be equal to the amount of oxidized activator formed (assuming that we added all the activator in the reduced form to begin with). In other words:

$$A' = (E + C) - E_0 \tag{8.109}$$

We are now in a position to determine all the variables, and we find for C:

$$(E_t - QC)(A_t + E_0 - QC) = (QC - E_0)K_1K_3C/S \tag{8.110}$$

with the abbreviation $Q = 1 + K_1/S$.

As S approaches infinity, Q approaches 1. The right side of 8.110 is zero, so one of the factors on the left must vanish. The solution obviously is that then $C = A_t + E_0$ if $A_t + E_0 < E_t$, while $C = E_t$ if $A_t + E_0 > E_t$ (a titration type of relation between C and A_t). We say "obviously," because C cannot exceed the initial amount of active enzyme E_0 plus the total of added activator when there is not enough activator to reduce all the enzyme present; when activator is in sufficient excess to reduce all enzyme, it can all be in the active form, and saturation with substrate will push it all into the form C.

Since the maximum value of C is E_t (in the presence of excess substrate and excess activator), the maximum rate will be $k_2 E_t$. The relative rate r is therefore $r = k_2 C/k_2 E_t = C/E_t$. Eliminating C in favor of r, we can solve for A_t, thus evading the solution of the quadratic for r:

$$A_t = (QE_t r - E_0)[1 + (K_1K_3/S)r/(1 - Qr)] \tag{8.111}$$

If we set $r = 0$, we get $A_t = -E_0$. This simply means that the rate r always starts at a level greater than zero because of the initial amount of active enzyme E_0. The starting level is given by setting the first factor on the right equal to zero, giving $r = E_0/QE_t$; this is the value of r for $A_t = 0$. For large S (saturation), this value is simply E_0/E_t (since Q is then equal to 1); let us call it r_i—it gives the initial fraction of active enzyme.

Setting the expression in square brackets equal to zero also gives an r for which A_t vanishes, $r = 1/(Q - K_1K_3/S)$. We reject this solution because it has no relation to E_0, which r must have when there is no activator present. Not only is this solution $r = 1/(Q - K_1K_3/S)$ greater than the other one, $r = E_0/QE_t$ (since $E_0 < E_t$, while $Q > Q - K_1K_3/S$); but it becomes equal to 1 when S becomes very large. The implication would be that the rate can

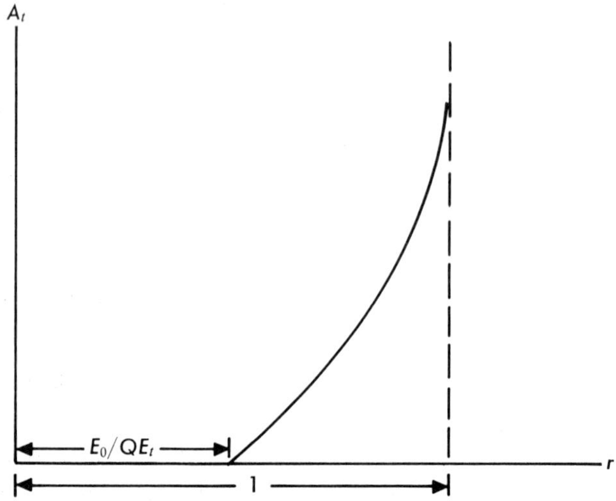

Fig. 8.31. Relation between Activator Concentration and Relative Rate in a System where the Activator Forms a Potential Inhibitor When it Activates the Enzyme (sulphydryl compound or other redox system).

have its maximum value at substrate saturation, even though there is **no** activator to get all the enzyme into the active state—a conclusion which cannot be accepted.

As r increases, it approaches $1/Q$, and A_t approaches infinity. Since $Q \geqslant 1$, $1/Q \leq 1$, and therefore lies within the physically meaningful range of r. This asymptotic value of r, then, is the value for saturation with A_t.

The activation curve (Fig. 8.31) is quite similar to that of an endogenous activator, the activity beginning with a rate determined by the initial amount of reduced enzyme in the preparation in this case, and a rate determined by the amount of enzyme activated by the endogenous activator in the previous case.

If the added activator were actually a mixture of the reduced and the oxidized forms, the increase in active or reduced enzyme would be to the increase in oxidized activator:

(8.112) $$A' - A_0' = (E + C) - E_0,$$

where A_0' is the amount of oxidized form present in the material added. The resulting equations would seem to give the same relation as 8.111, except that E_0 is changed to $E_0 - A_0'$:

(8.111′) $$A_t = (QE_t r - E_0 + A_0)'[1 + (K_1K_3/S)r/(1 - Qr)]$$

However, we must remember that A'_0 is not independent of A_t, from the way in which it is defined. For any preparation of activator, the oxidized form is some definite fraction of the total, which means that $A_0' = bA_t$. If we substitute this value into 8.111′, it is obvious that the equation must be rewritten to collect terms in A_t, and we have finally:

$$(8.113) \quad A_t = (QE_t r - E_0)[1 + (K_1K_3/S)r/(1 - Qr)]/ \{1 - b[1 + (K_1K_3/S)r/(1 - Qr)]\}$$

The initial value of r is E_0/QE_t as before. But if we let r approach $1/Q$, the denominator would become negative, since the term in square brackets in the denominator would be extremely large. This would give a negative value for A_t, and so would be senseless. Actually, as the term in square brackets in the denominator becomes larger with the growth of r, the whole denominator must be getting smaller, since the large term multiplied by b is being subtracted from 1. Hence, **before** r can closely approach the value $1/Q$, the **whole denominator** will approach **zero**, making A_t approach infinity. Setting the denominator equal to zero and solving for r, we get $(1-b)/[(1-b)Q + bK_1K_3/S]$. This is a smaller value than $1/Q$, as can be seen by factoring $1 - b$ from numerator and denominator, giving $1/[Q + K_1K_3b/S(1-b)]$, which has a larger denominator than $1/Q$ but the same numerator. The effect of adding activator contaminated with what we might call **anti-activator**, therefore, is to lower the rate which can be obtained by activator saturation. What this means physically is that, for every mole of total activator added, there is actually $1 - b$ mole tending to push reaction 3 to the right, while a fraction b of a mole of anti-activator is also added, tending to push it to the left. Since $E'/E = K_3 A'/A$, we will have (when A_t becomes so large that the slight shifts in its components due to transformation by reaction 3 can be neglected) $E'/E = K_3 bA_t/(A_t - bA_t) = K_3 b/(1 - b)$. Inasmuch as $E'/E = (E_t - QC)/(K_1C/S) = (1 - Qr)/(K_1r/S)$, setting this equal to $K_3 b/(1 - b)$ and solving for r, we get the result just found.

We may ask, finally, what would happen if we added an inhibitor which would oxidize the activator (for instance, if our activator were cysteine, we might add ferricyanide, iodine, or a trace of heavy metal). This would add the reaction:

$$(8.\mathrm{P}') \qquad A + I \underset{-4}{\overset{4}{\rightleftharpoons}} A' + I'$$

where I is the oxidant and I' the reductant form of the inhibitor. We must add an extra equation based on stoichiometric considerations, analogous to 8.109 and 8.112, inasmuch as the equation for reaction 4 again determines only the ratio I'/I and not either one separately:

$$(8.114) \qquad I' = A_0 - A = A_t - A_0'' - A$$

where A_0'' and A_0 are the amounts of oxidized and reduced activator present in the absence of inhibitor (imagine that we add activator to enzyme, let the oxidized and reduced forms come to equilibrium, then add inhibitor to shift the proportions). What the equation states is that the amount of inhibitor converted to reductant (assuming it all starts in the oxidant form) is equal to the loss in reduced activator due to oxidation by inhibitor.

Finally, we have the extra conservation equation:

$$I_t = I + I' \tag{8.115}$$

We find after tedious but elementary manipulations (introducing $r' = 1 - i$ by $C = r'C_u$ as usual):

$$I_t = \{1 + K_4(E_t - QC_u + QC_u i)/\{K_1K_3C_u(1-i)/S]\}. \quad \{A_t(E_t - QC_u + QC_u i)/[E_t - (Q - K_1K_3/S)C_u(1-i)] - A_0''\} \tag{8.116}$$

We can eliminate A_0'' from 8.16 if desired.[5] By definition, A_0'' is the A' of 8.109—the oxidized activator present **before** inhibitor was added. Hence, in terms of our present notation (C_u for uninhibited C value), $A_0'' = QC_u - E_0 = QC_u - r_i E_t$, where r_i is the (measured) initial rate of the uninhibited system.

When $i = 0$, the second factor of 8.116 vanishes by virtue of 8.110, the equation that determines C_u (remembering that $A_0'' = QC_u - E_0$). As i approaches 1, the first factor approaches infinity, while the second factor remains finite; hence I_t approaches infinity. The inhibition curve would thus seem to be conventional, the fractional inhibition going from 0 to 1 as I_t goes from 0 to large values.

Somewhat more information can be gathered if we introduce once again the variable $u = i/(1-i)$. From this relation, $i = u/(1+u)$, and $1 - i = 1/(1+u)$. Making these substitutions in 8.116, we have:

$$I_t = \{1 + K_4[(1+u)E_t - QC_u]/(K_1K_3/S)C_u\} \quad \{[(1+u)E_t - QC_u]A_t/[(1+u)E_t - QC_u + (K_1K_3/S)C_u] - A_0'' \tag{8.117}$$

The curve of I_t against u passes through the origin. It is linear for small u values. Applying the binomial expansion for small u, and using 8.110 and definition of A_0'' to cancel some terms, we get the initial slope

$$[1 + K_4(E_t - QC_u)/(K_1K_3/S)C_u](K_1K_3/S)C_u E_t A_t/[E_t - QC_u + (K_1K_3/S)C_u]^2$$

[5] Alternatively, we could eliminate QC_u in favor of A_0'', provided we now measure A_0'' directly in the uninhibited system. Whichever we do, it should be obvious that all relevant factors (E_t, S, A_t, and so on) must be the same in the inhibited and uninhibited system experiments whenever we are going to use the C_u of A_0'' of the uninhibited system in the formulas for the inhibited system.

When u is very large, I_t is again linear, with a slope

$$K_4 E_t(A_t - A_0'')/(K_1K_3/S)C_u.$$

These somewhat complex expressions simplify if we let A_t become rather large, as it usually is in activations by reducing agents. Now QC_u approaches E_t (8.111). Then the initial slope approaches

$$E_t A_t Q/(K_1K_3/S)E_t = (K_1 + S)A_t/K_1K_3$$

(definition of Q!), while the asymptotic slope is

$$K_4 E_t A_t Q/(K_1K_3/S)E_t = K_4(K_1 + S)A_t/K_1K_3.$$

The two slopes differ by just the factor K_4, which can therefore be estimated from their ratio. The intercept of the asymptote on the I_t axis is A_t.

The only two cases with which this might be confused are 8.103 and 8.106 (activator-mediated inhibition, which is of course closely related to the mechanism of this section). But the initial slope of 8.103 is proportional to A_t^2 for large A_t. In 8.106, the final slope is independent of A_t for large A_t. Both cases, therefore, can be distinguished from the present instance.

REFERENCES

General Background

Segal, H. L., Kachmar, J. F., and Boyer, P. D., Enzymologia *15*, 187 (1952).
Straus, O. H., and Goldstein, A., J. Gen. Physiol. *26*, 559 (1943).
Goldstein, A., J. Gen. Physiol. *27*, 529 (1944).

(These are excellent systematic studies, probably the best published during the forties and fifties.)

My warmest recommendation goes to:

J. L. Webb, "Enzyme and Metabolic Inhibitors", 3 vols., Academic Press, New York, 1963.

Aside from the merits of this work as a review of the topic of inhibitors, the first, volume strikes me as an exceptionally careful and lucid survey of enzyme kinetics, including linked systems.

Textual References

1. Elliott, W. H., J. Biol. Chem. *201*, 661 (1953); Kachmar, J. F., and Boyer, P. D., J. Biol. Chem. *200*, 669 (1952); Kuby, S. A., Noda, L., and Lardy, H. A., J. Biol Chem. *210*, 65 (1954).
2. Reiner, J. M., "The Organism as an Adaptive Control System", Prentice-Hall, Inc., Englewood Cliffs, New Jersey, 1967.
3. Monod, J., Changeux, J. P., and Jacob, F., J. Mol. Biol. *6*, 306 (1963).

4. Monod, J., Wyman, J., and Changeux, J. P., J. Mol. Biol. *12*, 88 (1965).
5. Sanwal, B. D., Stachow, C. S., and Cook, R. A., Biochemistry *4*, 410 (1965); Maeba, P., and Sanwal, B. D., Biochemistry *5*, 525 (1966); Sanwal, B. D., and Cook, R. A., Biochemistry *5*, 886 (1966); Atkinson, D. E., and Walton, G. M., J. Biol. Chem. *240*, 757 (1965); Atkinson, D. E., Hathaway, J. A., and Smith, E. C., J. Biol. Chem. *240*, 2682 (1965).
6. Koshland, D. E., Némethy, G., and Filmer, D., Biochemistry *5*, 365 (1966).
7. Noda, L., Nihei, T., and Morales, M. F., J. Biol. Chem. *235*, 2830–2834 (1960).
8. Robbins, E. A., Stulberg, M. P., and Boyer, P. D., Arch. Biochem. Biophys. *54*. 215 (1955).
9. Stumpf, P. K., Loomis, W. D., and Michelson, C., Arch. Biochem. *30*, 126 (1951).
10. Lehninger, A. L., Physiol. Revs. *30*, 393 (1950).
11. Greenberg, G. R., J. Biol. Chem. *181*, 781 (1949).
12. Singer, T. P., Muntz, J. A., Meyer, J., Gasvoda, B., and Barron, E. S. G., Metallurgical Laboratory Report CH-3739 (1947).
13. Reif, A. E., Arch. Biochem. Biophys. *47*, 396 (1953).
14. Town, B. W., Wills, E. D., Wilson, E. J., and Wormall, A., Biochem. J. *47*, 149 (1950).
15. Altman, K. I., J. Biol. Chem. *166*, 149 (1946).
16. Mackworth, J. F., Biochem. J. *42*, 82 (1948).
17. Kachmar, J. F., and Boyer, P. D., J. Biol. Chem. *200*, 669 (1952).
18. Reiner, J. M., Dissertation, University of Minnesota, Minneapolis (1946).
19. Berry, W. K., Biochem. J. *49*, 615 (1951).
20. Elliott, W. H., Biochem. J. *49*, 106 (1949).
21. Wilson, P. W. in "Respiratory Enzymes" (Lardy, H. A., ed.), Burgess Publishing Co., Minneapolis, 1949.
22. Monod, J., Cohen-Bazire, G., and Cohn, M., Biochem. Biophys. Acta *7*, 585 (1951).
23. Anson, M. L., in "Advances in Protein Chemistry" (Anson, M. L., and Edsall, J. T., eds.), vol. II, Academic Press, New York, 1945.
24. Berger, L. Slein, M. W., Colowick, S. P., and Cori, C. F., J. Gen. Physiol. *29*, 379 (1946).
25. London, M., McHugh, R., and Hudson, P. B., Arch. Biochem. Biophys. 55, 131 (1955).
26. Stearn, A. E., in "Advances in Enzymology" (Nord, F. F., ed.), vol. IX, Interscience Publishers, New York, 1949.
27. Evans, C. L., "Principles of Human Physiology", Lea and Febiger, Philadelphia, 1952.
28. Reiner, J. M., Tsuboi, K. K., and Hudson, P. B., Arch. Biochem. Biophys. *56*, 165 (1955).
29. Hildebrand, J. H., "Principles of Chemistry", Macmillan, New York, 1947.
30. Lohmann, K., Biochem. Z. *203*, 172 (1928).
31. Anagnastopoulos, C., Bull. Soc. Chim. Biol. *35*, 575 (1953).

9

Linked Reaction Systems

Any attempt to analyze cell behavior at the molecular level is bound to involve some consideration of the properties of linked reaction systems (1)[1]. Some of the more rudimentary problems were taken up in Chapter 6; in the present chapter we attempt a more general approach, and at least an introduction to some of the more complex situations.

OPEN SYSTEMS AND STEADY STATES

Consider the reaction chain of Fig. 9.1 (a). The differential equations are:

$$dX_i/dt = R_i - R_{i+1} \qquad (i = 1, 2, \ldots, n) \tag{9.1}$$

where we use R_i to represent the net rate of reaction i, and where we are temporarily ignoring compounds X_0 and X_{n+1}—but only temporarily!

A genuine steady state for this system of equations is at least conceivable. If it exists, the condition is:

$$(Ai)[R_i = R_{i+1}] \tag{9.2}$$

where "(Ai)" is a symbol of formal logic meaning "For all i", with the understanding that "all" encompasses precisely the set of admissible values already specified; this is therefore a convenient abbreviation for expressions like "$(i = 1, 2, \ldots, n)$."

In order that 9.2 be satisfied, the equations 9.2 must be consistent; and we may sometimes demand that they also possess non-trivial solutions. Thus,

[1] The topic is familiar from the early work of Dixon (2) under the cognomen of "multi-enzyme systems". I have selected the present terminology because it seems more suggestive.

$$X_0 \underset{}{\overset{1}{\rightleftharpoons}} X_1 \overset{2}{\rightleftharpoons} X_2 \overset{3}{\rightleftharpoons} \cdots \overset{n-1}{\rightleftharpoons} X_{n-1} \overset{n}{\rightleftharpoons} X_n \overset{n+1}{\rightleftharpoons} X_{n+1} \qquad \text{(a)}$$

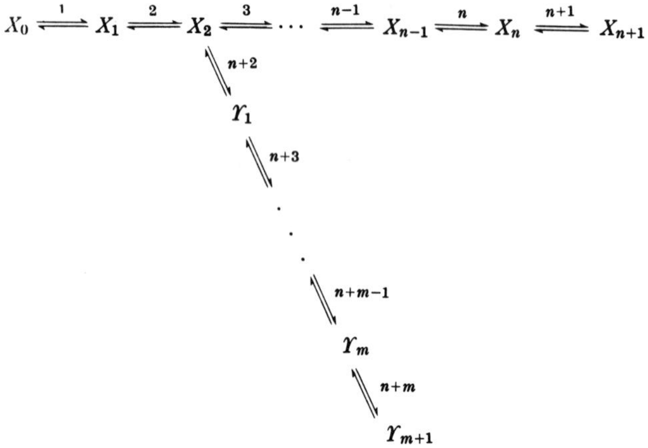

(b)

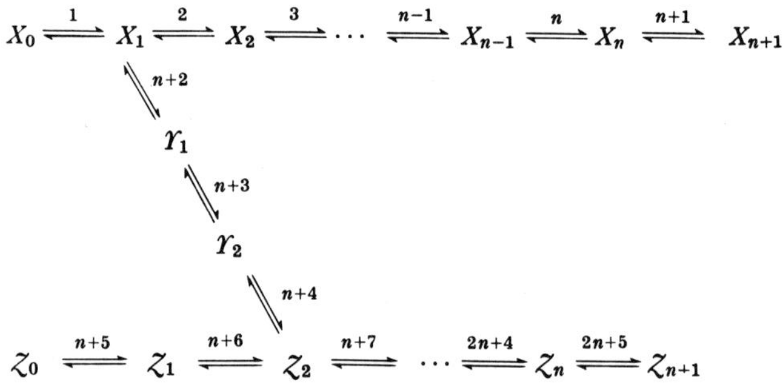

(c)

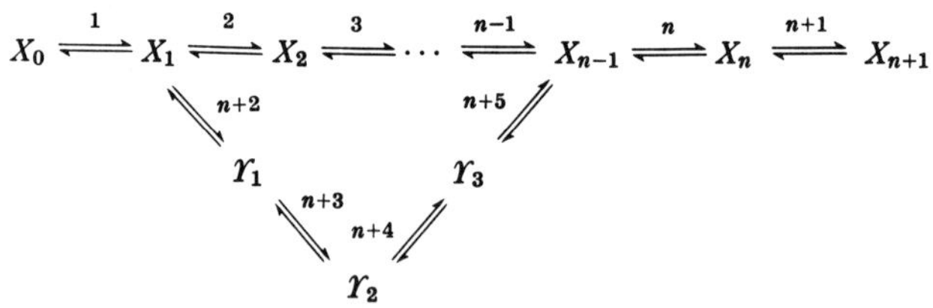

(d)

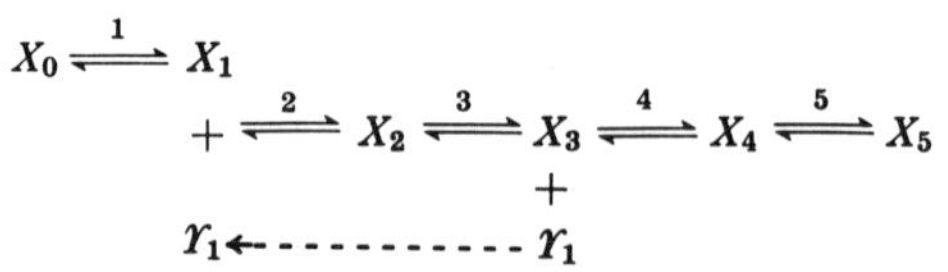

(e)

returning to the kinetic equations 2.1′ and 2.2′, it is evident that their right-hand sides cannot vanish simultaneously except for $C=0$, $S=0$; and we deem these values trivial because they represent the state of affairs when the entire reaction is finished and nothing more can occur.

It is conceivable that some of the equations 9.1 may be stationary but not all. In general, this would require their right sides to be free of the variables specified by the remaining equations. We shall consider some concrete examples later.

Assuming that all is well with 9.2, let us now ask about the neglected variables X_0 and X_{n+1}. Reactions 1 and $n+1$ being reversible as written, it is unlikely that 9.2 can be satisfied unless X_0 and X_{n+1} are also stationary (though not impossible in special cases). There are two possibilities (other than simply making the **terminal compounds** stationary by fiat): (1) We may arrange experimentally for the concentrations of these compounds to be kept stationary or virtually stationary; or (2) we may let $R_1=0$ and $R_{n+1}=0$. The second case is uninteresting: By virtue of 9.2 it implies $(Ai)[R_i=0]$, which means that each reaction separately is in equilibrium —no net chemical reaction can occur—the system is dead.

The first eventuality may be achieved directly by the experimenter as *deus ex machina* (perhaps more properly *deus per machina*): X_0 may be added from a suitably controlled burette as in an apparatus for automatic titration, and X_{n+1} may be removed by an adsorbent or a precipitant, or may be volatile like CO_2. Alternatively, we may consider X_0 and X_{n+1} to lie outside our system (which is a partial model of a cell), so that reactions 1 and $n+1$ are **transport reactions** (whether simple diffusion or active transport is immaterial for the moment). We may then arrange for the external volume to be so large that the concentrations of X_0 and X_{n+1} do not change significantly during a long period of time (long, that is, compared with the time of the observation, or with the life span of the cell). In the limit, the external volume may become the world, or the universe.

Here we have the model of an **open system** in a **steady state**. What we have found so far is that a closed system is incapable of a steady state that is not also an equilibrium state. (We will come to an apparent exception momentarily.) For the open system, the exterior (suspension fluid, tissue fluid, nutrient medium, ecological environment, universe) serves as a **source** of all compounds that enter the system and react in it, as a **sink** for all compounds produced in the system that leave it.

Let me note in passing that enzyme-catalyzed reactions possess an adaptive feature that serves for the easier provision of sources. We have known since Chapter 2 that such reactions may be "saturated" with substrate.

Fig. 9.1. Schemes for a Variety of Reaction Chains. (a) straight chain; (b) straight chain with branch; (c) two straight chains linked by branch; (d) straight chain with parallel loop; (e) straight chain with cycle.

Moreover, although the condition for saturation formally is "S approaches infinity," the practical condition is that S exceeds K by an order of magnitude or so. Under such conditions, the reaction rate is independent of S. Thus a **quasi-source** exists whenever, in a system like that of Fig. 9.1(a), reaction 1 is saturated with respect to X_0. This helps to explain why certain *in vitro* systems (for example, a yeast extract fermenting glucose) show steady rates for appreciable periods.

An important peculiarity of certain reactions (relevant also to the occurrence of quasi-sinks) is irreversibility or near-irreversibility. Thus in the system under consideration, if reaction $n+1$ is irreversible, the concentration of X_{n+1} is immaterial to the system (unless the compound is in some way toxic at sufficient concentration).

These two mechanisms furnish a clue to the way in which a subset of the reactions of a system may be stationary while the balance of the system is not. Suppose in the first place that one of the reactions, say 3, is irreversible. Then the reactions to the left of 3 depend only on compounds X_0, X_1, and X_2, while those to the right of 3 depend not only on the remaining compounds but on the first three as well. We may speak of the first three reactions as an **independent subset**, of the rest as a **dependent subset**. The first three reactions, the independent reaction subset, may be stationary while the others may or may not be. The reverse is not the case: If, for example, X_2 varies with time, the dependent subset of reactions cannot be stationary as long as the rate of reaction 3 is a function of X_2.

Saturation of an enzyme will of course permit a right-hand subset of reactions to be independent; thus in the case of the preceding paragraph saturation of reaction 3 with respect to X_2 releases the right-hand set from their dependence on the left-hand set. This independence is not absolute, however, but **contingent**: It holds only while X_2 (whether stationary or not) is sufficiently great to saturate.

One final observation about the straight-chain system is that 9.2 implies $R_1 = R_{n+1}$. With the interpretation of the first and last reactions as transport reactions, this amounts to the commonsense assertion that, in a stationary system, the rate of input must equal the rate of output. To be more specific, the number of molecules entering must equal the number of molecules leaving. We shall consider shortly whether—and if so, how—this statement must be modified for systems whose **connectivity** is more complex than that of a simple chain.

Branched Chains

Let us examine the system of Fig. 9.1(b). We specify that the terminal compounds of the main chain are outside, and are provided with source and sink respectively. The conditions for stationarity become:

$$(9.3) \qquad R_1 = R_2 \qquad R_3 = R_4 = \ldots = R_{n+1} \qquad R_{n+2} = \ldots = R_{n+m}$$
$$R_2 = R_3 + R_{n+2}$$

The last equation of the set gives the stationarity condition for the **nodal compound** X_2, which is at the place where the branch joins the main chain (**node** or **branch-point**).

These equations neglect Y_{m+1}. If this **secondary terminal** compound is inside the system, stationarity demands that $R_{n+m} = 0$, in which case all the reactions of the branch are in equilibrium by 9.3. The **input-output relation** then is $R_1 = R_{n+1}$ as before. If Y_{m+1} is outside the system and is provided with a sink, $R_{n+m} \neq 0$, and the input-output relation is: $R_1 = R_{n+1} + R_{n+m}$.

A special case occurs if the branch consists of just one step, which would represent the situation where X_2 can enter or leave the system. An obvious generalization is the addition of more branches; the conditions are left as an exercise for the reader.

It should be noted that the designation of the main chain is arbitrary. We may just as well consider that the main chain terminates with X_2, at which point it **diverges** in two branches. The formal relations will also be exactly the same if we want to speak of two chains **converging**. We would merely treat the reactions of the branch **into** X_2 as having positive rates—i.e., the rate of step $n+2$ from X_2 to Y_1 for instance, would be $-R_{n+2}$. The last equation of 9.3 would therefore read: $R_2 = R_3 - R_{n+2}$ or $R_2 + R_{n+2} = R_3$.

Fig. 9.1 (c) represents the common situation in which two main chains are linked by a common branch. The stationarity conditions are left as an exercise for the reader.

Loops and Cycles

Fig. 9.1(d) represents a novel kind of branching—a re-entrant branch which rejoins the main chain, forming a loop. The equations may be left for the reader to write down. It should be noted that the input-output condition obtained from the stationarity equations is simpler than the conditions for the branched systems (b) and (c). It is: $R_1 = R_{n+1}$. The loop simply divides the flow of material temporarily into two parallel streams (from X_1 to X_{n-1}).

An isolated loop, such as would be obtained if we dropped reactions 1, n, and $n+1$ from (d), is an apparent exception to the rule that stationarity requires an open system. There are thermodynamic objections to the system, into which I will not enter. The most significant objection, however, is that it is totally uninteresting as a model of a cell or part of a cell (or organism), precisely because it can serve as a model (assuming one ignores the thermodynamic snag) only while it is not connected to anything else in the universe.

Up to this point, we have treated all reactions as monomolecular; and for much of the formalism it was really immaterial what the order of the reactions was. (Why?) But, as has been obvious throughout most of this volume, most reactions are bimolecular, and especially most of the interesting ones. We must now take cognizance of this.

Fig. 9.1(e) illustrates one of the most significant situations of this sort. Compound Y_1 is used in reaction 2, regenerated in reaction 3 (the numbering is a matter of convenience; adjacent reaction steps need not be involved). The dashed arrow completes what I have termed a **cycle** (1), in contradistinction to the **loop** of system (d); the arrow does not represent a reaction path, but a connection between two reactions by virtue of a common compound not occurring trivially at a node.

The stationarity conditions for the X_i are just as in 9.2. The differential equation for Y_1 is:

$$dY_1/dt = R_3 - R_2 \tag{9.4}$$

It is immediate, then, that Y_1 is stationary if the X_i are.

Suppose, however, that Y_1 is involved in other reactions—for example, hydrolysis according to $R' = -kY_1$, or transport according to $R'' = -k'Y_1 + k''$. Then R' or R'' would be added to the right side of 9.4. If the X_i were stationary, the term $R_3 - R_2$ would drop out of 9.4, leaving the R' or R'' term. It would appear, then, that the X_i may be stationary while Y_1 is not. But this is not true.

Suppose we had an equation such as $dY_1/dt = R'$ or $dY_1/dt = R''$. The solution, regardless of the form of the right side, is $Y_1 = f(t)$. But R_2 and R_3 depend on Y_1—in a form, moreover (by virtue of the Law of Mass Action) which makes it virtually impossible for Y_1 to cancel out of $R_2 - R_3$. Consequently, the stationarity condition $R_2 = R_3$ will boil down to: $f(t) = \text{constant}$ —a flat contradiction.

The only escape from this dilemma is to insist that Y_1 shall also be stationary. This can be achieved, setting $R'' = 0$, at the cost of barring net transport of Y_1 (which may or may not be acceptable in a given concrete exemplification of our model). If we must set $R' = 0$, even this escape hatch is blocked; for the condition implies $Y_1 = 0$, which in turn means the collapse of our Y_1-dependent system of reactions.

The example illustrates two problems: that of consistency of stationarity conditions ($Y_1 = 0$ forces a trivial solution), and that of the difficulty of getting stationarity in a subset of a reaction set. Aside from these cautionary aspects, however, we shall see later in the chapter that systems involving cycles have specially interesting and significant features.

Before leaving the example, let us note that combining the two reactions of Y_1 removes all the difficulty. That is, if both hydrolysis and transport occur, the differential equation which concerns us is $dY_1/dt = R' + R''$ (when the X_i are stationary). But there is no difficulty about setting $R' + R'' = 0$; it means that both transport and hydrolysis go on, but at such a rate as to compensate one another. Y_1 has a perfectly good nonvanishing stationary value, $k''/(k + k')$.

This little problem illustrates the essence of the kind of stationarity we want as a model of a living system. Forcing stationarity into single reaction steps results in triviality (equilibrium or zero conceotrations); combining distinct steps (whether all chemical, or chemical and transport) permits nontrivial stationarity.

Pure Bimolecular Systems

Let us finally observe that bimolecular systems free of cycles present exactly the same sort of problem, but more acutely. Suppose we erase the dashed arrow in (e), and denote the second occurrence of Y_1 by Y_1' (a distinct compound). Now, if Y_1 and Y_1' are involved in no other reactions, stationarity demands $R_2 = 0$ and $R_3 = 0$, and the whole system collapses into equilibrium.

The only way out of this situation is for Y_1 and Y_1' to be involved in other reactions. Moreover, these "outside" reactions are not entirely free: A constraint is imposed by consistency. Thus, suppose that $Y_1 \rightleftharpoons Z_1$ and $Y_1' \rightleftharpoons Z_1'$, with rates R_6 and R_7 respectively (Fig. 9.2(a)). Then the differential equations are: $dY_1/dt = -R_6 - R_2$ and $dY_2/dt = R_3 - R_7$, whence the stationarity conditions are: $R_2 = -R_6$ and $R_3 = R_7$. But the stationarity of the X_i imposes $R_2 = R_3$. Accordingly, the entire system can be stationary only if $-R_6 = R_7$.

This is not the only restriction, moreover. Suppose Y_1, for instance, is a member of a chain parallel to the X_i chain: $Y_0 \overset{a}{\rightleftharpoons} Y_1 \overset{b}{\rightleftharpoons} Y_2 \ldots$. (Fig. 9.2(b)) Then the second chain is stationary if $R_a = R_b = R_c = \ldots$. But in this case $dY_1/dt = -R_2 + R_a - R_b$, and when $R_a = R_b$ the right side is simply $-R_2$. Hence stationarity of Y_1 demands $R_2 = 0$, or equilibrium of the first main chain. For the entire system to be stationary, therefore, either Y_1 must initiate only a **simple secondary branch** rather than being interdigitated in a parallel chain, or Y_1 must participate in **still another** reaction, which would then constitute a branch of the second chain (Fig. 9.2(c)). (By "secondary branch" we mean one not involving the X_i of the main chain, which do initiate the branches in systems like (b) and (c) Fig. 9.1.)

Let me define an **isolated bimolecular reactant** as any compound which appears, as reactant or product, exactly once in a chain. (Every X_i

appears twice: as product of i and reactant of $i + 1$.) Let me define a **simple branch** as a branch one of whose termini is free, and none of whose elements is a member of more than one main chain. Let an **intersecting branch** of two chains be one which has a member in two chains, and which does not

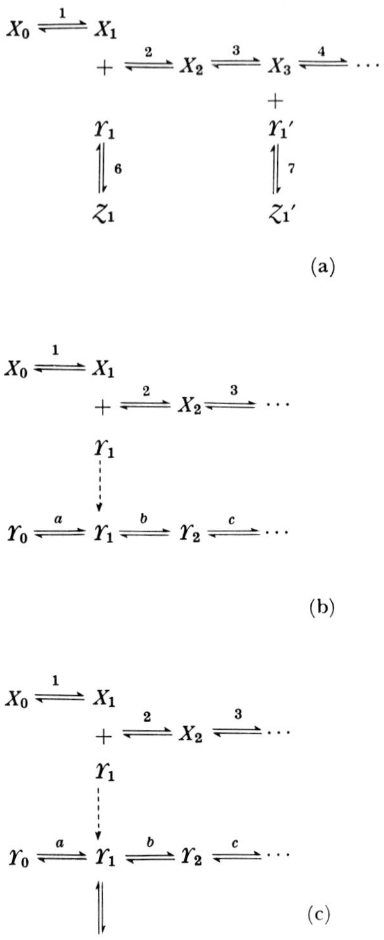

Fig. 9.2. Bimolecular Schemes.

terminate in both chains. Then our conclusions may be summarized as follows:

A system involving bimolecular reactions can be nontrivially stationary only if any isolated bimolecular reactant in a chain either initiates a simple branch or belongs to an intersecting branch of that chain. (Fig. 9.2,(a) or (c).)

STABILITY OF THE STATIONARY STATE: MODE OF APPROACH

If a reaction system admits of a stationary state, two further questions are of interest: (a) Is the steady state **stable**?—that is, will the system return to that state after it has been moved out of it by some external disturbance? and (b) Do all the components of a system approach their stationary values unidirectionally from above or below, or do they sometimes exhibit the phenomena of overshoot or false start which we encountered in the first section of Chapter 8 (with X as a variable, rather than t as here)?

The first of these questions is a classical problem in the theory of differential equations; and I will merely outline the approach to it here. It was first considered with respect to biological and chemical systems by Alfred Lotka (3); and any biologist who has not already read all of Lotka's charming and interesting book should take this as a good excuse to do so. (To those who find Lotka's references to the mathematical literature on this problem discouraging because they are in French and German, I will say first that they should be ashamed of themselves, and second that they can find the material in any reasonably good textbook on the subject, such as Ross (4).)

Given a system of first-order differential equations, such as 9.1, the values of the X_i for which the right-hand sides all equal zero are the **critical points**. These critical points can be classified according to their stability and certain other properties.

To understand what is involved, imagine that the differential equations have been solved, so that we have $X_i = X_i(t)$. For any value of t, the set of numbers $X_1(t)$, $X_2(t)$, ..., $X_n(t)$ can be thought of as coordinates of a point in an n-dimensional space; and as we vary t we get a set of points—the **path** or orbit of the system. (There is actually a system of paths, infinite in number; which path the system in fact follows will be determined by the initial conditions $X_i(0) = a_i$.) What we ask about the critical points of the system is how they are related to the system of possible paths. If all the paths of a system **converge** on a critical point with increasing t (either unidirectionally or, in some cases, with damped oscillation), the point is stable (stable **node** or stable **spiral point**). If all paths **diverge** with time, the point is unstable. If all paths are **closed curves** around the critical point, it is a **center**; and the system is **periodic** in its behavior. Finally, there are **metastable** or **saddle** points: One or two paths converge on the point, all other paths slide past the point (for a two-dimensional example, visualize two diagonal lines passing through the origin of a Cartesian plane, and the system of rectangular hyperbolae which have these lines as asymptotes). The idiosyncrasy of these points is that disturbances which displace the system precisely along one of the diagonal paths will leave the system stable—it will return to the critical point; but any other perturbation will put the system on a path which removes it from the critical point.

The way one actually gets this kind of information, at least for systems such as interest us here, is fairly straightforward. Let us begin by agreeing that the right-hand sides of our chemical equations will be polynomials of degree two in the X_i.[2] Thus:

$$dX_i/dt = \sum_j \sum_k a_{ijk} X_j X_k + \sum_j a_{ij} X_j + \alpha_i \tag{9.5}$$

The critical values of the system are of course given by the set of $\bar{X}_i$ which satisfy:

$$\sum_j \sum_k a_{ijk} \bar{X}_j \bar{X}_k + \sum_j a_{ij} \bar{X}_j + \alpha_i = 0 \tag{9.6}$$

It will simplify matters if we introduce the variables:

$$u_i = X_i - \bar{X}_i \tag{9.7}$$

the differences between the values of the X_i at any time t and their critical values. Introducing the u_i into 9.5 by 9.7, and making use of 9.6 to simplify the result, we have:

$$du_i/dt = \sum_j \sum_k a_{ijk} u_j u_k + \sum_j b_{ij} u_j \tag{9.8}$$

with the abbreviation:

$$b_{ij} = a_{ij} + \sum_k (a_{ikj} + a_{ijk}) \bar{X}_k \tag{9.9}$$

The critical point is now, in terms of the u_i, given by:

$$(Ai)[u_i = 0] \tag{9.10}$$

Now! Virtually all the information we need can be obtained if we drop the quadratic terms from 9.8 and write:

$$du_i/dt = \sum_j b_{ij} u_j \tag{9.11}$$

To be precise: Every statement we can make about a critical point of 9.11 remains true for a critical point of 9.8, except when the 9.11 critical point is a center.

[2] This may be viewed as an approximation, in which we neglect the concentration-dependent terms in the denominator of a typical bimolecular rate expresssion (Chapter 7) on the ground that most metabolites occur at rather low concentrations in cells. The form will be exact, however, if we elect to include enzymes in their various free and bound forms among the components of the system.

The linear system 9.11 has of course the well-known solution:

$$u_i = \sum_m A_{im}\, e^{\lambda_m t} \tag{9.12}$$

where the A_{im} are constants of integration, to be fixed by the initial conditions, and the λ_m are the (distinct) roots of the characteristic equation. The characteristic equation is obtained by subtracting λ from each diagonal entry in the determinant of the b_{ij}, and setting the determinant equal to zero. Formally stated:

$$|b_{ij} - \lambda\, \delta_{ij}| = 0 \tag{9.13}$$

where δ_{ij} is the "Kronecker delta" which equals 1 when $i = j$ and 0 otherwise.

The properties of the critical points that interest us in connection with the idea of the stationary state can now be summarized roughly as follows: If all the λ_m are negative real numbers, or are complex numbers with negative real parts, the critical point is stable.

It is worth while to try to see why this routine works. First of all, as to the solutions of 9.11: It is obvious from 9.12 that, if each λ_m is negative or has a negative real part, every term of the sum will approach zero with increasing t, so that each of the u_i does so. Since the critical point is represented by $u_i = 0$, it follows that under such conditions the system approaches the critical point, and will return to if it the transient state is started up all over again by some disturbance that removes the system from the critical point. On the other hand, if even one of the λ_m is positive, the corresponding term approaches infinity with t; and so the system will run away from the critical point (unless the corresponding A_{im} is identically zero, in which case λ_m will not influence the behavior of the system).

Why can the analysis of 9.11 be applied, as a rule, to 9.8? In the neighborhood of the critical point, all the u_i are small by definition (9.7). The quadratic terms in 9.8 are therefore small compared with the linear terms (recall discussion of power series in Chapter 1); and 9.11 is therefore a good approximation to 9.8, which is better the closer we are to the critical point. Hence, for any path that passes through the critical point of 9.8, there will be a path (perhaps slightly different in shape) that passes through the critical point of 9.11 and behaves essentially the same way in the neighborhood of the critical point. The qualification made earlier about the method in the case of a center is now obvious: If a critical point is a center, the paths of the system are closed curves that are described **around** the center, but do not pass **through** it. Hence we cannot confidently say that 9.11 will have all the essential properties of 9.8 near the critical point, because most paths will not come close to that point, and the u_i values on those paths will not be small.

Obviously the practical problem of finding the λ_m may be vexatious when $n > 2$. But there are numerical methods, suitable for digital computation, for getting the roots of a polynomial to any desired accuracy. It should be remembered, moreover, that in practice many if not most of the model systems we are likely to be investigating involve only two or three (independent) concentrations, so that the characteristic equation will be amenable to ordinary methods.

The problem of overshoot and false start does not have so general a solution. We can say the following:

1. Both forms of behavior imply the existence of an extremum. But not all the components of a system can have such an intermediate extremum. For this would imply $dX_i/dt = 0$ for all i, which is the condition for the critical point(s).

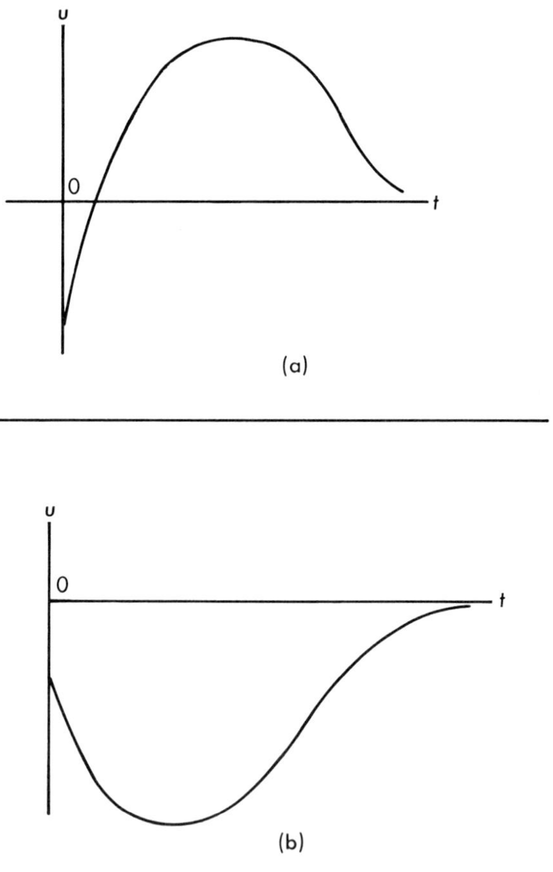

Fig. 9.3. Time Curves for (a) Overshoot and (b) False Start.

2. As is evident from Fig. 9.3(a), for a component i with simple overshoot the function u_i must change sign at some finite value of t; and du_i/dt must be positive at that point. (The corresponding figure for undershoot may be supplied by the reader.)

3. From Fig. 9.3(b) it follows that, for a component i with simple false start, the initial values (at $t = 0$) of u_i and du_i/dt must have the same sign. (The figure is drawn for a component going from a negative to an eventual zero value; the corresponding figure for a component starting with positive u_i and eventually sinking to zero can readily be supplied by the reader.)

As stated, these conditions do not seem very stringent. It is only in working out specific examples that their significance can become evident. Denbigh (5) has worked out some simple cases, and his paper should be consulted.

PROBLEMS: 1. Demonstrate that $S = 0$, $C = 0$ is the critical point of 2.1′ + 2.2′, and that it is stable.

2. For $n = 2$, prove that u_1 can overshoot only if A_{11} and A_{12} differ in sign. Find the conditions for overshoot in this case, for the initial condition where $u_i(0) > 0$. Specifically, prove that, if u_1 changes sign, there is an extremum. (Hint: Prove that, if there is a $t = T > 0$ such that $u_1(T) = 0$, then there is a $t = T' > 0$ such that $du_1/dt = 0$ at T'. The expressions for T and T' are related, and so are the conditions that they be positive.)

3. For $n = 2$, find the conditions on the constants to get false start of u_1.

SOME ILLUSTRATIVE EXAMPLES

1. If a substrate S enters a cell by passive permeation, and is converted to product by an enzyme inside, how does the observable reaction rate depend on the external concentration of S?

The scheme is:

(9.A) $$S_e \rightleftharpoons S \qquad S + E \xrightarrow{1} C \qquad C \xrightarrow{2} E + P$$

We express the rate of the transport reaction (from left to right) as $hS_e - h'S$. For the steady state of S and C, we readily find, recalling that $v = k_2 C$:

(9.14) $$S_e = [v + Kh'v/(V - v)]/h$$

where V stands for $k_4 E_t$ as usual. (We could invert 9.14 and solve the quadratic equation for v; but it is simpler to examine the inverse relation.)

Following the usual procedures, we find that S_e is linear near the origin, with initial slope $(1 + Kh'/V)/h$, and that it has a pole (see Chapter 1) at $v = V$. The graph therefore looks very much like Fig. 2.5. It is distinguishable, however, by an elementary trick. In the equation $v = VS/(K + S)$, which

Fig. 2.5 illustrates, $S = K$ when $v = \frac{1}{2}V$. But the value of S_e for $v = \frac{1}{2}V$, according to 9.14, is $(\frac{1}{2}V + Kh')/h$. If the enzyme has been or can be studied *in vitro*, so that V and K are known, the two cases should be readily distinguishable.

PROBLEM: In the same system, suppose that S is transported inward by a carrier mechanism, and leaks out at the rate $h'S$ as before. Assume that the transport rate by the carrier mechanism follows the relation: Rate = $V_m S_e/(K_m + S_e)$, where V_m and K_m are constants of the carrier system. Analyze the relation between v and S_e as above. Is the rate at saturating S_e values still V? If the results are simplified by neglecting the outward leak ($h' = 0$), how does the maximal rate behave?

2. Given the scheme:

(9.B)

$$S + E_1 \overset{1}{\rightleftharpoons} C_1 \qquad C_1 \overset{2}{\longrightarrow} E_1 + P_1 \qquad P_1 + E_2 \overset{3}{\rightleftharpoons} C_2 \qquad C_2 \overset{4}{\longrightarrow} E_2 + P_2,$$

how does the **metabolic intermediate** P_1 behave? What determines its steady-state value?

The kinetic equation for P_1 is of course:

(9.15) $$dP_1/dt = V_1 S/(K_1 + S) - V_2 P_1/(K_2 + P_1)$$

where the constants V_1, V_2, K_1, and K_2 have the usual significance.

Let us assume that there is a source to maintain S constant. Then the first term on the right side of 9.15, which we shall refer to as v_1, is constant.

Setting $dP_1/dt = 0$, we get the stationary value of P_1:

(9.16) $$\bar{P}_1 = K_2 v_1/(V_2 - v_1) = K_2 V_1 S/[K_1 V_2 + (V_2 - V_1)S]$$

This is positive, hence meaningful, only if $v_1 < V_2$, and for saturating levels of S only if $V_1 < V_2$.

The most interesting feature of 9.16, best seen by dividing numerator and denominator by V_2, is that $\bar{P}_1$ is an increasing function of V_1/V_2, hence also of E_{1t}/E_{2t}. That is, insofar as the chemical composition of our cell model may be viewed as given by the distribution of stationary values of the various metabolic intermediates, it is not independent of enzyme levels, as a true closed system would be, but depends on the distribution of the enzymes, and also of their turnover rates. This very simplified example indicates how the cell makeup is fixed by the activity of the genome and its various controls (repression, derepression), which fixes, under given conditions, the level of all the enzymes.

PROBLEMS: 1. Solve 9.15 for $P_1(t)$. Show that $\bar{P}_1$ is an asymptotic value of $P_1(t)$ (as is the case for stationary values in general). Prove that, if $v_1 > V_2$, there is no stationary state, and that P_1 increases indefinitely with t; find the asymptotic form of $P_1(t)$.

2. Repeat the analysis in the foregoing section, assuming that the enzymes in 9.B are reversible. Is the conclusion about the effect of enzyme distribution valid?

3. Fermentation involves cycles (Fig. 9.1(e)), such as the reduction and reoxidation of DPN. How do these affect the activity of the main chain (the transformations of the carbon compounds)?

Consider a highly telescoped version of a fermentation scheme:

(9.C) $$S + B_o \xrightarrow{1} P + B_r \qquad P + B_r \xrightarrow{2} P' + B_o$$

In this scheme, S is the carbon source (glucose or triose phosphate), and B_o and B_r are DPN and DPNH$^+$ respectively, while P is an oxidized intermediate, such as pyruvate or acetaldehyde.

It is immediately obvious from the scheme that we have a conservation equation for co-substrate:

(9.17) $$B_t = B_o + B_r$$

Similar considerations show that $dP/dt + dB_o/dt = 0$, so that:

(9.18) $$P + B_o = M$$

a constant. Since this relation holds for all t, it must hold for $t = 0$; hence the constant has the value $P_o + B_{o0}$, where P_o is the initial value of P and B_{o0} the initial value of B_o.

Treating the reactions of the scheme as approximately simple bimolecular (see footnote 2), we may write

(9.19) $$dB_o/dt = k_2 PB_r - k_1 SB_o$$

We can get the stationary value of B_o by setting the right side equal to zero, eliminating P and B_r by 9.17 and 9.18, and solving the resulting quadratic:

(9.20) $$B_o = \tfrac{1}{2}\{(B_t + M + k_1S/k_2) - [(B_t + M + k_1S/k_2)^2 - 4MB_t]^{\frac{1}{2}}\}.$$

We select the root with the negative sign before the radical because $B_o = 0$ if $B_t = 0$ is an obvious requirement.

Since cellular concentrations of co-substrates like DPN are usually quite low, we may simplify this result by applying the binomial approximation for small B_t to the radical:

(9.21) $$B_o = MB_t/(M + B_t + k_1S/k_2)$$

Since $R_1 = R_2$ in the stationary condition, we may multiply 9.21 by k_1S to get the overall rate of fermentation in the steady state:

(9.22) $$v = k_1SMB_t/(M + B_t + k_1S/k_2).$$

The rate behaves in the expected fashion with respect to S and B_t. It is of interest to note the occurrence of the ratio k_1/k_2, which shows how the overall rate of fermentation is limited by the capacity of the system to reoxidize DPNH$^+$; and indeed the rate at saturation with S is simply $k_2 MB_t$, proportional to k_2.

The most interesting feature of the rate expression, however, is the dependence on $M = P_o + B_{o0}$. If one looks at scheme 9.C, one realizes that the system will not start, even for $B_t \neq 0$, if no B_o is initially available —i.e., if $B_{o0} = 0$. But the interchangeability of B_{o0} for P_o, evident from the expression for M and from consideration of the scheme, is impressive. In effect, the scheme can be primed with oxidant either at the beginning or in the middle. The experimental counterpart of this, of course, is Meyerhof's classical observation that the induction period of fermentation by yeast juice could be eliminated by a trace of acetaldehyde.

PROBLEMS: 1. If the expressions for R_1 and R_2 were more complete (e.g., 7.14) instead of approximate as in 9.19, would the essential features of the dependence of B_o and v on M be altered?

2. Demonstrate that the steady state of B_o is stable—that is, that any perturbation will cause activity tending to restore the steady state.

4. The fact that a cell's ability to metabolize depends on its previous history, as the preceding example suggests, raises some interesting questions. Offhand, one would suppose that a cell carrying on a fermentative regime would always have some sort of balanced supply of B_o and B_r, since every molecule of S that goes through the mill first uses B_o and then regenerates it. In fact, if we return to 9.20 to see what exhaustion of substrate would do, we find that, for $S = 0$, B_o will equal B_t or M, whichever is smaller. Correspondingly, B_r would become equal to 0 or to $B_t - M$. (There is no contradiction in $B_t < M$, by the way, though obviously $B_t \geqslant B_{o0}$; the relation depends on the magnitude of P_0.) Removal of S stops reaction 1, while reaction 2 continues to reoxidize B_r.

Thus the balance may not be very impressive. Moreover, we know that cells contain enzymes which break down and others which synthesize DPN (6). Some of the enzymes that split DPN are in fact specific for the oxidized form (7). The synthetic enzymes all require energy in the form of ATP. Thus we would expect that a starved cell might have no DPNH$^+$ at the onset of starvation, and might then lose its DPN to breakdown processes. Eventually, such a cell would be totally unable to utilize a new supply of substrate without priming.

It therefore seems worth while to investigate the kinetics of the processes whereby the cell maintains its supply of co-metabolites. Let us consider a highly simplified version. We ignore the oxidation-reduction aspects, reducing the scheme 9.C to a single reaction. We add reactions for breakdown and resynthesis of DPN (B), and we include in the first reaction the conversion of ADP (A') to ATP (A):

$$S + B + A' \xrightarrow{1} P + B + A \qquad B \xrightarrow{2} X \qquad X + A \xrightarrow{3} B \qquad A \xrightarrow{4} A' \tag{9.D}$$

where the fourth reaction takes into account any endogenous utilization of ATP in biosynthesis or enzymatic hydrolysis. We also include a conservation equation for the adenylic system:

$$A_t = A + A' \tag{9.23}$$

and one for the B system (on the assumption that the breakdown product X is not further converted and does not leak out of the system:

$$B_t = B + X \tag{9.24}$$

Assuming, as usual, constancy of S, there are two independent variables, say A and B. Again approximating rate expressions by simple bimolecular formulations, and eliminating A' and X with the help of 9.23 and 9.24, the kinetic equations can be written:

$$\begin{aligned} dB/dt &= k_3 B_t A - k_3 AB - k_2 B \\ dA/dt &= k_1 SB(A_t - A) - k_3 B_t A + k_3 AB - k_4 A \end{aligned} \tag{9.25}$$

Some care is required in the manipulations that lead to the stationary values of A and B. If we set the right sides of 9.25 equal to zero, solve both algebraic equations for $\overline{B}$, and put the results equal to each other in order to get an equation for $\overline{A}$, we obtain an expression (set equal to zero) with a factor $\overline{A}$. One solution, then, is $\overline{A} = 0$; and it is readily found, by putting this back into either expression for $\overline{B}$, that the corresponding value is $\overline{B} = 0$. In one steady state, therefore, the system is in a state of total collapse—it is devoid of the DPN required for fermentation and of the energy source required to synthesize the DPN.

If we set this solution aside, and use the other factor of the $\overline{A}$ equation, we find eventually:

$$\begin{aligned} \overline{A} &= [k_3 B_t(k_1 SA_t - k_2) - k_2 k_4]/k_3(k_1 SB_t + k_4) \\ \overline{B} &= [k_3 B_t(k_1 SA_t - k_2) - k_2 k_4]/[k_3(k_1 SA_t - k_2) + k_1 k_2 S] \end{aligned} \tag{9.26}$$

An immediately obvious condition for the meaningful character of these expressions is that the common numerator be positive, for which of course a necessary but not sufficient condition is that $k_1 SA_t - k_2 > 0$. If $k_4 = 0$ (no diversion of ATP from the system), the necessary condition is also sufficient.

If $k_2 = 0$, the condition is trivially satisfied (there is no breakdown of DPN, hence no problem).

Confronted with two possible stationary states, we must ask whether only one is stable, and if so which one, and under what conditions. We can apply the methods discussed in the preceding section. We introduce difference variables ($u_1 = B - 0 = B$, $u_2 = A - 0 = A$, in the first case, $u_1 = B - \overline{B}$, $u_2 = A - \overline{A}$, in the second case), use the condition of stationarity to cancel some terms, and drop nonlinear terms. We then set up the characteristic equation for the solution to the linear system in each case, and investigate the properties of the λ_m.

Omitting some tedious but not at all difficult alegbraic manipulation, I will come at once to the results. If the criterion for physical significance of the nontrivial steady state (9.26) is fulfilled, that state is stable—both roots of the characteristic equation are real and negative. If the condition is not fulfilled, then of course the expressions 9.26 are meaningless—but also, leaving that question aside, we find that one of the roots is positive and one negative, so that the critical point would be a saddle point (metastable).

For the "dead" stationary condition, precisely the reverse holds. The roots are always real; but they are both negative only when the condition just discussed does **not** hold. If it does hold, then one root is positive, and the (0, 0) critical point is metastable.

Thus it appears that the expression

$$k_3 B_t(k_1 S A_t - k_2) - k_2 k_4$$

has a crucial significance for our model. If this is positive, a nontrivial stationary state is possible. If anything happens to depress it to or below zero, not only will the nontrivial stationary condition no longer be possible, but the system will in fact drop all the way to the (0, 0) condition and remain there. The system will fail to function at a sufficiently low level of carbon source, for instance, even though it is not being starved. A starved system, on the other hand, will not recover at any arbitrary level of renewed nutrition, but only at a sufficiently high level. In short, while an ideal system can adjust its metabolic rate to its supply of nutrient, the real system, which has to pull up its socks all the time, can function at all only above a certain level.

Any of the variables in the system may be manipulated. It must be remembered that the apparent velocity constants, for example, in fact embody enzyme concentrations. Thus one may read off from the parts of the critical expression such obviously sensible conclusions as: A more active DPNase may be compensated by less endogenous ATP consumption, or by more active phosphorylation; a lower substrate level may be compensated by a higher level of the adenylic system, and in part by a higher total concentration of DPN; and so on.

It is of interest, in the light of these results, to speculate about their relevance to the high sensitivity of brain to oxygen deficit. The anaerobic metabolism is certainly not weak (8); but it is obviously inefficient in generating ATP compared with its rather vigorous aerobic metabolism (9). Moreover, brain is known to have a quite vigorous DPNase (10). In this tissue, therefore, even more than in others, the stage is set for anaerobiosis to depress our critical expression to the danger point.

One may also speculate, somewhat more tentatively, about the application of our conclusions to other pathological conditions. The significant feature, which I suspect may apply, is that the homeostatic mechanisms of real systems are not infinitely elastic, and that, under suitable circumstances, limited cell damage may result in complete collapse.

5. Our treatment of the adenylate cycle has been somewhat sketchy. It may be of interest to see whether it can be improved. To avoid getting into a hopelessly complex even though not very interesting mathematical mess, we carry out this analysis neglecting the DPN cycle. The scheme is:

$$(9.E) \qquad S + A \xrightarrow{1} P + A' \qquad P + A' \xrightarrow{2} P' + A \qquad A \xrightarrow{3} A'$$

$$A' \xrightarrow{4} A'' \qquad A + A'' \underset{}{\overset{5}{\rightleftharpoons}} 2A'$$

where A'' represents AMP, and reaction 5 is catalyzed by ATP-AMP phosphotransferase (myokinase) (11, 12). We also have an extended conservation equation for the adenylate system:

$$(9.27) \qquad A_t = A + A' + A''$$

In setting up the kinetic equations, we must bear in mind that two molecules of ATP are regenerated for every S (half-glucose) that passes through the glycolytic system. In the equations for A and A', therefore, the rate of reaction 2 appears multiplied by a factor 2. The rest of the stoichiometry is self-explanatory. (If we elected to include oxidative phosphorylation, the factor 2 would become n, a number of the order of 15.) We use 9.27 to eliminate A'' where it appears.

The kinetic equations are therefore:

$$(9.28) \qquad dA/dt = -k_1SA + 2k_2PA' - k_3A - k_5A(A_t - A - A') + k_{-5}A'^2$$

$$dA'/dt = k_1SA - 2k_2PA' + k_3A - k_4A' + 2k_5A(A_t - A - A') - 2k_{-5}A'^2$$

$$dP/dt = k_1SA - k_2PA'$$

The formidable-appearing equations for the steady state present no difficulty. We eliminate $\overline{P}$ with the help of the third equation. Adding twice the first equation to the second cancels the quadratic terms, and permits us to solve for $\overline{A}'$ in terms of $\overline{A}$. Eliminating $\overline{A}'$, with the help of this result, from the sum of the first two equations gives an equation for $\overline{A}$. There is a factor $\overline{A}$, and a trivial solution $\overline{A} = 0$, $\overline{A}' = 0$, in addition to the desired solution.

This may be written, using the abbreviation $w = k_1S - k_3$:

$$\overline{A} = k_4(k_5 A_t - w)/[k_5(w + k_4) + k_{-5} w^2/k_4] \tag{9.29}$$

$$\overline{A}' = \overline{A}w/k_4$$

From the proportionality of $\overline{A}'$ to w and from the numerator of $\overline{A}$, it follows that we must have $k_5 A_t > w > 0$, so that in this case the level of carbon source has an upper as well as a lower limit. The mechanism of this is simple: Reaction 1 (proportional to k_1S) consumes A and produces A', which is converted to A'' by reaction 4. The regeneration of A'' to a higher level of phosphorylation by reaction 5 also consumes A. and produces only A'. Thus increasing w tends to decrease the supply of A for the essential first reaction.

The rate at which S is moved through the system in the steady state, the stationary glycolytic rate, is $k_1S\overline{A} = (w + k_3)\overline{A}$. Inasmuch as the first factor increases with w while the second decreases, we may reasonably inquire into the possibility of an optimum. Setting dv/dw equal to zero, we find after a little algebra:

$$w_m = [-F \pm (F^2 + GH)^{\frac{1}{2}}]/G \tag{9.30}$$

with the abbreviations $F = k_4 + k_3 K_5 k_5 A_t/k_4$, $G = 1 + K_5(k_5 A_t - k_3)/k_4$, and $H = (k_4 - k_3)k_5 A_t - k_3 k_4$, and $K_5 = k_{-5}/k_5$.

It follows from 9.30 that the requirement $w_m > 0$ can be satisfied whenever G and H have the same sign, provided we take the positive sign before the radical when they are both positive, and the negative sign when they are both negative. The remaining possibility is $G < 0$, $H > 0$; but this can be ruled out by the simple proof that $G < 0$ entails $H < 0$. ($G < 0$ means $k_5 A_t < k_3 - k_4/K_5$, hence $k_5 A_t < k_3$; but $k_5 A_t - k_3 < 0$ means $H < 0$, by inspection of the expression for H.)

The condition $w_m < k_5 A_t$ is satisfied for G and H both positive (use 9.30, multiply the inequality by the positive number G, transpose F, and square to remove the radical sign, after which a couple of elementary cancellations make the left side negative and the right side positive, thus proving the inequality, inasmuch as all the steps are reversible). If G and H are both negative, this is not the case; multiplying by the negative G reverses the direction of the inequality, and requires the negative of the radical to be **greater than** $F + k_5 A_t G$, which can easily be shown to be positive, so that the condition cannot hold.

Inspection of dv/dw at $w = 0$ immediately demonstrates that the initial slope is positive for $H > 0$ and negative for $H < 0$. It is also easy to show that

$d^2v/dw^2 < 0$ for $w = w_m$ in the case $G > 0$, $H > 0$, so that the extremum in that case is obviously a maximum.

Drawing the graphs with S itself as abscissa, rather than w, we get Figs. 9.4 and 9.5 for the cases $H > 0$ and $H < 0$ respectively. From the definition of H, it follows that the dividing line between the two shapes is determined by $k_5 A_t = k_3 k_4/(k_4 - k_3)$, provided $k_4 > k_3$. If $k_4 < k_3$, $H < 0$ trivially (i.e., for all values of A_t). In the former case, it would be possible to shift the shape of the v-S curve from one form to the other by varying A_t; in the latter case, only the shape of Fig. 9.5 should be observed.

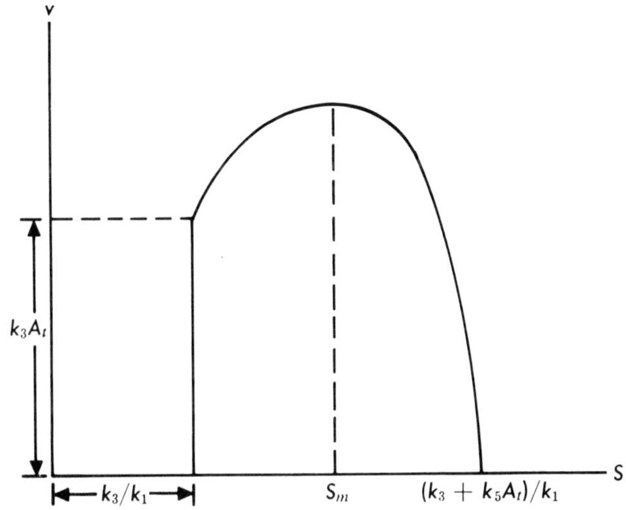

Fig. 9.4. The Phosphorylative Cycle: Metabolic Rate as a Function of Substrate Concentration ($H > 0$).

The restrictions on this system are striking, and make sense in terms of the assumed model. It would be of interest to have experimental confirmation of some of the features of this imaginary but not really improbable simplified glycolytic system. It should be noted that the model ignored the metabolism of inorganic phosphate—i.e., it is implicitly assumed that the supply of phosphate is sufficient to saturate those enzymes at which phosphate uptake occurs (reaction 2 in the scheme). It should also be kept in mind, in thinking about the implications of our results, that each of the apparent velocity constants k_i is proportional to the concentration of the corresponding enzyme. The true turnovers of the enzymes cannot be varied, except in a Pickwickian sense by a comparative study of different glycolytic systems; but the enzyme concentrations are in principle accessible parameters.

PROBLEMS: 1. Analyze the critical point $\bar{A}=0$, $\bar{A}'=0$, $\bar{A}''=A_t$, for stability. What is the value of $\bar{P}$ at this critical point? Does it matter what value you assume or calculate?

2. Can you say anything about the stability of the nontrivial critical point?

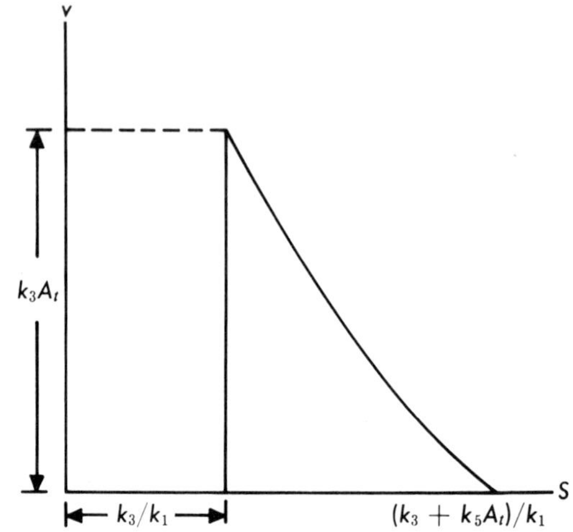

Fig. 9.5. The Phosphorylative Cycle: Metabolic Rate as a Function of Substrate Concentration ($H<0$).

6. When the first edition of this volume was in manuscript, the mechanism of biosynthesis was just being unravelled, especially in the case of macromolecular syntheses. A number of simple models for the relation between energy metabolism and biosynthesis were considered.

We are currently far better informed about synthetic mechanisms. Moreover, biosynthesis as a part of the growth process presents special complications which were not given any consideration in the previous simple models. For the study of these systems, which is beyond the scope of this volume, the reader is referred to the author's analysis (13.).

There are, however, some cases of macromolecular biosynthesis in the absence of growth, such as the synthesis of serum albumin by mammalian liver and of various enzymes by prostate and by other secretary tissues. It is recommended that the reader experiment with simple models, following the lines indicated in the foregoing. The essential connections between energy metabolism and biosynthesis, which must be included in such models, are the following:

a. The metabolic system generates ATP and uses ADP and inorganic phosphate. It also uses ATP as a primer (hexokinase, phosphofructokinase)

b. The biosynthetic system uses ATP, and generates ADP and inorganic phosphate. (It sometimes generates AMP and pyrophosphate, so that the behavior of the interlocked systems will depend on myokinase, and also on the activity of pyrophosphatase.)

c. The metabolic system forms intermediates, such as keto acids, which are converted to precursors of the biosynthetic system, at some cost in ATP.

d. Some precursors enter certain systems from without. The transport reactions are generally active, and require expenditure of ATP, directly or indirectly.

THE STUDY OF METABOLISM BY MEANS OF ISOTOPIC TRACERS

The virtual equivalence of the various isotopes of many elements with respect to the chemical reactions of compounds containing those elements has led to some of the most remarkable and significant research in the history of biochemistry, as well as to some of the damnedest nonsense.[3] The use of isotopes as tracers in the study of metabolism has naturally divided itself into the pursuit of two classes of problems: (1) the analysis of the connectivity of metabolic networks, and (2) the measurement of metabolic rates. The separation is pragmatic and not fundamental: The analysis of hypotheses about the form of metabolic networks has from time to time demanded a careful study of rates; and, as this section will show, the study of rates invariably demands a detailed knowledge of connectivity.

I will have almost nothing to say about the first class of problems. The achievements of investigators like H. G. Wood, M. F. Utter, and Lester Krampitz in the study of carbohydrate metabolism, or of others such as G. R. Greenberg and John Buchanan in the study of purine biosynthesis, have become classics of biochemistry, and are the commonplaces of any textbook worth opening.

[3] The pioneers of metabolic tracer work, even if they had not been exceptionally intelligent and scrupulous, could not have afforded badly designed and slovenly experiments; the heavy labor invested in every experiment is particularly vivid in the mind of one who has watched Harland Wood keep a thermal diffusion column and a mass spectrometer going, and listened to Martin Kamen's account of what was involved in pre-war research with carbon 11 from the Berkeley cyclotron. After the war, when isotopes became cheaply available and counters could be purchased rather than built from scratch, tracer work became attractive to the kind of people who jump on every bandwagon of technique as a substitute for ideas. To judge from some of the publications appearing at about the time the present analysis was undertaken, many believed that tracer research had been reduced to stuffing isotope in one end of an animal and collecting a neatly typed paper out of the other.

In the study of metabolic rates by means of tracers, the theoretical problem is to relate kinetically meaningful rates to observable changes in the tracer content of the interesting molecules.[4]

Isotopically Unimolecular Systems

If we are dealing with reactions where a labelled molecule has only a single type of labelled immediate precursor, the analysis is quite simple.

Consider any reaction network, with the components numbered as elsewhere in this chapter. Let us concentrate on a compound i, with concentration C_i. We denote by v_{ji} the rate of formation of i from j (e.g., moles of j converted to i per liter per second), where j is any other compound directly convertible to i in the network. Similarly, let v_{ik} be the rate at which i is converted to any other molecule k in a reaction of the network.

Let us define the specific activity[5] of any compound k in terms of its total concentration C_k and of the concentration of labelled molecules of k, $C_k{}^*$ (in the same units, obviously):

$$s_k = C_k{}^*/C_k \tag{9.31}$$

The number of labelled molecules of i formed from j in unit time is obviously $s_j v_{ji}$—the total number of j molecules converted times the fraction of them that is labelled. In the same way, the number of labelled i molecules converted to k per unit time is $s_i v_{ik}$. The total rate of change of labelled i per unit time is the difference of such terms, summed over all the j and k from which i can be formed or to which i can react:

$$dC_i{}^*/dt = \sum_j s_j v_{ji} - \sum_k s_i v_{ik} \tag{9.32}$$

The kinetic equation for total i is of course:

$$dC_i/dt = \sum_j v_{ji} - \sum_k v_{ik} \tag{9.33}$$

[4] Some of the intellectual confusion that still persists in this field is attributable to the fact that even quite capable investigators succumbed to the temptation of using picturesquely inaccurate or semi-empirical terms such as "half-life" or "percent replacement". The curious reader will find a critical analysis of these terms in the original papers (14). I consider them very dead mutton, and will not invest more ink and paper in discussing them here.

[5] Various interpretations of "specific activity" are current, some of them best described as laboratory slang. The earlier workers with stable isotopes were strict: They used "atom percent excess", which is exactly the specific activity defined in the text, but expressed as a percentage instead of a fraction. The word "excess" appears because the "natural" element frequently includes some of the isotopes used as tracers. Workers with radioactive isotopes usually express specific activity as counts per minute in a sample divided by total labelled material in the sample (frequently in any old units, such as milligrams or micrograms). Since the rate of disintegration of a sample of radioactive material is proportional to the number of atoms of isotope present, the transformation to meaningful units is easy.

Using 9.31 to eliminate C_i^* and 9.33 to eliminate dC_i/dt, we can rewrite 9.32 in a form that is frequently more convenient:

$$C_i \, ds_i/dt = \sum_j v_{ji}(s_j - s_i) \tag{9.32'}$$

Equations 9.32 and 9.33, or 9.32′ and 9.33, jointly specify the kinetics of the system, together with any other independent constraints known to hold.

To ask for the metabolic rates with respect to a particular compound i is to ask for the values of the various v_{ji} and v_{ik}. The equations involve these quantities together with a number of quantities which are, at least in principle, experimentally measurable: the s_i, the C_i, and their first derivatives with respect to t. We have therefore taken the first step in the analysis—to relate the quantities of theoretical interest to the measurable quantities.[6]

If there are n compounds in the network, we have $2n$ linear algebraic equations in the v_{ji} and v_{ik}. For any moment t, the measurable quantities in these equations are parameters—if their numerical values have been ascertained, we can proceed to try to solve the equations for the metabolic rates. What we must ask is whether the equations suffice to fix the unknowns. We can assume that the equations are consistent from the method of derivation, provided that we have not committed an error in our picture of the network. But are there enough of them?

If every compound were reversibly connected to every other compound in the network, the number of rates to be found would be $n(n-1)$. This can equal the number of equations, $2n$, only for $n = 3$ (trivially for $n = 0$). More realistically, let us note that the number of immediate neighbors N_k of a compound k is 2 if it is inside a straight chain, 1 if it is terminal, and 3 if it is nodal (either at a simple branch or at one end of a loop). If there are b branches, c loops, and d termini, there are $n - b - 2c - d$ compounds in interior simple straight-chain positions. The total number of connections of all compounds, which is half the number of unknown metabolic rates if all reactions are reversible, is:

$$\sum_K N_k = 2n + b + 2c - d \tag{9.34}$$

Recalling the discussion of the earlier parts of the chapter, it is clear that some of the unknown rates are transport rates, in or out. If we have reason

[6] It is obvious that I have been writing as if the tissue or organism under investigation were strictly homogeneous. The problems associated with words like "compartments" and "pools" present no new features. Compounds common to various compartments or pools are assigned distinct numbers (e.g., plasma glucose might be C_1 while muscle glucose is C_{23}), and may be connected by transport reactions between contiguous pools. The formal aspects of distribution of matter among spatial compartments and among chemical compartments are identical, provided the compartments are so defined that we may assume perfect mixing within each one.

to know the values of any of these independently (as would be the case, for instance, if we were perfusing an organ with a fluid of known composition at a controlled rate), we may deduct their number (say I) from the $2\sum_k N_k$ unknowns that must be determined. If we know that certain reactions of the network are irreversible, we may include the number of these in the deduction I. We require therefore:

$$2(2n + b + 2c - d) - I = 2n \tag{9.35}$$

It may seem plausible to add another condition to the right side of the account in certain cases. For, if we add all the tracer label used in an experiment to the organism at once (say by injection), it is certainly true that the total label at any later time, distributed among whatever compounds of the network, must equal the amount initially added, corrected for excretion if any. But in fact the equations 9.32 take care of this problem. Let us examine them more carefully.

We have supposed that labelled material does not enter the system except at the moment $t = 0$, at which, in effect, a supply of a certain compound m, with specific activity $s_m(0) = s_{m0}$, has been generated. Thus the first set of terms in 9.32, which represent input into any compound i, contains no transport terms. The second set of terms, the outputs from any compound i, contains an excretion term, however. We may separate this from the others, and write it as $s_i v_{i0}$ (since the rate is not to another compound of the system, but to the outside).

Now let us sum 9.32 over i, getting:

$$\sum_i dC_i^*/dt = \sum_i \sum_j s_j v_{ji} - \sum_i \sum_k s_i v_{ik} - \sum_i s_i v_{i0} \tag{9.36}$$

From the discussion of variable subscripts in Chapter 1, it follows that we may change i to j and k to i in the second double summation—i.e., that the two double sums are identical, and will cancel:

$$\sum_i dC_i^*/dt = d\left(\sum_i C_i^*\right)/dt = -\sum_i s_i v_{i0} \tag{9.36'}$$

Integrating:

$$\sum_i C_i^* - \sum_i C_i^*(0) = \int_0^t \left(-\sum_i s_i v_{i0}\right) dt \tag{9.37}$$

The first term on the left is the total label (per unit volume) at time t, the second term is the total label per unit volume at time 0 (actually one term, under our assumptions), and the right side is the total excretion of label per unit volume up to time t, since the integrand is the **rate** of excretion of label. The condition of conservation of label is therefore implicit in our kinetic equations. (If label is not added at time 0, but continuously, we may separate the input terms from the first summation of 9.32, just as we did with the output terms in the second summation; and the procedure just outlined will then give a general condition for the conservation of label: Label at any time equals total input less total output up to t.)

Equation 9.35 is therefore complete. If it is satisfied, the set of rates is exactly specified by 9.32′ and 9.33, the measurable quantities being known at any time t. If the left side is less than the right, the unknowns are overdetermined. Under our assumption that the equations are consistent, this presents no problem: It means that we do not need all the available information to fix the unknowns. If the left side is greater than the right side, the system is underdetermined, and we need more information than one set of values of the measurable quantities will afford. This problem is easily solved, at least in principle, by using measurements at other times, thus gaining, for each set of measurements, $2n$ additional equations.[7]

Steady State and "Turnover"

None of the relations arrived at in the preceding paragraphs requires stationarity of the metabolic network.[8] So long as we view the system 9.32′–33 as a set of algebraic equations specifying unknowns v_{ji}, with parameters C_i, s_i, and ds_i/dt, the simplification produced by the condition $(Ai)(dC_i/dt = 0)$ is not very striking. The metabolic rates will be strictly constant under stationary conditions, instead of depending on t—that is, not only their velocity constants but their concentration factors will be constant, so that application of the equations to measurements made at any time whatever should give the same numbers.

If we view the equations 9.32′ as differential equations in t, to be solved for the specific activities s_i as functions of t, with the v_{ji} as parameters, then the simplification achieved by the condition of stationarity is enormous. For under that condition the equations constitute a linear system with constant coefficients, whose solution can be written down at once (see equations 9.11 and 9.12). In the absence of stationarity, the equations would constitute a system with variable coefficients, these coefficients being themselves the solutions of another set (9.33) of differential equations (in general nonlinear); and it would be a happy accident if one could obtain a usable solution.

[7] It is obvious that one can always obtain more information by measuring at more points in time; and I hope it is obvious that this is a good idea even when one set of measurements would suffice, since the consistency of independent estimates of a rate parameter is a check on the theory as well as on the arithmetic. One set will not, in any event, give a good ds/dt value at a point. It is also important to remember that v may be of the form kC, and that kC need not be the same for different t if C varies—but of course k must be the same within experimental error.

[8] Contrary to the apparent assumptions of earlier theorists (15).

It is also true that stationarity is the condition that one expects to find in a non-growing and non-differentiating organism, such as the adult forms of many species. Whatever importance we care to attribute to it for its mathematical convenience is therefore emphasized by its biological significance.

Closely associated (historically if not logically) with the concept of stationarity is that of turnover. The intuition underlying this often undefined or badly defined term is obvious: If a substance exhibits no net change in unit time, this may be because nothing happened, or because the rate of synthesis was precisely balanced by the rate of breakdown or removal. This balanced process with no net consequences is what is thought of as "turnover."

That the connection to stationarity is trivial may be grasped with minimal effort. Thus, if we now suppose that a substance exhibits net gain dC/dt in unit time, this may be because there was a gain dC/dt and no loss, or because there was in fact a gain greater than dC/dt and some (corresponding) loss. A portion of the synthesis is not measurable as **net** synthesis because it is offset by breakdown or removal. That portion, it would seem, is what one thinks of as turnover. A similar argument defines turnover if there is net loss rather than net gain.

Examination of 9.33 in connection with the arguments of the last two paragraphs indicates that we may most conveniently define the rate of turnover of a substance as $\sum_i v_{ji}$ or $\sum_k v_{ik}$, **whichever is smaller**. For, if there is net gain, the portion of the total gain that is camouflaged, as it were, is measured by the total loss; and so forth. The stationary condition is the simplest; gain and loss are equal, and either of them is equal to the rate of turnover.

That turnover rate can be precisely and consistently defined is thus obvious. Whether it is worth while to define and use the term is arguable. Certainly no mathematical simplification is achieved thereby: As a glance at 9.32′ will show, it is impossible to express this simply in terms of turnover rate (except for the massive and improbable coincidence whereby all s_j are the same). If we are to learn about metabolic rates, we must calculate all of them. The sum of a subset of these rates embodies only a fraction of the information that will have been gained; and it seems to me that gratuitously discarding the rest of the information is silly. Granted, there is something intuitively appealing about the notion of a single number that measures the metabolic activity in which a compound is involved when you cannot detect any activity by ordinary means; but this number is of very little assistance in the causal analysis of the system.

It remains only to ask whether there is anything substantial to be gained by solving 9.32′, under the simplifications afforded by the stationary condition. This seems to me to resolve itself into a matter of taste. As the reader may see by consulting the second half of my original paper, simple systems show fairly characteristic patterns of radioactivity-time curves under specific experimental conditions; and the ease of recognition is an indubitable advantage. Unfortunately, this perspicuity diminishes rapidly as the system becomes more complicated; and there is a large and boring[9] literature concerned with the resolution of experimental curves into combinations of exponentials. There is a genuine need for a way to see whether our theoretical assumptions about a given system are correct and above all complete; but I do not believe that slope-splitting and curve-fitting supply the required information reliably. Solving 9.32′ and 9.33 for the rates at more than one time, and applying the criterion that the rates estimated from measurable quantities at different times should be the same in the steady state, or that the rate parameters should be the same in the nonsteady state, seem much more likely to do the trick.

PROBLEM: Suppose we assume that a compound X has just one precursor A, and that the X portion of the total metabolic network looks like: $A \xrightarrow{1} X \xrightarrow{2} \ldots$. We solve the appropriate equations for the rates v_1 and v_2 under the further assumption of stationarity. Demonstrate that we will be able to detect our error if in fact:

(a) X has a second precursor;

(b) reaction 2 is reversible;

(c) the immediate precursor of X is actually another compound B, lying between A and X on the pathway.

Precursor-Product Relations

A well-established notion that fares even worse under analysis than that of turnover is the assertion that the specific activity-time curve of the immediate precursor of a compound crosses the curve of that compound at its maximum. This was first proposed (15) as a property of stationary systems, but has nothing special to do with them.

Suppose that the summation in 9.32′ has just one member—i.e., that i has just one immediate precursor, so that the equation becomes:

$$C_i ds_i/dt = v_{ji}(s_j - s_i) \tag{9.38}$$

Then obviously, on the trivial assumptions $C_i \neq 0$ and $v_{ji} \neq 0$, $ds_i/dt = 0$ if and only if $s_j = s_i$. If the extremum of i is a maximum, the s_j curve must be decreasing at the point of intersection. (Why?)

Suppose now that k is the sole immediate precursor of j, so that it is a precursor but not the immediate precursor of i. It is easy to see that the k

[9] I do not mean to imply that it is bad work of its kind. There are certain kinds of routine domestic virtue that are boring in the nature of the case. Some people have a higher threshold for boredom than others.

curve must cross the i curve before the maximum. Thus: (a) if the j curve is decreasing at the i maximum, the maximum of the j curve must precede the maximum of the i curve (i.e., occurs at a smaller t value). But (b) by the reasoning of the preceding paragraph, at the j maximum the k curve must cross it, and be decreasing, so that s_j becomes greater than s_k. Hence (c), when $s_i = s_k$ the right side of 9.38 is positive, so that the k-i intersection occurs on the rising part of the i curve. (Fig. 9.6.)

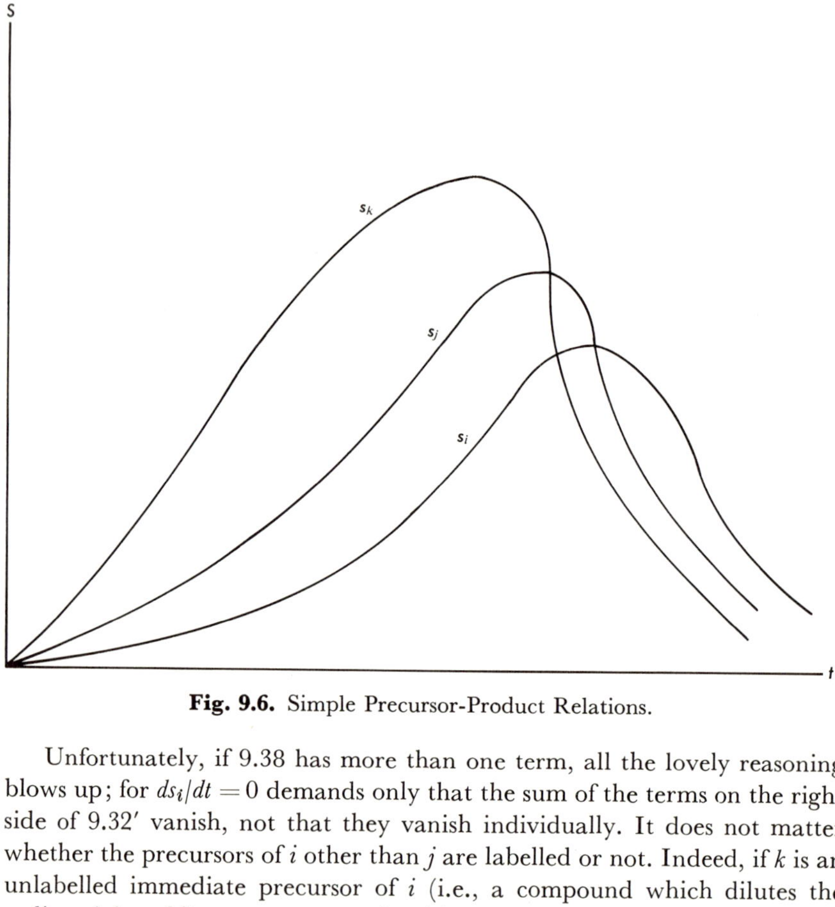

Fig. 9.6. Simple Precursor-Product Relations.

Unfortunately, if 9.38 has more than one term, all the lovely reasoning blows up; for $ds_i/dt = 0$ demands only that the sum of the terms on the right side of 9.32′ vanish, not that they vanish individually. It does not matter whether the precursors of i other than j are labelled or not. Indeed, if k is an unlabelled immediate precursor of i (i.e., a compound which dilutes the radioactivity of i), a term $-v_{ki}s_i$ is added to the right side of 9.38, and the maximum of s_i occurs when $s_i = v_{ji}s_j/(v_{ji} + v_{ki}) < s_j$, so that the intersection of the i and j curves occurs after the i maximum (16). The same argument applies to any number of diluent compounds.

The criterion is therefore useless for the most part, except in a negative sense.

Isotopically Bimolecular Systems

The preceding subsections apply only under the presumption of isotopic unimolecularity—i.e., under the presumption that each compound is formed from no more than one molecule of labelled immediate precursor. Unfortunately for the interests of simplicity, there are many common and important systems where this is not true—e.g., the tricarboxylic acid cycle.

Suppose we consider compound X, formed from A and B. We denote by $X_1{}^*$ the concentration of X molecules formed from labelled A and unlabelled B; by $X_2{}^*$ the concentration of X molecules formed from unlabelled A and labelled B; and by $X_3{}^*$ the concentration of X molecules formed from both labelled A and labelled B.

We shall ignore, for the moment, reactions other than the formation of X from A and B; the expressions can be appropriately corrected for such reactions later. Let the basic rate of the reaction be $v = kAB$. Then:

(9.39) $dX_1{}^*/dt = kA^*\,(B-B^*) = vs_a\,(1-s_b) \quad dX_2{}^*/dt = k(A-A^*)B^* = vs_b\,(1-s_a)$

$$dX_3{}^*/dt = kA^*B^* = vs_a s_b$$

The obvious next step is to combine the various labelled forms to define $s_x = (X_1{}^* + X_2{}^* + X_3{}^*)/X$. From this, 9.39, and the obvious $dX/dt = v$, we can get an expression analogous to 9.32′:

$$Xds_x/dt = v(s_a + s_b - s_a s_b - s_x) \tag{9.40}$$

There are two evident drawbacks to 9.40. First of all, it is nonlinear in s_a and s_b. Secondly, while our expressions count the varieties of labelled molecules correctly, they will not give the right answer when we seek to convert these numbers into counts (or indeed into mass-spectrometric assays), since a doubly labelled molecule will give twice as many counts as a singly labelled one.

Taking up the second problem first, there are two ways to proceed. One is to leave 9.40 alone, with a mental note to insert a factor 2 in the right place when dealing with experimental results. If anyone is willing to do this, he is welcome; I confess it makes my flesh creep!

There is a much better alternative available, which would occur naturally to anyone schooled in the classical isotopic approach to metabolic pathways. Let us treat X chemically so as to separate the part derived from A from that derived from B. Calling the concentrations X_a and X_b, it is immediately obvious that:

$$(9.41) \qquad X_a = X_b = X \qquad X_a{}^* = X_1{}^* + X_3{}^* \qquad X_b{}^* = X_2{}^* + X_3{}^*$$

The relations in 9.41 are immediately translatable into counts per minute or atoms of an isotope without any correction factor. If we define the obvious $s_{xa} = X_a{}^*/X_a$, etc., and make use of 9.41, we get:

$$(9.42) \quad X_a ds_{xa}/dt = v(s_a - s_{xa}) \quad X_b ds_{xb}/dt = v(s_b - s_{xb})$$

as the appropriate analogue of 9.32′. From 9.41 also, the analogue of 9.33 is simply $dX_a/dt = v$ (plus terms for any other reactions we may have neglected in this discussion), and so on.

Splitting the molecule and the equations obviously involves the experimental investigator in extra labor; but it is labor which yields valuable extra information, and which in any case is indispensable for an adequate analysis of the system.

The reasoning of the foregoing paragraphs is readily extended to reactions in which X is converted to other compounds. For instance, if the A-derived group is split off (as in the decarboxylation of oxalosuccinate to α-ketoglutarate), we must treat this as a reaction of X_a. The balance of the molecule, containing the B-derived group, will figure as a reaction product of X_b.

PROBLEMS: 1. Do the considerations of the last subsection apply to double labelling experiments (e.g., C^{14} and P^{32} or C^{14} and H^3)?

2. Formulate the problem of studying the synthesis of a specific protein by labelling with one amino acid which appears several times in the protein.

Enzymatic Equilibrium Studies

I have already pointed out in an earlier section that the reactions involved in enzymatic catalysis, being either unimolecular or bimolecular, constitute special cases of the reaction networks discussed there. The same is true here; and equations 9.32 and 9.33 (or 9.42) are applicable.

For example, a simple reversible system, such as that represented by scheme 3.B, involves three reversible reactions, hence six rates to be ascertained. Equations 3.8 and 3.9 (used jointly with 3.10) are analogous to 9.33; and four more equations analogous to 9.32′ are readily written down. Six of the eight equations are independent, there being two constraints (conservation of substrate and conservation of isotope).

If no further restriction is placed on such a system, we are embarrassed by a technical difficulty—the requirement that we be able to measure separately C_1 and C_2 and the corresponding specific activities. However, technical difficulties are made to be overcome; and good enzymologists are prodigies of ingenuity. I recommend this problem to them as being not only amusing but significant.

If we are willing to ignore the distinction between C_1 and C_2, and revert to the simpler scheme 3.A, we must still be able to separate C from E. There are times when a large miracle is just as easy to produce as a small one!

Stationarity improves matter appreciably. For then we know that (in the notation of Chapter 3) $C = E_t(S/K + P/K')/(1 + S/K + P/K')$; and, armed with this, we can calculate the specific activity $s_c = C^*/C = (C^*/E_t)/(C/E_t)$. The numerator is the specific activity of total enzyme, so it is now only necessary for the exploited enzymologist to separate enzyme from the rest of the incubation mixture without significant dissociation of C. Not necessarily easy, but surely much less difficult than fractionating the enzyme.

A theoretically sound but impractical way to get around the difficulties of estimating s_c (which would be valid even in the nonstationary condition) is to use the condition of conservation of isotope, which may be written:

$$s_c C + S^* + P^* = S_0^* \tag{9.43}$$

where S_0^* is the total amount of radioactive material added initially (as substrate or product or both). Similar to this is the condition for conservation of substrate:

$$C + S + P = S_t \tag{9.44}$$

The trouble with both of these relations is that they give C in one case, and $C^* = s_c C$ in the other, as small differences of much greater numbers. They are likely to be quite inaccurate, therefore. In situations like those of Chapter 4, where a great deal of substrate is bound, they would work, and would be welcome checks on other methods.

When the enzyme-catalyzed reaction is at equilibrium, $P = K_{ee} S$, and the stationary form of C simplifies to $C = \alpha S E_t/(\alpha S + K)$, where $\alpha = 1 + K_{ee} K/K'$. Since $dS/dt = 0$ and $dP/dt = 0$, the four rates to be estimated boil down to two pairs of equal rates. Moreover, since S and P are constant, only their specific activities need be measured as functions of time; thus aliquots, if they are radiochemically pure, will do. It is easy to see why this case is a favorite, since it permits the estimation of all velocity constants individually with a minimum of labor.

Boyer has derived a number of relations for the equilibrium state (17); these are not immediately recognizable, however, as cognate with those I have been discussing. The method is ingenious: Boyer proposes that C be considered as made up of molecules most recently derived from S (C_s) and of molecules most recently derived from $P(C_p)$. Using the equilibrium expression for the E-S reaction and the equality of forward and backward rates (S to P and P to S) in the equilibrium state, he obtains expressions for C_s and C_p.

How this approach fits into a more orthodox scheme becomes evident from Boyer's footnote 3. He asks the reader to visualize beginning with an equilibrium mixture of labelled S and unlabelled P, and adding a small amount of enzyme. When stationary conditions are reached, the specific activity of C (neglecting labelling of and back reaction from P) should depend on the fraction of C molecules most recently derived from S, and should be equal to $(C_s/C)s_s$. Now if we take the activity equation for C:

$$Cds_c/dt = v_1(s_s - s_c) + v_{-2}(s_p - s_c)$$

and discard all terms in s_p and s_c as being initially negligible compared to s_s, we have approximately:

$$Cds_c/dt = v_1 s_s$$

If we also assume that s_s is virtually constant during the period of application of this relation, we may integrate from 0 to t, getting after division by C:

$$s_c = (v_1 t/C)s_s$$

Comparison of this with Boyer's expression for s_c identifies his fraction most recently derived from S, C_s, as $v_1 t$. This makes a certain kind of sense, for indeed v_1 is the rate of formation of C from S, and so $v_1 t$ should be the C formed from S during t. If t is sufficiently short, and s_s is constant during t, all is well.

If I have any tendency to be critical of this extremely neat performance, it is based on the fact that Boyer's discussion only tangentially indicates the critical role of timing in the applicability of his equations. It is all in there somewhere (since Boyer is far too bright ever to say anything silly); but the style of the cautionary remarks is somewhat bland. What is really implied is that we add enzyme, mix, and almost immediately take our sample for analysis. The reader might well get the idea that the only limitation on timing is on how **soon** one may take the sample. The sole indication to the contrary is the comment that "it may not be convenient to neglect the back reaction." Since this is followed in the next paragraph by "measurements for more than one time period are valuable," the net effect is somewhat misleading.

Correction for Decay

It is customary for careful workers to correct radioactivity figures for radioactive decay when time studies are conducted, since otherwise activities at different times will not be strictly comparable. The correction is simple. If the decay constant of the isotope is k, the radioactivity at time t is related to that at time 0 by: $N = N_0 e^{-kt}$. Activity of a sample taken at time t (assuming it is counted immediately) can therefore be expressed as the undecayed activity by multiplying by e^{kt}.

Let us see how this problem affects our equations, in which decay was neglected. Our equations for activity are all of the form:

$$dX^*/dt = F$$

where F is a homogeneous function of the various activities. Suppose now that we write $X^* = X_e^* e^{kt}$, where X_e^* is the experimentally measured radioactivity, and that we correct all activities this way. Then, if F is linear in the activities, every term of F will have a factor e^{kt}, and the equations will read:

$$d(X_e^* e^{kt})/dt = e^{kt} dX_e^*/dt + kX_e^* e^{kt} = F_e e^{kt}$$

where F_e is F expressed as a function of the experimentally measured radioactivities. For finite values of t, e^{kt} is finite and nonvanishing, and, since the equation is valid for all t, may be cancelled. We get:

$$dX_e^*/dt = F_e - kX_e^*$$

which is precisely the equation we would have obtained if we had corrected our original equation for decay.

The same result will be obtained if the sample taken at time t is not counted immediately, but T time units later. The theoretical quantity X^* (the value which one would have found if there had been no decay until counting was initiated) will now be related to the measured activity by a factor $e^{k(t+T)} = e^{kt}e^{kT}$. We get a t-independent factor e^{kT} in every term of the equation, and this cancels as before.

In short, our theoretical equations, if otherwise correct, will properly predict experimental results corrected for decay.

Summary and Conclusions

The equations derived in this section are universally applicable; but their application in any given case requires a good deal of prior information. To state the problem concretely: The general equations, applicable to any metabolic network (or network of compartments or pools), can be applied to a specific network only if the connectivity of that network is known. This may not sound like a very harsh restriction; but it contradicts a great deal of past practice in the field of tracer studies of metabolic rates. An enormous amount of published work is based on the implicit premise that the metabolic rates of formation and/or breakdown of a compound can be calculated if incorporation of isotope into that compound with time is measured, regardless of our knowledge of the network in which that compound is imbedded. This section has demonstrated unarguably that such a premise is nonsense.

Simply by inspection of 9.32 or 9.32′ (or 9.42) and 9.33 it is possible to specify just what information is required:

1. We must know the reactions in which any compound of interest is formed or removed. In particular, we must be able to identify every immediate precursor of that compound.

2. The activity and concentration of the compound and of each of its immediate precursors must be measured. Activity alone, or specific activity of an aliquot of the unknown total, will not suffice.

3. Time curves must be filled in adequately. Not only is it of great value to check consistency by comparing rates calculated from measurements at different times, but it is indispensable to obtain adequate estimates of rates of change of specific activity—that is, of slopes of time curves. The all-too-frequent "curve" consisting of a zero-time measurement and one other measurement is worthless.

These statements may sound as if they were designed to discourage tracer work. But this is so only if one holds the view that tracer methods give the easy and effortless way to study metabolism—that they are an effective substitute for good chemistry and hard work. Such a view can scarcely be held by anyone with adequate experience in tracer techniques. The fact is that preparative and analytical techniques which would more than suffice for a problem not involving isotopes may be too crude for a tracer study, simply because of the ever-present problem of radiochemical purity: A trivial but "hot" contaminant, which might not in the least affect the assay of quantity of material, may be catastrophic for the assay of radioactivity of a specific compound. No one who knows anything about such complications will ever undertake a tracer study on the assumption that it will be easy.

The most obvious candidate for tracer studies of metabolic rates lies in the *in vivo* pursuit of problems already far along *in vitro*. Thus it makes good sense to ask how the network of carbohydrate and/or fat metabolism actually operates in a living animal. The reactions are known, the substrates and coenzymes have been identified, the enzymes have been isolated and their properties have been studied. Any freshman medical student could write down the requisite information. We are not so far along in the study of protein and nucleic acid metabolism, but nevertheless we are quite close to the point where a rigorous *in vivo* analysis would be limited chiefly by one's willingness to invest the labor demanded.

I do not want to be understood as saying that no tracer study of metabolic rates is legitimate unless the system is already completely known. But I want to stress that studies based on incomplete knowledge of the system must be regarded as exploratory and speculative rather than conclusive. There is no reason why one should not postulate a set of pathways, and analyze adequate data in the light of such a postulate. If the results pass the consistency tests mentioned earlier in this section, one might legitimately say, "There is a fighting chance that the proposed network is in fact the correct one." But one may not say, "These numbers **are** the metabolic rates of substances *X*, *Y* and *Z*." I have tried throughout this volume to stress that a major role of theory is to go in advance of experiment rather than behind it, to inspire and help design novel experiments rather than merely to preserve previous results. But to design experiments based on a theoretical postulate, and eventually thereby perhaps to gain confidence in those postulates, is not the same thing as to design experiments based on theory and then interpret the data on the unproved assumption that the theory is correct.

In view of the tremendous complexity of metabolism in the animal body, the extremely high degree of linkage among the various metabolic systems, we should not leave this topic without considering briefly the problem of partial analysis. To what extent can subsystems, even though tied to the whole, be analyzed in themselves?

Suppose we are, for whatever reasons, interested in just one compound i. If i has just one precursor j, and just one product k, the equations for i (9.32′ and 9.33) boil down to:

$$(9.45) \qquad C_i\, ds_i/dt = v_{ji}(s_j - s_i) \qquad dC_i/dt = v_{ji} - v_{ik}$$

and the system is completely solvable, if j and k are known.

The simplest deviations from this ideal state would be that (a) there is actually a compound j' (the true immediate precursor) between j and i or (b) there is a second diluent precursor m. If we have a little luck, and the two deviations do not occur simultaneously and exactly compensating one another, we may find out about them by applying the criteria mentioned earlier—the location of the point where the s_j curve and the s_i curve cross relative to the s_i maximum.

If we find indications of (a), one can only say with resignation: Back to the drawing board! Some chemistry and a lot of fumbling will be required to identify j'. Once it is identified, however, we may be in good shape.

If we find indications of (b), there is a simple approach to the problem, provided there is only one significant diluent m. For then:

$$(9.46) \qquad C_i\, ds_i/dt = v_{ji}(s_j - s_i) - v_{mi}\, s_i \qquad dC_i/dt = v_{ji} + v_{mi} - v_{ik}$$

The strategy in this case is to accumulate points on the specific activity curves as close as possible to zero time. Assuming we started with i unlabelled, s_i should decrease as we move closer to $t = 0$. If we plot ds_i/dt against s_i, and extrapolate to $s_i = 0$, we will have $C_i \, ds_i/dt = v_{ji} s_j$ in the limit. From this we get v_{ji}, and from points at later times we now get v_{mi} using 9.46.

The case of more than one labelled immediate precursor is more difficult to deal with in all its generality—but fortunately also more unusual. If we try to think about concrete cases, we come up with something like glucose plus ATP and glucose-1-phosphate as precursors of glucose-6-phosphate, or the various precursors of CO_2 from pyruvate on. In the first instance, the nature of the pathways helps; for, if labelled glucose is the ultimate source of isotope, glucose-6-phosphate will be labelled before glucose-1-phosphate and hence the approach of the preceding paragraph should be possible at early times, whereas with glycogen as the ultimate source of isotope glucose should be labelled only very slowly. In the case of CO_2, with uniformly labelled glucose as the eventual source of isotope, naturally everything will become labelled; but the use of C_1-labelled or C_6-labelled glucose will largely confine tracer to the methyl group of pyruvate, and the one labelled immediate precursor of CO_2 will be oxalosuccinate.

A closely related case is that of a small active pool on the main pathway, connected to a much larger and less active dead-end pool (18). It is easy to show (try it!) that the duality is easily overlooked, and that confounding the two pools will lead to incorrect estimates of the rates. Arthur Koch, who has undertaken a detailed analysis of some of the boobytraps that await the unwary (19), has investigated this case rather fully (personal communication).

We may conclude that partial analysis is by no means out of the question, with some careful consideration of the sources of isotope and of sources of error like the dual-pool case, and of course a little luck. The apparently depressing conclusions of preceding paragraphs were not intended as an invitation to all workers in this field to jump off a bridge, but as a prophylactic against the occupational disease of the field, overconfidence.

REFERENCES

General Background

There is a delightful discussion of the stability of dynamic systems in "Design for a Brain," by W. Ross Ashby, Wiley, New York, 1952. The scope of this work is much broader than the title suggests. The mathematically tougher reader, who likes to eat raw meat for breakfast, may care to look at "Non-linear Oscillations," by Nicholas Minorsky, D. Van Nostrand, Princeton, 1962.

The most elegant and powerful analysis of the kinetics of linear systems from the biological point of view is that of John Z. Hearon (Annals of N. Y.

Acad. Sci. *108*, 36–68, 1963). The reader is advised to brush up on at least the elements of matrix theory before trying to read it.

My own papers on tracer kinetics were preceded by exactly one that deserves serious consideration. (This is a personal evaluation; and I intend to ignore all cries of outrage. My reaction to most of the stuff written both before and after this period is like that of a notoriously outspoken expert in intermediary metabolism who sat in the row ahead of me at a session on brain metabolism at a national meeting. As paper after paper fumbled about the periphery of problems whose solution had been published several years earlier (8), this gentleman reached his rather low boiling point. He turned around, leaned over confidentially, and observed in a hoarse and contemptuous whisper that reverberated through the auditorium, "A bunch of goddam amateurs!") The paper by Sheppard and Householder, written in terms of compartmental analysis, is a masterpiece; unfortunately, it was published in a journal that was unlikely to reach the eyes of most of the people who should have been reading the paper (J. Appl. Phys. 22, 510, 1951). C. W. Sheppard has since published an excellent book, "Basic Principles of the Tracer Method" (Wiley, New York, 1962).

The late Jane Russell, who probably knew as much about intermediary metabolism in the animal body as anyone in the world, wrote a paper (Perspectives in Biol. & Med. *1*, 138, 1958) which is a veritable museum of sins committed through ignorance of the principles of tracer kinetics. Jane collected such specimens with something of the zest with which Henry Mencken collected the antics of Bible-belt elders and New York Times pundits, and like Mencken she did not treat them with superfluous delicacy. She had fully earned her right to be impatient with nonsense—she had no more special mathematical training than most biochemists or physiologists, but she had brains and rather enjoyed using them.

Textual References

1. Reiner, John M., "The Organism as an Adaptive Control System", Prentice-Hall, Englewood Cliffs, New Jersey, 1967.
2. Dixon, M., "Multi-Enzyme Systems", Cambridge, 1949.
3. Lotka, Alfred, "Elements of Physical Biology", Williams & Wilkins, Baltimore, 1925.
4. Ross, S., "Differential Equations" Blaisdell, New York, 1964.
5. Denbigh, K. G., Hicks, M., and Page, F. M., Trans. Faraday Soc. *44*, 479 (1948).
6. Kornberg, A., and Pricer, W. E., Jr., J. Biol. Chem. *182*, 763 (1950).
7. Kaplan, N. O., Colowick, S. P., and Nason, A., J. Biol. Chem. *191*, 473 (1951).
8. Utter, M. F., Wood, H. G., and Reiner, J. M., J. Biol. Chem. *161*, 197 (1945).
9. Reiner, John M., Arch. Biochem. *12*, 327–338 (1947).

10. Handler, P., and Klein, J. R., J. Biol. Chem. *143,* 49 (1942).
11. Sable, H. Z., Proc. Soc. Exp. Biol. Med. *75,* 215 (1950).
12. Colowick, S. P., and Kalckar, H. M., J. Biol. Chem. *148,* 117 (1943).
13. Reiner, John M., Bull. Math. Biophys. *29,* 349–361 (1967).
14. Reiner, John M., Arch. Biochem. Biophys. *46,* 53–99 (1953).
15. Zilversmit, D. B., Entenman, C., and Fishler, M. C., J. Gen. Physiol. *26,* 325 (1943).
16. Barnum, C. P., and Huseby, R. A., Arch. Biochem. *29,* 7 (1950).
17. Boyer, P., Arch. Biochem. Biophys. *82,* 387–410 (1959).
18. McCarthy. B. J., and Britten, R. J., Biophys. J. *2,* 35–47 (1962).
19. Koch, A. L., J. Theoret, Biol. *3,* 283–303 (1962).

Added in press: An interesting application of tracer theory, for the special case of an isotopic steady state, to the study of pathways of carbohydrate metabolism, is:

Landau, B. R., Bartsch, G. E., Katz, J., and Wood, H. G., J. Biol. Chem. *239,* 686-696 (1964).

Despite the steady-state restriction, the paper admirably illustrates my point about using the rate equations as algebraic equations for rate parameters rather than as differential equations for specific activities.

10

Effects of pH and Temperature on Enzyme Action

THE EFFECT OF pH

It has been known for many years that the activity of most enzyme-catalyzed reactions varies with pH (1). The curve of activity against pH usually though not invariably exhibits a maximum, whose location varies widely from enzyme to enzyme.

The importance of a theory of pH transcends the possibility of explaining why enzymes have characteristic pH curves. The study of pH effects can and should give information about certain aspects of the reaction mechanism, and particularly about the structure of the so-called "active centre" of the enzyme. This area therefore constitutes one of the major bridges between the study of kinetics and that of reaction mechanisms proper.

The Restricted Theories

The most obvious and elementary approach to a theory of pH curves is based on a familiar principle of physical chemistry. If a compound that can yield hydrogen ions by dissociation does so in several stages, one can proceed in a familiar fashion to calculate the concentrations of the various ionic forms at equilibrium. For example, if there are two dissociable hydrogens one can find the concentrations of the undissociated form and of the two dissociated forms, expressed as functions of the hydrogen ion concentration. It then appears that the undissociated form increases monotonically with

H^+, the most dissociated form decreases monotonically with H^+, and the intermediate (singly ionized) form has a maximal value as a function of H^+. The argument then goes readily: Enzymes, which are proteins, are polyvalent electrolytes, and will exhibit stages of dissociation similar to those just mentioned; if only certain ones of these stages are enzymatically active, the activity will follow the concentration-pH curves of those, and the various forms of activity-pH curves are thus explained (2).

This plausible approach runs aground on several counts. It was T. H. Huxley, I think, who wisely said, "There is nothing so sad as a beautiful theory destroyed by a brute fact." If we take the two-stage dissociation model as at least an approximation, the maximizing value of h (which is how I shall refer to H^+ for convenience henceforth) is $(K_1K_2)^{\frac{1}{2}}$, where K_1 and K_2 are the two dissociation constants; and similar relations may be inferred for the more general case. This expression is what, in the case of an electrolyte like a protein, would be called the isoelectric point (the pH at which the molecule carries an equal number of positive and negative charges). However, the optimal pH values of many if not most enzymes markedly fail to coincide with their isoelectric points.

Moreover, Northrop, in studies on pepsin and trypsin (3, 4), observed that the activity curve coincided remarkably well with the titration curves of a number of substrates, and that maximum combination of these substrates with the enzyme occurred close to the pH optima. It was evident from these data, if it had not been evident on logical grounds, that the various charged forms of the substrate might have different probabilities of combining with the enzyme, and that the contribution of the substrate to the pH-activity relation could not be neglected.

The most complete quantitative treatments of the problem, restricted to a few ionic states, are those of Waley (5) and Laidler (6).[1] I shall indicate the nature of the theories rather briefly, since the procedure is a fairly obvious one, and let the reader fill in details for himself.

Consider first the scheme in which substrate ionization is ignored:

$$(10.A)\quad H_2E^+ \overset{e_1}{\rightleftharpoons} H^+ + HE \qquad HE \overset{e_2}{\rightleftharpoons} H^+ + E^- \qquad H_2ES^+ \overset{c_1}{\rightleftharpoons} H^+ + HES$$

$$HES \overset{c_2}{\rightleftharpoons} H^+ + ES^- \qquad HE + S \overset{1}{\rightleftharpoons} HES \qquad HES \overset{2}{\longrightarrow} HE + P$$

The ionic reactions have been labelled e_1, e_2, c_1, and c_2 in self-explanatory fashion to distinguish them readily from the combination and turnover reactions. I call the reader's attention to the fact that some superfluous

[1] Laidler raises the superfluous problem of what would happen if the almost instantly equilibrated dissociation reactions were thrown into the pseudostationary pot along with the reactions of enzyme and substrate. To no one's great surprise, he finds that the resulting equations are a terrible mess, and carries out his further analysis with what are essentially Waley's equations.

reactions have been ignored; specifically, direct combination of S with the forms H_2E^+ and E^- has been omitted, and the hypothetical compounds treated as obtained from ionizations of HES. The justification for this is that H_2ES^+ is a dead-end compound; hence a reaction like $H_2E^+ + S \rightleftharpoons H_2ES^+$ would necessarily, if stationary, be in true equilibrium—and the equilibrium concentration of H_2ES^+ would have to agree with that obtained from the equilibrium of reaction c_1. What we would get, if we included such reactions in the scheme, would be two new equilibrium constants which would, under the consistency conditions, be expressible in terms of the other constants.

With K_{e1}, etc., as the dissociation constants of the respective ionization reactions, and $K = (k_{-1} + k_2)/k_1$ in the familiar fashion:

$$v = k_2 SE_t/[K(1 + h/K_{e1} + K_{e2}/h) + S(1 + h/K_{c1} + K_{c2}/h)] \tag{10.1}$$

Inasmuch as comparisons with experimental data will usually be carried out by obtaining apparent K and V values from Lineweaver-Burk reciprocal plots at various h values, it is convenient to get this into the conventional form $v = V'S/(K' + S)$ by dividing through by the coefficient of S in the denominator, whence:

$$\begin{aligned} V' &= V/(1 + h/K_{c1} + K_{c2}/h) \\ K' &= K(1 + h/K_{e1} + K_{e2}/h)/(1 + h/K_{c1} + K_{c2}/h) \end{aligned} \tag{10.2}$$

If we maximize 10.1 with respect to h, we get:

$$h_m = [K_{e1}K_{c1}(KK_{e2} + SK_{c2})/(KK_{c1} + SK_{e1})]^{\frac{1}{2}} \tag{10.3}$$

which is a function of S, and varies from $(K_{e1}K_{e2})^{\frac{1}{2}}$ to $(K_{c1}K_{c2})^{\frac{1}{2}}$ as S varies from zero to infinity.

The variation of pH optima with S is a well-known phenomenon. One of the most dramatic examples is the case of the alkaline phosphatases (7), whose pH optima at the conventional substrate saturation are at ridiculously alkaline levels (ca. pH 10). As the S level is dropped to the much lower values at which one might reasonably find phosphate esters in the cell, the pH optima drop toward neutrality. Thus we make sense out of the otherwise grotesque paradox of an intracellular enzyme that functions best under conditions where the cell would at once be mortally damaged.

Laidler argues from this model (and its extension to the case of substrate ionization) that the occurrence of an optimum rate implies that both an acidic and a basic group must somehow be involved in the enzymatic reaction. As h is varied, the forms of enzyme will shift from maximally dissociated to minimally dissociated, hence from negatively charged to positively charged (or less negatively charged) forms. The probability of finding two groups of opposite ionic character on one molecule will of course be maximal somewhere in between the extremes of the h range.

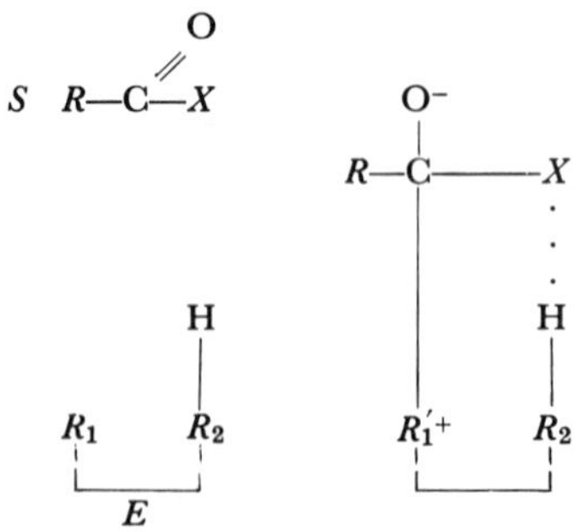

Fig. 10.1. Illustration of the Theory of Acidic and Basic Pairs of Catalytic Groups. See text for details.

An example (Fig. 10.1) is the binding of a carbonyl compound. Group R_1, a nucleophilic group, is seen as forming a bond which permits "ionization" of the carbonyl group, while the acidic group R_2H forms a hydrogen bond with group X, which in turn is nucleophilic (e.g. —OH, —NH_2, —OR′, —NHR′).

This view is a plausible one. We shall see in the next chapter that it is a shade too simple; but the essence of it is not too far off the mark.

Problems: 1. Derive 10.1 and 10.3.

2. Suppose that the substrate is actually a compound HS capable of a single dissociation, that the S of scheme 10.A is actually the ionized form S^-, and that the reaction $HS + E^- \rightleftharpoons HES$ can also take place. Derive the rate expression and find the optimum pH.

3. Returning to scheme 10.A: Suppose the form of enzyme that combines with S and turns it over is ES^- rather than HES. Find and characterize the rate as a function of pH. (See (8), dealing with serum cholinesterase, for an example of such a system.)

A General Theory of Enzyme Activity and pH

A more general formulation of the problem is by no means as difficult as it may once have seemed. I have therefore carried out such a generalization, and present it in this subsection.

We may (although we need not) begin by assuming that all the dissociable groups of the enzyme, substrate, and enzyme-substrate complex are distinct in their properties. Let the groups be numbered in some fashion. We assign to each such group a suitable dissociation constant, say K_{ei} for the ith group on free enzyme, etc. When several groups on a single molecule are dissociated, they no doubt influence one another; and the constants may not be the same as they would have been if the dissociated groups had remained isolated, or had been arranged in different ways (as in Chapter 5). However, in view of the fact that we are here dealing with equilibrium reactions, the **order** in which the dissociations of a given set of groups were carried out will not affect the values of the equilibrium constants. (The thermodynamic argument is the familiar one: The difference in free energy between two states of a systems, accessible from one another through a variety of reversible paths, is independent of the actual reversible path followed.)

We shall in fact approximate by neglecting the mutual influences of the dissociated groups, and assume that each dissociation constant is first of all characteristic of the group itself, and next characteristic of its (fixed) immediate neighborhood in the molecule. In the case of protein molecules, the justification for this as a first approximation is the well-known decrease of such effects with distance, illustrated by the case of the chlorinated straight-chain carboxylic acids (9). Moreover, if the protons are as mobile in their distribution as Kirkwood suggests (Cohn and Edsall (10)), a time-averaged dissociation constant is what we actually observe. In the case of low-molecular substrate molecules, the approximation might be very bad, but is also superfluous, as will become evident shortly.

We shall use the familiar symbols E, S, and C, and employ subscripts to denote the number of protons that have been removed. Thus the reference forms, with maximal number of protons, will be labelled with index 0.

We shall denote specific **sets** of dissociable groups by Greek letters. Appending such letters as subscripts will signify that we are referring some such specific group. Thus α might represent the set of groups 1, 4, 5, 7, 13, 22; and $E_{6\alpha}$ would signify the number of free enzyme molecules with six protons removed, namely from groups 1, 4, 5, 7, 13, 22.[2]

We may then readily express a typical dissociation equilibrium. If, for instance, β is the set formed by adding j to α:

$$E_{i+1,\beta} = K_{ej}\, E_{i\alpha}/h \tag{10.4}$$

with similar expressions for the S and C dissociations. This type of equation obviously (see Chapters 1 and 5) defines a set of recursive functions, and it is easy to write the general solution:

$$E_{i\alpha} = P_{i\alpha}\, E_0/h^i \qquad P_{i\alpha} = \prod_{j \varepsilon \alpha}^{i} K_{ej} \tag{10.5}$$

where the limit i on the product means that there are i factors, and the expression "$j \ \varepsilon \ \alpha$", employing a standard symbol of set theory, states that every index j in the product is a member of the set α (read "j is in α").

[2] Since sets such as α are characterized among other things by their measure (for such finite sets, the cardinal number of their elements), they should properly be labelled with an appropriate suffix, as: α_i. I have omitted the suffix in the interests of keeping the notation no more complicated than it absolutely has to be.

The total number of enzyme molecules with i groups dissociated is the sum of 10.5 over all sets α containing i members:

(10.6) $$E_i = \sum_\alpha E_{i\alpha} = E_0 \sum_\alpha P_{i\alpha}/h^i$$

with similar expressions for the S and C series. Finally, the total number of molecules of free enzyme is:

(10.7) $$E = \sum_i^{r_e} E_i = \sum_i \sum_\alpha E_{i\alpha} = E_0 \sum_i \sum_\alpha P_{i\alpha}/h^i$$

and so forth, where r_e is the total number of dissociable groups on a molecule of free enzyme. It will be convenient to introduce an abbreviation for the double summation:

(10.8) $$g_e = \sum_i \sum_\alpha P_{i\alpha}/h^i$$

with similar abbreviations g_s and g_c for the double sums in the S and C series. We may then write:

(10.9) $$E = g_e E_0 \qquad S = g_s S_0 \qquad C = g_c C_0$$

It is reasonable to suppose that only certain complexes included in the total C will actually be enzymatically active. Let us denote one such complex by $C_{q\gamma}$. Let us also agree that this complex may be formed in more than one way, according to the scheme:

(10.B) $$E_{r\alpha} + S_{p\beta} \overset{1}{\rightleftharpoons} C_{q\gamma} \qquad E_{s\delta} + S_{t\eta} \overset{1'}{\rightleftharpoons} C_{q\gamma} \text{ etc.}$$

$$C_{q\gamma} \xrightarrow{2} \text{products}$$

The triples of subscripts (r, p, q), (s, t, q), etc., are subject to the condition of **electroneutrality**: The net charge in the system must be zero, or, neglecting ions that do not participate in any of the reactions of interest to us, the balance of the net charge must not be changed by dissociation and/or complex formation. Thus if the charge of an E_0 molecule is e_0, that of an S_0 is s_0, and that of a C_0 is c_0, it must be true that the initial charge $e_0 + s_0$ before dissociation and complex formation must be equal to charge $c_0 - q + r + p$ that results when $r + p$ protons are released by dissociation and then the dissociated E_r and S_p molecules are combined to a complex with net charge $c_0 - q$. The same holds for the E_s and S_t dissociations, and any similar ones. Thus:

(10.10) $$e_0 + s_0 - c_0 = r + p - q = s + t - q = \ldots$$

Moreover, if there were another enzymatically active complex $C_{m\lambda}$ distinct from $C_{q\gamma}$, the same reasoning would hold. The change in the charge of the system $(E + S)$ that occurs when C is formed is therefore a constant.

Let it also be observed that, as in the preceding subsection, I have neglected with impunity the E-S reactions which do not lead to enzymatically active complexes; for the inactive ones are dead-end compounds, and their (equilibrium) concentrations are already fixed by the dissociation equilibria.

The pseudostationary equations corresponding to scheme 10.B give:

$$C_{q\gamma} = \sum E_{r\alpha} S_{p\beta}/K_i \tag{10.11}$$

where the summation is over all the pairs of reactants in the scheme in reactions 1, 1′, etc., and

$$K_i = (k_{-1} + k_{-1}' + \cdots + k_2)/k_i \tag{10.12}$$

Using 10.5 in 10.11, we have:

$$C_0/S_0 E_0 = \sum Q/K_i \tag{10.13}$$

where the range of the summation is as in 10.11, and where we have introduced the abbreviations:

$$Q = P_{r\alpha} P_{p\beta}/P_{q\gamma} h^{r+p-q} \tag{10.14}$$

and so on for the other combinations (s, t, q, etc.).

The result 10.13 can be simplified by a neat trick, which shows up a necessary relation between Q, Q', etc. Suppose reactions 1, 1′, etc., were actually in equilibrium rather than in the pseudostationary state. Then we would have $[C_{q\gamma}/E_{r\alpha} S_{p\beta}]_{eq} = k_1/k_{-1}$, and similar relations for reaction 1′ and the others. Using 10.5 again, we would get $[C_0/E_0 S_0]_{eq} = Qk_1/k_{-1} = Q'k_1'/k_{-1}' = \ldots\ldots$ (Note that, from the consequences of the electroneutrality law, $h^{r+p-q} = h^{s+t-q} = \ldots$, so that the powers of h cancel from these relations, which are thus independent of pH as they should be.) Introducing these relations into 10.13, we can now write simply:

$$C_0/E_0 S_0 = Q/K \qquad K = k_{-1}K_1/(k_1 K_1 - k_2) \tag{10.13'}$$

The same line of reasoning of course guarantees that, if there were an active complex other than $C_{q\gamma}$, leading to an analogue of 10.13′ with a ratio Q^*/K^*, this would have to equal Q/K.

The usual conservation equation holds for enzyme:

$$E_t = E + C \tag{10.15}$$

We shall take $S \simeq S_t$ for simplicity as usual. Using 10.9 and 10.13′ in 10.15:

$$E_0 = E_t/(g_e + g_c SQ/Kg_s) \tag{10.16}$$

Since the analogue of 10.5 for the C series gives each member in terms of E_0, h, and the constants, the problem is now completed. Ignoring for the time being the existence of active complexes other than $C_{q\gamma}$, and taking therefore $v = k_2 C_{q\gamma}$, we get a rate expression which, after slight rearrangement, reads:

$$(10.17) \qquad v = [k_2 S E_t P_{q\gamma}/h^q g_c]/[(K g_e g_s/Q g_c) + S]$$

From 10.17 we can at once express the apparent parameters of a Lineweaver-Burk plot as:

$$(10.18) \qquad V' = k_2 E_t P_{q\gamma}/h^q g_c \qquad K' = K g_e g_s/g_c Q$$

If such expressions can be made amenable to analysis, we may be able to make something out of experimental data. What we would like to know is: (a) something about the character of the active forms of enzyme and (b) the values of the various parameters, including the dissociation constants.

Such an achievement depends on factors similar to those entering the analysis of titration curves of simpler compounds. As everyone knows, the architectural features of such curves (plateaux, inflections) depend on and are simply related to the values of the various dissociation constants. The curves depend, in fact, on the logarithms of expressions like $1 + K_a/h + K_a K_b/h^2$ (of which 10.8 is an obvious generalization). As we titrate from acid to alkaline pH values, $1/h$ increases, and the successively higher powers of $1/h$ will come to dominate the expression. But a simple interpretation of the curve and ready estimate of the parameters depends on the possibility that discriminable regions of the curve depend chiefly on one of the terms at a time: We would like to find an h range in which the term 1 is an order of magnitude greater than the others, then a range in which K_a/h is an order of magnitude greater than either 1 or $K_a K_b/h^2$, and so on. It is easy to see that this demands a difference of two orders of magnitude between K_a and K_b; and it follows that we will have to titrate through two orders of magnitude of h (2 pH units) in order to pass from the region in which the $1/h$ term is the only significant one to the region in which the $1/h^2$ term dominates.

Roughly speaking, then, the pH analysis of a molecule with n distinct dissociable groups would demand titration over $2n$ pH units. The practical pH range for such titrations is perhaps 10 to 14 pH units, allowing us to analyze for 5 to 7 distinct groups.

Consultation of suitable tables (9) shows two things: (a) proteins tend to contain over 20 dissociable groups per 100 amino acids, so that a typical enzyme of molecular weight ca. 60,000 would have over 100 dissociable groups; (b) the dissociable groups are not distinct, but fall into a very limited number of chemical classes. It is the second circumstance that permits any kind of analysis.

It is known that the influence of the neighborhood, when an amino acid becomes part of a protein, affects the value of the dissociation constant (10, 11). It is nevertheless true that the shifts in pK values are rarely so radical and unaccountable as to make titration of proteins wholly pointless; nor do such effects vitiate the hopeful observation that the titratable groups fall into a small number of sets, and that in general the dissociation constants of the various members of a set are fairly close to one another, and surely closer than they are to those of any members of other sets. Sometimes a chemical set must be split into two sets because some members have an atypical environment; this is true of hemoglobin, where some histidines are associated with the iron-porphyrin prosthetic group.

I shall therefore assume that the r_e dissociable groups of free enzyme fall into a small number (say w) of **dissociation sets**, such that the ith set has r_{ei} members, all with dissociation constant K_{ei}, and that $\sum_i^w r_{ei} = r_e$. Let the enzyme-substrate complex have w' sets with similar properties (symbols: r_c, r_{ci}, K_{ci}). (Substrate can be treated the same way when it is macromolecular, so that such treatment is both necessary and admissible.) (We call numbers like r_{ei} the **dissociation group indices** of the molecule.)

Expressions like $P_{i\alpha}/h^i$ (10.5, 10.7, 10.8) are products of contributions from the various sets. A typical product might be $\prod_j (K_{ej}/h)^{m_{ej}}$, where the groups are represented by m_{ej} members of group j, and of course $\sum_j m_{ej} = i$.

The summation over α (10.7, 10.8) must contain more terms with the same set of m_{ej} values. The number of ways to pick m_{ej} objects out of a set of r_{ej} is in fact the familiar combinatorial coefficient $\binom{r_{ej}}{m_{ej}}$, and so (writing the combinatorial temporarily as j_e for typographical convenience) there must be altogether $\prod_j j_e$ such terms in the summation; and they can all be grouped together. There must also be other sets of values (say m'_{ej}) which are distinct from the set $\{m_{ej}\}$, but which also add up to the correct number i ($\sum_j m'_{ej} = i$). The entire summation, then, can be described as a multiple summation of products:

$$\sum{}' \prod_j j_e (K_{ej}/h)^{m_{ej}},$$

where $\sum'$ stands for the sum over all sets of m_{ej} values such that $\sum_j m_{ej} = i$.

Now according to 10.7 and 10.8 we must now further sum over i, whose possible values range from 0 to r_e. But this simply amounts to removing the restriction $\sum_j m_{ej} = i$ on the numbers m_{ej}. If we let m_{ej} range from 0 to r_{ej}, the multiple summation will now give us all the desired cases. At one extreme, when every $m_{ej} = 0$, their sum is 0, which corresponds to the value $i = 0$; at the other extreme, when every m_{ej} is equal to its maximal value r_{ej} (the dissociation index of set j), their sum will equal the sum of the r_{ej}, which corresponds to the case $i = r_e$. For every selection of m_{ej} values between these two extremes, they will add up to some number i lying between 0 and r_e; and if we were to group together all terms with the same sum i, we would get precisely the restricted summation ($\sum'$) shown in the preceding paragraph.

Now the **unrestricted** summation of products:

$$\sum_{m_{e1}}^{r_{e1}} \cdots \sum_{m_{ew}}^{r_{ew}} \prod_{j}^{w} je(K_{ej}/h)^{m_{ej}}$$

can now be regrouped as a product of sums (from the laws of multiple summation, Chapter 1):

$$\prod_{j}^{w} \sum_{m_{ej}}^{r_{ej}} je(K_{ej}/h)^{m_{ej}}.$$

But each of these sums is familiar: It is the binomial expansion of $(1 + K_{ej}/h)^{r_{ej}}$. We have therefore achieved the desired simplification:

$$\text{(10.19)} \qquad g_e = \prod_{j}^{w} (1 + K_{ej}/h)^{r_{ej}}$$

$$= (1 + K_{e1}/h)^{r_{e1}}(1 + K_{e2}/h)^{r_{e2}} \ldots (1 + K_{ew}/h)^{r_{ew}},$$

with a similar expression for g_c (and for g_s when needed).

Finally, we must ask how well we can specify, in terms of our dissociation sets, the nature of the active complex $C_{q\gamma}$. Ideally, we would like to assign a number (and a set of coordinates) to every dissociable group in the complex, and to say that activity will be found if carboxyl group #11 is dissociated, carboxyl groups #4 and #20 are not dissociated, imidazole #3 is dissociated, and so forth.

This is beyond our powers at the moment. Our knowledge about the active groups of enzymes continues to increase; but it is still far too fragmentary. We may, however, try to approximate by specifying **how many** members of each dissociation set must be dissociated and how many must be undissociated for activity to be found. Let us suppose that τ_j members of set j must be dissociated and that σ_j members must not be dissociated; we will refer to these numbers as **activity indices.**

In these terms, we may write, for the factor $P_{q\gamma}/h^q$ which appears in V' (10.18):

$$\prod_{j} \varphi_j (K_{cj}/h)^{\tau_j}$$

where the factor φ_j represents the contribution from those members of set j, out of the total of $r_{cj} - \tau_j - \sigma_j$ unspecified (free) ones, which may in fact be dissociated.

Putting the matter in this way suggests that we should assume the occurrence of more than one active complex—that all complexes which have the right set of activity indices are active, no matter how the remaining free groups are or are not dissociated. If this is true, and if we make the reasonable assumption that the turnover constant k_2 will be the same for all the forms of active complex (i.e., that all free groups are by definition irrelevant to the behavior of the complex), we may sum the product of the φ_j over all states of the free groups. Since each φ_j is of the form

$$j_c(K_{cj}/h)^{m_{cj}},$$

where

$$j_c = \binom{r_{cj} - \tau_j - \sigma_j}{m_{cj}}$$

where m_{cj}, the number of free groups dissociated in a given form of complex, may range from 0 to $r_{cj} - \tau_j - \sigma_j$, the total number of free groups, it is evident that the summation is exactly like the one we carried out previously, except for the upper limit of the m_{cj}. We have simply hung a "reserved" sign on a certain number of groups, and the others may range through all the possibilities. It follows that:

$$\sum_{free} \prod_j \varphi_j = \prod_j (1 + K_{cj}/h)^{r_{cj} - \tau_j - \sigma_j}$$

Hence, taking into account the definition of g_c (the C analogue of 10.19), we may write for V' (10.18):

$$V' = k_2 E_t \prod_j [(K_{cj}/h)^{\tau_j}/(1 + K_{cj}/h)^{\tau_j + \sigma_j}] \tag{10.20}$$

In carrying through the foregoing derivation, we have obviously dropped the previous limitation of the active complex to a total dissociation of q. This seems reasonable enough: If we have been able to select the activation indices properly, so that free groups are really irrelevant to the activity of the complexes, then the limitation is meaningless.

We are now in a position to attempt an analysis of K' and V'. Writing $m_0 = e_0 + s_0 - c_0$ and $Q = A/h^m$, taking the negative logarithm of K', we get:

$$\text{pK}' = \text{pK} + \log A + \log g_c - \log g_e - \log g_s + m_0 \,\text{pH} \tag{10.21}$$

We also note for convenience that:

$$\log g_e = \sum_j r_{ej} \log (1 + K_{ej}/h) \tag{10.22}$$

with similar expressions for $\log g_c$ and $\log g_s$.

The graph of pK′ against pH has a linear term with slope m_0 (in general a positive integer). The other terms are of the form $\log (1 + K_i/h)$, where K_i

is one of the dissociation constants. This may be written as $\log(1 + 10^{\text{pH}-\text{pK}_i})$. When $\text{pH} < \text{pK}_i$, the exponent of 10 is negative and the power is a small fraction. Hence, expanding the logarithm as usual (Chapter 1), $\log(1 + 10^{\text{pH}-\text{pK}_i}) \sim 10^{\text{pH}-\text{pK}_i}$, which is itself a small number. When $\text{pH} > \text{pK}_i$, the logarithm may be written as

$$\log[10^{\text{pH}-\text{pK}_i}(1 + 10^{-\text{pH}+\text{pK}_i})] = \log 10^{\text{pH}-\text{pK}_i} + \log(1 + 10^{-\text{pH}+\text{pK}_i}).$$

The first term is exactly $\text{pH} - \text{pK}_i$, while the second is approximately $10^{-\text{pH}+\text{pK}_i}$, a small fraction. The expression $\log(1 + K_i/h)$, considered as a function of pH, thus begins an exponential rise like $10^{\text{pH}-\text{pK}_i}$ for pH values below pK_i, and approaches the linear asymptote $\text{pH} - \text{pK}_i$, as illustrated in Fig. 10.2. The pH value of maximum curvature (sharpest bend) is not easy to locate visually from such a graph; but it can be proved to lie below the intercept pK_i [3] by about half a pH unit. Applying the same analysis to one of the terms of 10.22, the foregoing remarks still hold, except that the point of maximum curvature lies even further below pK_i ($\log r_{ej}$ units below, in fact).

There is no loss of generality if we suppose that the dissociation sets are so numbered that $\text{pK}_{e1} < \text{pK}_{e2} < \ldots$, and similarly for the pK_{cj}, etc. Then, if the pK values are fairly well separated, the first term of 10.22 will have emerged from the region of maximum curvature and be approaching the asymptote while the second term is still small and the remaining terms are negligible. The graph of 10.22 will therefore—roughly stated—consist, after the first low-level exponential phase, of linear regions of ever-increasing slope, the gain in slope from one such phase to the next being equal to the corresponding r_{ej} value.

If we now return to 10.21, and for the moment neglect the role of g_s, we see that $\log g_c$ makes a positive contribution to the value of pK′ while $\log g_e$ makes a negative contribution. Once we emerge from the first low-level exponential phase, indeed, we may expect the following: Assuming that K_{cj} and K_{ej} are not too different (i.e., assuming that combination with substrate has not radically altered the dissociation behavior of the various sets), the asymptotically linear phases of the g_c and g_e contributions will combine into asymptotically linear phases with slopes $r_{cj} - r_{ej}$, and intercepts on the

[3] The curvature of $y = f(x)$ is defined by $K = (d^2y/dx^2)/[1 + (dy/dx)^2]^{3/2}$. If in the present case we set up and maximize the curvature, we obtain a cubic equation in the variable K_i/h. This can be solved approximately for K_i/h small (pK_i large compared to pH) by neglecting the cubic term, and for K_i/h large (pK_i small compared to pH) by neglecting the constant term. In both cases, the approximate solution is a pH less than pK_i, so that only the first approximation is self-consistent. Using the corresponding result, the conclusion in the text is readily obtained.

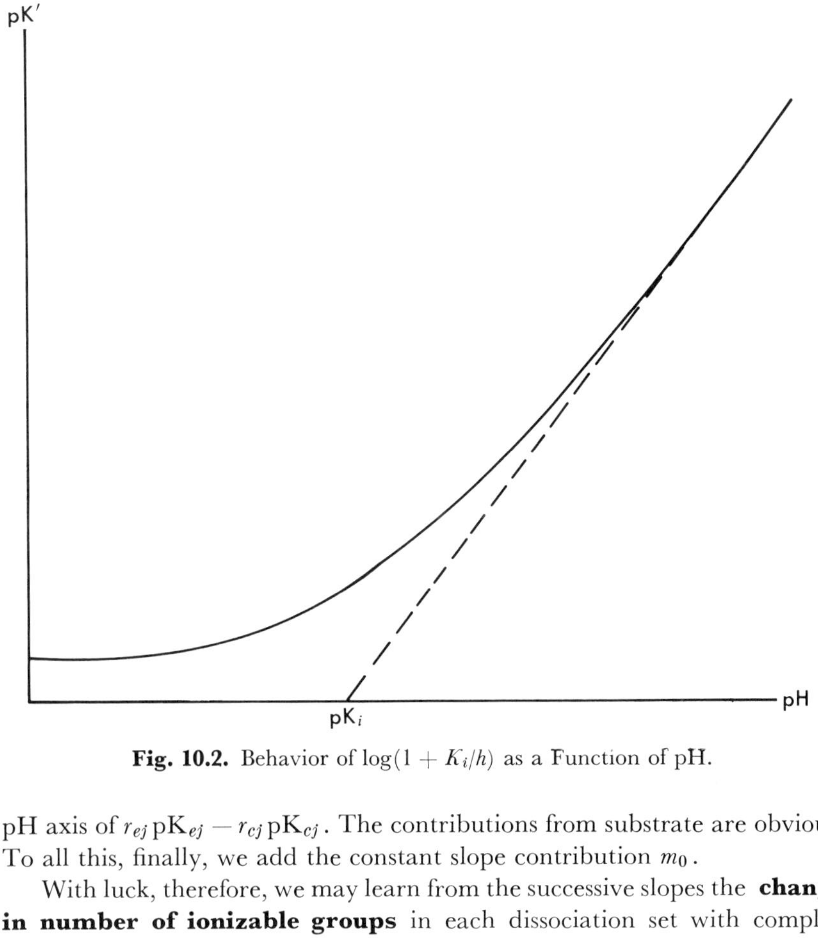

Fig. 10.2. Behavior of $\log(1 + K_i/h)$ as a Function of pH.

pH axis of $r_{ej}\,\mathrm{pK}_{ej} - r_{cj}\,\mathrm{pK}_{cj}$. The contributions from substrate are obvious. To all this, finally, we add the constant slope contribution m_0.

With luck, therefore, we may learn from the successive slopes the **change in number of ionizable groups** in each dissociation set with complex formation. The constant slope contribution m_0 may hopefully be obtained by extrapolation at the low pH end of the curve; we shall consider later another method of measuring it.

Both the luck already mentioned and a good deal of care are demanded by this sort of analysis. If happily the pK value of a group has changed very little on complex formation, so that pK_{ej} and pK_{cj} have, approximately, a common value pK_j, then the intercept of the asymptotically linear portion of a dissociation wave on the pH axis is approximately $(r_{ej} - r_{cj})\mathrm{pK}_j$. Hence, measuring the slope contribution of the dissociation wave (slope of linear portion less slope of previous linear portion), we may calculate pK_j from the intercept, and so hopefully identify the dissociating groups. The contribution, if any, of substrates of low molecular weight to the expressions can easily be anticipated and deducted: We will know the r_{sj} and pK_{sj} values in advance, or be able to get them with minimal difficulty.

The fact that the slope method gives only differences in the census of dissociation sets is not a serious obstacle. For, assuming that the nature of the dissociating groups can be discerned, the amino acid analysis of the free enzyme will give r_{ej}, so that the value of r_{cj} may now be found (as r_{ej} plus the slope contribution). By the same token, if the estimate of pK_j as a common value for enzyme and substrate is not very good, but close enough to permit guessing the identity of the dissociation set, the knowledge of r_{ej}, r_{cj}, and (known or measured) pK_{ej} will permit us to use the more exact expression for the intercept ($r_{ej}\,\mathrm{pK}_{ej} - r_{cj}\,\mathrm{pK}_{cj}$) to obtain pK_{cj}.[4]

Up to this point we have been considering a titration curve of activity against pH for the analysis of K'. Further information and a check on information already obtained come from a conventional titration. The basis of this method is the observation that the base bound per mole at any pH by any of the components of the system is equal to the slope of the corresponding graph of log g against pH. For it follows from 10.7 and 10.8: (a) $g = \sum_i A_i/h^i$, where the A_i are constants, so that $d(\log g)/dpH = (\sum_i iA_i/h^i)/(\sum_i A_i/h^i)$; and (b) the numerator of $d(\log g)/dpH$ is proportional to the base bound, i being the number of protons removed in titrating to get the i-fold dissociated form, while the denominator represents the total moles of component, with the same constant of proportionality. Thus, denoting the base bound per mole of each component by B_e, B_s, and B_c respectively:

$$\text{slope of pK}' = B_c - B_e - B_s + m_0 \tag{10.23}$$

Obviously, we cannot get B_c alone except at saturation of the enzyme with substrate; and then $B_c - B_e - B_s$ may be a small difference of large numbers, hence not a very accurate estimate. But we may perform what I have called (by analogy with the well-known method of differential spectrophotometry (13)) a **differential titration.** We may titrate a solution of substrate and one of enzyme separately under suitable conditions, and then titrate a mixture in which the same amounts of E and S are present. Since the total base bound by, say, E is EB_e, etc., we can use 10.9 and 10.15 to show that the difference between the titre of the mixture and that of the separate components is:

$$\Delta B = SE_t(B_c - B_e)/(K' + S), \tag{10.24}$$

[4] Dixon (12) has proposed a theory of pH curves which has some superficial resemblance to the present one. The theory has three patent defects: (a) It concentrates on what I have called K' and ignores V'; (b) it therefore misses the constant-slope pH term $m_0\,\mathrm{pH}$; (c) it falsely and inconsistently assumes that the Haldane constant for the reaction (misnamed "Michaelis constant") is given by $K = ES/C$, and that there is a similar constant for the reference forms, $K_0 = E_0S_0/C_0$, and that both constants are independent of h.

whence we may get $B_c - B_e$ at any convenient S value. Introducing the results into 10.23, we can compare the slope at various pH values with the slope obtained from the activity curve, or we can use the results to get an independent estimate of m_0, since the extrapolation to low pH is not always accurate. (Note that the independently obtainable values of e_0 and s_0, given m_0, can be used to get a value for c_0.)

Finally, let us consider the properties of V' and the information obtainable from the maxima of pH curves. To begin with, we may show at once that the correlation of a maximum rate with the essential role of both acidic and basic groups is sound. We can rewrite 10.20 as follows:

$$V' = k_2 E_t / \prod_j (1 + h/K_{cj})^{\tau_j} (1 + K_{cj}/h)^{\sigma_j}] \tag{10.20'}$$

Only the denominator of 10.20′ involves h. The denominator is a product of binomials in h by binomials in $1/h$; hence if expanded it would take the form $\sum_n b_n h^n + \sum_n B_n/h^n$. Its derivative will therefore be of the form $\sum_n nb_n h^{n-1} + \sum_n (-nB_n/h^{n+1})$; dividing through by the highest power of $1/h$ will give a polynomial, some of whose roots may represent real and positive (hence physically meaningful) minima of the denominator and maxima of V'.

Two points are immediately obvious: (a) If either all the τ_j or all the σ_j vanish, all terms of the polynomial are of the same sign, and there exists no real and positive root; (b) if such is not the case, there is exactly one maximum of V', since the polynomial consists of a series of terms of one sign followed by a series of terms of the opposite sign, and such a polynomial is known to possess exactly one real positive root.

The detailed analysis of V' follows the same lines as that of K', using the expression:

$$\log(1/V') = \log(1/k_2 E_t) + \sum_j \tau_j \log(1 + h/K_{cj}) + \sum_j \sigma_j \log(1 + K_{cj}/h) \tag{10.25}$$

For pH $<$ pK_{cj}, the expansion of the logarithms in the first summation gives τ_j(pK_{cj} − pH) for each of them, so that the total contribution to the slope is $-\tau = -\sum_j \tau_j$. The second summation gives small terms, which we may neglect by comparison.

When pK_{c1} $<$ pH $<$ pK_{c2}, the contribution of the first term of the first summation becomes small, while there is a significant (asymptotically linear) contribution from the first term of the second summation. The pH-dependent part of the expression now becomes, to the usual approximation, $-(\tau - \tau_1 - \sigma_1)$pH. As the pH increases and passes other pK_{cj} values, similar changes take place. Thus the initial slope of the pH curve gives τ, the next phase gives (by difference) $\tau_1 + \sigma_1$, and so on. The intercepts of the linear phases on the pH axis give first $\sum_j \tau_j \mathrm{pK}_{cj}$, then (by difference) $(\tau_1 + \sigma_1)\mathrm{pK}_{c1}$, and so forth.

The slope of the curve is initially negative. It will change sign at a pH value corresponding to the maximum of V'. This value will be attained when $\mathrm{pK}_{cm} < \mathrm{pH} < \mathrm{pK}_{c,m+1}$, where m is such that $\sum_{j=m+1} \tau_j = \sum_{j=1}^{m} \sigma_j$. An approximation to the maximal pH will therefore be given by

$$\tfrac{1}{2}(\mathrm{pK}_{cm} + \mathrm{pK}_{c,m+1}), \text{ or } h_m = (K_{cm} K_{c,m+1})^{\frac{1}{2}}.$$[5]

We have demonstrated that the maximum of V' is unique if it exists. This does not exclude the possibility of multiple pH optima, however. If there should be more than one set of activity specifications (e.g., "t_1, u_2, **or** t_3, u_4"), we will obtain a new expression for V', a sum of expressions like 10.20. If the regions of optimal pH for the several terms do not overlap too strongly, each term will be at its maximum when the other terms are far away from theirs, and the sum will have several maxima.[6]

PROBLEMS: 1. Prove that the system with activity specification $\tau_1 = \sigma_2$ has the same maximum pH as the system $\tau_1 = 1$, $\sigma_2 = 1$.

2. Prove that V' is independent of pH if only active complexes are formed.

3. If a non-ionizable substrate does not affect the ionizable groups of an enzyme on combining with it, prove that K' is independent of pH, but not V'.

4. If the activity specification is: $\tau_1 = 2$, $\sigma_2 = 1$, find two approximate solutions to the cubic equation for h_m, and show that only one of them is self-consistent.

5. Analyze the pH dependence of an enzyme system containing an effector (e.g., a "competitive" inhibitor). May such an analysis afford information over and above the inhibitor-free study?

6. Discuss the plausibility of such activity specifications as "τ_1, σ_1" or "τ_2, σ_1" or more generally "τ_i, σ_j, $i \geqslant j$".

[5] Insofar as our activity specifications are valid at all, the situation is not as complex as the general equations might lead one to think. In a typical protein of molecular weight about 60,000, containing more than 100 dissociable groups, about 50 of these are the free carboxyls of aspartic and glutamic acids (estimated after correcting for amide content), and another 50 come about equally from the ε amino group of lysine and the guanidinium group of arginine (9). Histidine contributes only about 8, and in the case of a single polypeptide chain there would be at most a single terminal amino group and a single terminal carboxyl. The number of dissociation sets that must be considered (in the absence of additional special information) is really quite small; and there is essentially just one rather large set with pK in the acid range.

[6] Schwimmer (14) has also proposed certain inhibited systems as models for multiple optima. If this is to be treated seriously as a model for multiple optima occurring in systems that have not been deliberately inhibited, it must be argued that the system usually contains effectors (for example, specific buffer ions) which act as inhibitors. This is by no means implausible, and should also be quite readily tested by switching buffers.

THE EFFECT OF TEMPERATURE VARIATION

The analysis of thermal effects in chemical systems rests upon two familiar relations. One of these expresses the equilibrium constant of a reaction in terms of the standard free energy change:

$$K = e^{-\Delta F/RT} \tag{10.26}$$

where T is the absolute temperature and ΔF is the standard free energy change per mole for the reaction. (R is the usual gas law constant, equal to about 2 calories per mole deg.)

The other relation expresses the velocity constant of a reaction in terms of the free energy of activation for that reaction (15)

$$k_r = (kT/h)e^{-\Delta F^*/RT} \tag{10.27}$$

where k is the gas law constant per molecule (1.37×10^{-16} erg/deg), h is Planck's constant (6.55×10^{-27} erg sec), and ΔF^* is the activation energy of the reaction in the specified direction—that is, the difference in free energy between the reactants and the activated complex.

The distinction between ΔF and ΔF^* is evident in Fig. 10.3, where the energy of a reacting system is plotted against a suitable reaction coordinate (e.g., the length of a bond which is broken in the reaction). As the system

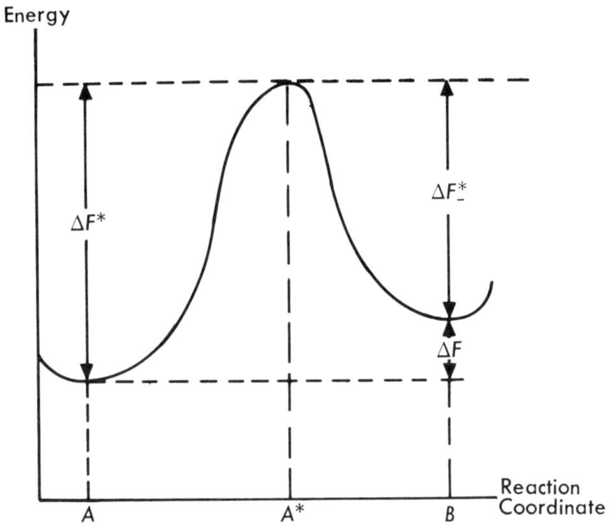

Fig. 10.3. Energy of a Reaction System as a Function of Reaction Coordinate.

goes from state A to state B, its free energy increases—in other words, energy is required to deform the bond which is to be broken. At the relative maximum of free energy, state A^*, the system has gained ΔF^* in free energy. As it now proceeds to the state B (reaction products), some free energy is regained, inasmuch as the reaction products are not under stress. The free energy regained, ΔF^*_-, is also the energy of activation for the reverse reaction (from B to A). The difference in free energy between state B and state A is ΔF.

Assuming the Law of Mass Action, the rate of the forward reaction is $k_r f_A$, where f_A is a suitable function of the reactant concentrations (see Chapter 2 and Chapter 9), and the rate of backward reaction is $k_{-r} f_B$. At equilibrium, when these are equal, the equilibrium constant $K = f_B/f_A = k_r/k_{-r}$, from which by 10.26 and 10.27:

$$\Delta F = \Delta F^* - \Delta F^*_- \tag{10.28}$$

It follows, as is well known, that the equilibrium of a reaction is predictable from the rates, but not conversely. In the figure, indeed, the free energy of the reaction is quite small compared with either of the activation energies.

Rate and Temperature

If we consider the primitive model of an enzymatic reaction (Chapter 2), there are two part reactions distinguishable: the formation of the complex C, and its conversion to E and P. The situation is illustrated in Fig. 10.4. Since

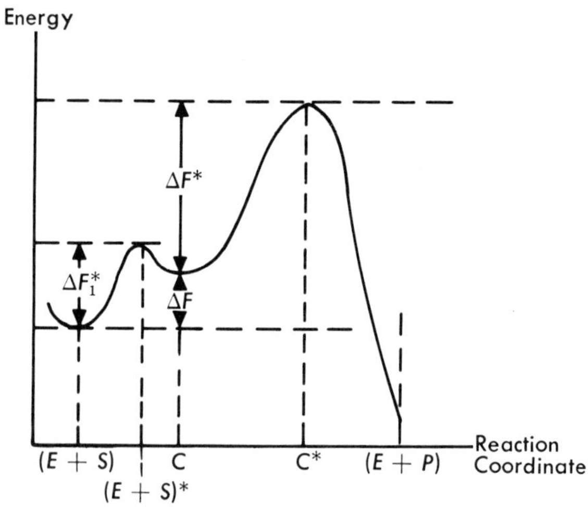

Fig. 10.4. Energy of a Simple Model of an Enzyme-catalyzed Reaction as a Function of a Reaction Coordinate.

only the first reaction is treated as reversible in this model, there is only one ΔF—the standard free energy of formation of C from E and S. Since $K = (k_{-1} + k_2)/k_1 = k_{-1}/k_1 + k_2/k_1$, and bearing in mind that the equilibrium constant for the formation of C from E and S is k_1/k_{-1}, we may introduce the correct forms from 10.26 and 10.27 into the usual $v = k_2 SE_t/(K + S)$, obtaining:

$$v = SE_t(kT/h)e^{-\Delta F^*/RT}/[S + e^{\Delta F/RT} + e^{-(\Delta F^* - \Delta F_1^*)/RT}] \quad (10.29)$$

Since the thermal variation due to the exponential terms is much stronger than any possible contribution from kT/h (which may be treated as practically constant over the usual temperature range for enzyme studies), it is convenient to think of v as plotted against $1/T$ rather than T. (This amounts to plotting against T from right to left instead of left to right, on a somewhat distorted scale.) With the relative values and signs of the energy parameters as in Fig. 10.4, it is easy to see that v decreases steadily as $1/T$ increases—that is, v increases with T.

In view of the fact that enzymatic reactions invariably show temperature optima, it is reasonable to ask whether there are any possible arrangements of the energy parameters that would admit of such optima. Setting the derivative of 10.29 equal to zero gives the condition:

$$(\Delta F + \Delta F^*)e^{\Delta F/RT} + S\Delta F^* + \Delta F_1^* e^{-(\Delta F^* - \Delta F_1^*)/RT} = 0 \quad (10.30)$$

This expression can evidently vanish only if one of the terms can become negative. Since the activation energies are intrinsically positive, this is possible only if $\Delta F < 0$.

We are not in a position to test 10.30 directly. However, a slightly different formulation permits us to make use of available data. First of all, we note a thermodynamic identity:

$$\Delta F^* = \Delta H^* - T\,\Delta S^* \quad (10.31)$$

where ΔH^* and ΔS^* have the usual significance of enthalpy and entropy differences. We also observe that 10.27 may be rewritten:

$$\ln k_r = \ln(kT/h) - \Delta H^*/RT + \Delta S^*/R \quad (10.27')$$

Over a moderate range of temperatures, it follows from this that the major temperature-dependent term is $\Delta H^*/RT$.

If we now differentiate $v = k_2 SE_t/(K + S)$, and set the derivative equal to zero, we may write the condition approximately (neglecting variation of kT/h) as:

$$(1 + S/K)d(\ln k_2)/dT - d(\ln K)dT =$$
$$(1 + S/K)\Delta H^*/RT^2 - d(\ln K)/dT = 0 \quad (10.32)$$

Gibson (16) has critically surveyed much of the available data; his equations are expressed in terms of derivatives with respect to $1/T$, so that all the terms are of opposite signs to those in 10.32. From Gibson's tabulated values, it appears that the terms of 10.32 are almost always of the same sign; the sole exception is fumarase acting on malate, from Massey's data (17). Thus it seems unlikely, as a general rule, that the temperature variation of the usual rate of an enzyme-catalyzed reaction will explain the occurrence of temperature optima.

A classical qualitative hypothesis about temperature optima in enzyme systems is that it is due to the occurrence of some denaturation; crudely put, rise in temperature increases the rate but increasingly puts some enzyme out of commission, and a balance between the two processes results in a temperature optimum (18).[7] If we add to the simple enzyme model just discussed the assumption that there is an equilibrium $E \rightleftharpoons E_d$, with equilibrium constant K_d, such that E_d does not combine with S, the rate equation becomes:

$$v = k_2 S E_t / [K(1 + K_d) + S] \tag{10.33}$$

Differentiating with respect to T and setting the derivative equal to zero:

$$(1 + S/K)\Delta H^*/RT^2 - d(\ln K)/dT - d[\ln(1 + K_d)]/dT = 0 \tag{10.34}$$

which differs from 10.32 by the extra (denaturation) term.

This, however, changes the situation. For K_d must vary with T according to the analogues of 10.26 and 10.31:

$$K_d = e^{-\Delta F_d/RT} = e^{-(\Delta H_d - T\Delta S_d)/RT} \tag{10.35}$$

The last term of 10.34 must therefore be $-(\Delta H_d/RT^2)K_d/(1 + K_d) < 0$.

The second term of 10.34, according to Gibson's tabulation (our ΔH^* is E in Gibson's notation, and our second term times RT^2 is Gibson's ΔH_M), is of the same order as or less than the first term for moderate or small S values. If we neglect S/K compared with 1 in 10.34, and write $y = \Delta H^* + \Delta H_M$, we can solve for $1/T_m$ with the help of 10.35:

$$1/T_m = [R \ln(\Delta H_d/y - 1) + \Delta S_d]/\Delta H_d \tag{10.36}$$

[7] The problem of denaturation will be considered in greater detail in Chapter 11. For the moment, it will suffice to consider the additional reaction as any reaction (not involving substrate) which immobilizes some fraction of the free enzyme. **Any competitive inhibitor should have precisely the same effect**; so should compounds that affect protein structure nonspecifically, such as guanidine, urea, and organic solvents. In principle, according to the reasoning which follows, in the text, it should be possible for an inhibitor to generate a temperature optimum at a T value below the usual optimum due to thermal variation alone, given suitable ΔH and ΔS values for the inhibitor binding equilibrium. Unfortunately—for the idea is a neat and appealing one—very few if any such equilibria are likely to have sufficiently large values of ΔH and ΔS (or even values of the right sign).

Substituting reasonable values into this equation (for example, for the denaturation of trypsin and the activation of trypsin with benzoyl-L-arginine amide as substrate (19)), $\Delta H_d = 67{,}600$ cal/mole, $\Delta H^* = 14{,}900$ cal/mole, $\Delta S_d = 213$ cal/mole/deg, and $\Delta H^* \leqslant y \leqslant 2\,\Delta H^*$, we find $T_m \sim 316°$ K or 43° C, which is a fairly reasonable value. It is important to notice that the range of variation allowed for y has little effect on the result, inasmuch as ΔS_d dominates the expression in brackets. Indeed, the solution will be real and positive as long as ΔH_d is just a little larger than y (e.g., $\ln(10^{-10}) \simeq -46$). But the numerical values used are fairly typical. However, I would warn the reader against laying too much stress on the precise value of T_m, for, as Bayliss already pointed out in 1914 (18), this will depend on the time of incubation, since the denaturation process eventually becomes irreversible.

The role of S raises an immediate problem. If we permit S to approach infinity, the effect of the denaturation equilibrium vanishes. It is difficult to find out whether this is really a valid prediction or an indication of a deficiency in the model. Studies on temperature **optima** (which constitute a small part of the material on thermal effects) have for the most part treated the question of substrate level with a casual vagueness and slovenliness that provoked explicit criticism from Gibson (16). It is "well known"[8] that substrate protects enzyme against various noxious influences, including denaturation; but I am unable to locate information sufficiently quantitative to settle the point at issue.

The problem, it would seem, may be sidestepped by assuming that substrate protection is not perfect, and that there is an equilibrium $C \rightleftharpoons C_d$ in addition to the previous $E \rightleftharpoons E_d$. It is then easy to show that the maximum rate of the enzyme is $k_2 E_t/(1 + K'_d)$, where K'_d is the equilibrium constant for the new denaturation equilibrium; and the possibility of a temperature optimum survives saturation with S. (The proof of this proposition, if it is not obvious, is left to the reader.) But it is embarrassing to note that the even more plausible equilibrium $C \rightleftharpoons E_d + S$ (denaturation renders the enzyme unable to hang on to previously bound substrate) does not work!

Let me observe, before leaving the topic of thermal optima, that the quite common assumption which treats activation energies, heats of reaction, and the like as constants is an approximation. If the study of the thermal behavior of enzyme systems is ever ready for extremely precise methods, it will be essential to test these points, and to correct them insofar as the approximation seems too rough. The approach to such corrections may be found in any good text of thermodynamics.

[8] "Well known that": standard phrase in the argot of biological science. Meaning: "I'm sure I've seen it somewhere, but for the life of me I can't find the reference."

Evaluation of Thermodynamic Parameters

With the reservations just mentioned, thermodynamic parameters of enzymatic reactions may in part be estimated from temperature curves (say $\ln v$ against $1/T$). The parameters are always **heats** of some kind, as is evident from the last subsection. An obvious extension would seem to give information about entropies. If we take the logarithmic form of 10.27 and use 10.31, the plot of $\ln k_r$ against $1/T$ will have an intercept (on the $\ln k_r$ axis) of $\Delta S^*/R$. Hence we should be able to learn something about entropies of reaction and activation.

There are problems, however. To make the issue clear, let us get rid of K in $v = k_2 SE_t/(K + S)$ by using a saturating level of S and studying the maximum rate $V = k_2 E_t$. We have:

$$\ln V = \ln k_2 + \ln E_t = -\Delta H^*/RT + \Delta S^*/R + \ln E_t \tag{10.37}$$

The intercept of the graph is contaminated by the term $\ln E_t$; and this term cannot in general be ignored. For example, if $\Delta S^* = 20$ cal/mole/deg, $\Delta S^*/R \cong 10$; and for $E_t = 10^{-6}M$, $\ln E_t \cong -14$. In order that temperature curves may be used to get entropies, then, E_t must be known in absolute units (and V must be expressed in the same units, of course). This difficulty was pointed out, in somewhat different language, by Stearn (19), who speaks of "an entropy of dilution or of concentration change from some unknown value to a rational standard state."

Thus most of the scanty data are not suitable for the desired calculation. The data of Butler (20) and of Laidler and Hoare (21) are exceptions; and they agree, for quite different enzyme systems, in finding ΔS^* values between -5 and -10 kcal/mole/deg. (See also refs. 22, 23, 24.)

The problem of E_t does not enter the study of K. We may write $\ln r = \ln(v/V) = \ln S - \ln(K + S)$, or even better, working at low S values, $\ln r = \ln S - \ln K$. But K is given (for the simplest enzyme model) by the last two terms of the denominator of 10.29. The analysis of K therefore demands a preliminary evaluation of ΔF^* and ΔF_1^* (or at least of k_2/k_1). If this is feasible (e.g., by using transient data as described in Chapter 2), one could plot $\ln(K - k_2/k_1)$ against $1/T$; the slope is $\Delta F/R$ and the intercept the corresponding $\Delta S/R$.

There is a large literature on the subject of the thermodynamics of enzymatic equilibria (e.g., refs. 25–31). With the proper technicalities (corrections for ionization and the like), the fact that $K_{ee} = K_e$ (Chapter 3) permits equilibrium data to be used to calculate the overall heat and entropy of various reactions, which must be the same as the heats and entropies that presumably would be found for the spontaneous reactions if they could occur fast enough for equilibrium to be attained in a reasonable time.

Although this kind of study, concerned chiefly with the properties of substrates rather than of enzyme systems, is somewhat aside from the main foci of interest in this book, it may be well to say just a word in passing, in view of the persistent disposition of many people to regard reaction free energies or enthalpies (especially those of the so-called "high-energy" compounds) as something like bond energies. What is usually the chief term (not necessarily the only term) in the ΔF or ΔH of reaction is the difference in energy between the ground states of products and of reactants. This, as consultation of Fig. 10.3 will remind you, tells nothing about the energy required to form or to break the relevant bonds, which quantities are more nearly measured by the activation parameters ΔF^* or ΔH^* of the back and forward reactions respectively. The reader who cares to play around with the relative magnitudes of the peak and the two valleys in the figure will immediately convince himself that large ΔF values are equally consistent with large and with small bond energies. If one is disposed to deem a figure like 15,000 cal/mole awe-inspiring, compare the ΔF for hydrolysis of the peptide bond (a moderate value lying between 1000 and 5000 cal/mole according to circumstances) with the energy of the C—N bond, 70,000 cal/mole!

If one wishes to interpret the ΔF of reaction as the energy released, and therefore the energy available for other processes, it is also essential to recall that the "standard" free energy change of a reaction is the actual free energy change only if the reactants and products are at unit concentrations, and that the actual free energy change involves concentration-dependent terms (the so-called free energy of dilution). It is for this reason that most textbooks write ΔF^0 for what we have been calling ΔF.

The reader who is interested in pursuing these points is advised to read suitable discussions, such as Chapters 11 and 12 of reference 32 and Chapters 2 and 3 of reference 33.

Finally, I will remark (even if it be superfluous) that the equilibrium properties of a reaction yield no information about the enzyme and its mode of activity. The constant K_{ee} is a product of three equilibrium constants, as already remarked in Chapter 3; and the thermodynamic parameters of the overall reaction are sums of the corresponding parameters for the three individual reactions. To disentangle these requires a procedure analogous to that pursued a couple of pages ago, when we approached the analysis of K for the simplest (and irreversible) enzyme model by getting k_1 and k_2 from transient data (if possible!), and then plotting the logarithm of $K - k_2/k_1 = k_{-1}/k_1$. In the case of the more nearly correct model of the latter half of Chapter 3, in which we take into account the reaction in which the *E-S* complex is converted to the *E-P* complex, the pursuit of the first part of this goal will be arduous indeed, as anyone may convince himself by inspecting equations 3.12 and 3.13.

What we would really like to know about this key reaction is its ΔF^* and ΔF^*_-, which means knowing the thermal variation of the velocity constants k_2 and k_{-2} of scheme 3.B. It is conventional to say that catalysis by enzymes is achieved by lowering the activation energy of a reaction; and indeed it is hard to see how this can fail to be true. But it is, deplorably, the case that we do not really have the necessary supporting evidence for this statement. To get such evidence requires finding systems which are unimolecular or quasi-unimolecular and reversible, and for which it is possible to distinguish and separately measure C_1 and C_2 in both the transient and the pseudostationary condition. The thermal variation of such systems would furnish the desired information.

REFERENCES

1. Haldane, J. B. S., "Enzymes", Longmans, Green, London, 1930.
2. Michaelis, L., and Davidsohn, H., Biochem. Z. *35*, 386 (1911).
3. Northrop, J. H., J. Gen. Physiol. *5*, 263 (1922).
4. Northrop, J. H., J. Gen. Physiol. 2, 113 (1919).
5. Waley, S. G., Biochem. Biophys. Acta *10*, 27 (1953).
6. Laidler, K. J., Trans. Faraday Soc. *51*, 528 (1955).
7. Neumann, H., Biochem. Biophys. Acta *3*, 117 (1949).
8. Hase, J. Biochem. (Japan) *39*, 259 (1952).
9. Tristram, G. R., and Smith, R. H., in "The Proteins" (Neurath, H. ed.), vol. 1, Academic Press, New York. 1963.
10. Cohn, E. J., and Edsall, J. T., "Proteins, Amino Acids and Peptides as Ions and Dipolar Ions", Reinhold, New York, 1943.
11. Tanford, C., Advances in Protein Chemistry, *17*, 69 (1962).
12. Dixon, M., Biochem. J. *55*, 161 (1953).
13. German, B., and Wyman, J., Jr., J. Biol. Chem. *117*, 533 (1937).
14. Schwimmer, S., J. Theoret, Biol. *3*, 102 (1962).
15. Glasstone, S., Laidler, K. J., and Eyring, H., "The Theory of Rate Processes", McGraw-Hill, New York, 1941.
16. Gibson, K. D., Biochem. Biophys. Acta *10*, 221 (1953).
17. Massey, V., Biochem. J. *55*, 172 (1953).
18. Bayliss, W. M., "The Nature of Enzyme Action", Longmans, Green, London, 1917.
19. Stearn, A. E., Advances in Enzymology *9* (1949).
20. Butler, J. A. V., J.A.C.S. *63*, 2971 (1941).
21. Laidler, K. J., and Hoare, J. P., J.A.C.S. *72*, 2487 (1950).
22. Ouellet, L., Laidler, K. J., and Morales, M., Arch. Biochem. Biophys. *39*, 37 (1952).
23. Snoke, J. E., and Neurath, H., J. Biol. Chem. *182*, 577 (1950).
24. Casey, E. J., and Laidler, K. J., J.A.C.S. *72*, 2159 (1950).
25. Borsook, H., and Huffman, H. M., in "Chemistry of the Amino Acids" (Schmidt, C.L.A., ed.), C. Thomas, Springfield, Ill., 1938.

26. Meyerhof, O., Ann. N. Y. Acad. Sci. *45*, 377 (1944).
27. Ogston, A. G., and Smithies, O., Physiol. Revs. *28*, 283 (1948).
28. Oesper, P., Arch. Biochem. *27*, 255 (1950).
29. Oesper, P., in "Phosphorus Metabolism" (McElroy, W. D., and Glass, B., eds.), vol. I., Johns Hopkins Press, Baltimore, 1951.
30. Hill, T. L., and Morales, M. F. J.A.C.S. *73*, 1656 (1951).
31a. Burton, K., and Wilson, T. H., Biochem. J. *54*, 86 (1953).
b. Burton, K., and Krebs, H. A., Biochem. J. *54*, 94 (1953).
32. Daudel, R., Lefebvre, R., and Moser, C., "Quantum Chemistry: Methods and Applications", Interscience, New York, 1959.
33. Ingold, C. K., "Structure and Mechanism in Organic Chemistry", Cornell, 1953.

II

Protein Structure and Stability and Mechanisms of Enzyme Action

The inclusion of a chapter on structure and mechanism in a book on kinetic theory is an unavoidable compromise. It is unavoidable because (as I have repeatedly stressed) the major function of theory is the disclosure of mechanism; and mechanism in the sense of stoichiometry is only half the story. It is a compromise because one chapter—even a long one—does not really suffice. The reader is warned, therefore, that this chapter does not purport to be exhaustive or authoritative. It is intended to be an introduction to a topic that really deserves its own book.

As I have just implied, most of kinetic theory is aimed at revealing mechanisms in a very specific and rather schematic sense: that of the stoichiometry of a single enzymatic reaction or a system of such reactions. The task has been accomplished when we can state what substrates, cosubstrates, and effectors participate in the activity of the system, and according to what stoichiometric schemes. This is the sense in which most papers entitled "The Mechanism of Action of ——ase" use the term.

In this chapter we are approaching very different questions, of which the following are perhaps typical:

1. What is the sequence of atomic and electronic events which enables an enzyme, as a catalyst, to accelerate chemical reaction?

2. Is it essential that an enzyme be a protein, or at least a macromolecule? If so, what does the macromolecular character contribute to the function?

3. Are there features of protein structure—for instance, certain sequences

of amino acids—common to all enzymatic activity? If not, are such sequences common to large classes of activity types, or to extremely narrow classes? Or may amino acid sequence be relevant to activity only indirectly, by way of its influence on conformation?

4. Can it be demonstrated that an enzyme is a perfect catalyst—i.e., that the molecule of enzyme released after the catalytic act is conformationally and energetically the same as it was when it entered the field?

5. Besides the stoichiometric generalization of Chapter 2—that all enzymatic catalysis occurs by way of compound or complex formation between the enzyme and the various other participants—is there any general law governing enzymatic catalysis, as distinguished from specific explanations of individual cases?

6. To what extent is the kinetic behavior of enzymes specifically related to the mechanism of catalysis? Put another way, is the currently accepted and experimentally supported theory a purely formal and abstract theory, compatible with many distinct catalytic mechanisms, or does it narrow down the possibilities, if not to one mechanism then at least to very few?

It seems clear that questions like these, though they are by no means logically independent, lead us in two different though eventually convergent directions. Question 3, for example, obviously invites us to do what in fact many have done—to investigate various specific cases in search of factual information. Some of the other questions, on the contrary, evidently demand the application of the usual deductive methods of theory, with the eventual development of testable and critical predictions.

What we shall try to do in the following sections is to sample what has been done, in the hope of giving the reader a picture of where we stand at the moment.

THE STRUCTURE OF THE BINDING SITE

It is obvious from the structure of the substrates, the nature of the reactions, and the specificity of the enzymatic functions that binding of substrate to enzyme almost certainly involves more than one group of each. Yet[1] most of our kinetic theory has treated the binding site as a featureless and indivisible entity. Let us see what can be learned by going further.

[1] For example, in studying the hydrolysis of a series of phosphate esters by a phosphatase, we find near identity of turnover numbers and wide variation in K values (1). This is inexplicable unless we assume that both phosphoryl group and alcohol residue participate in the binding. (To assume that the phosphoryl group is dangling out in space, of course, explains the binding specificity but makes the enzyme action inexplicable.)

Consider as the simplest example the situation illustrated in Fig. 11.1 and schematized thus:

(11.A) $$E + S \underset{-1}{\overset{1}{\rightleftharpoons}} C_1 \qquad C_1 \underset{-2}{\overset{2}{\rightleftharpoons}} C \qquad C \overset{3}{\rightarrow} E + P \qquad C_1 + S \underset{-4}{\overset{4}{\rightleftharpoons}} C_{12}$$

Here C_1 represents the **unipolar complex**, stage (b) of the figure, C the **bipolar complex**, stage (c), and C_{12} the enzymatically useless **double unipolar complex**, stage (d).

With the abbreviations $K_1 = (k_{-1} + k_3/K_2)/k_1$, $K_2 = (k_{-2} + k_3)/k_2$, $K_4 = k_{-4}/k_4$, we get:

(11.1) $$v = k_3 E_t S/[K_1 K_2 + S(1 + K_2 + K_2 S/K_4)]$$

The rate considered as a function of S has a maximum at $S_m = (K_1 K_4)^{1/2}$. It is distinguishable from the autoinhibition case of Chapter 5 (Fig. 5.1(c)) by the fact that the asymptotic value in the present case is zero while in the previous case it was not. Moreover, the occurrence of a maximum here is compulsory instead of contingent: It can fail only if K_4 approaches infinity.

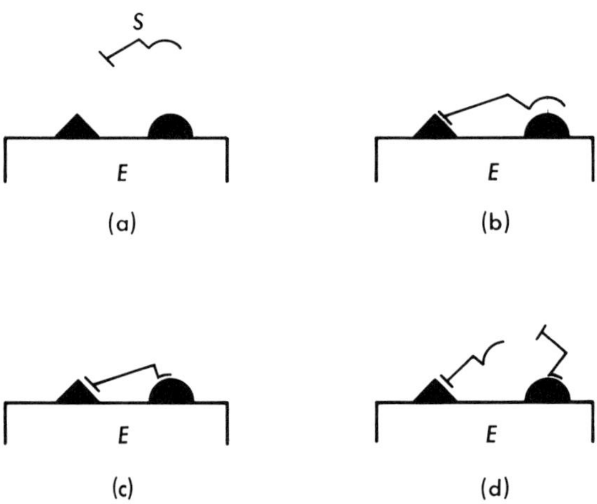

Fig. 11.1. Diagram Showing the Stages of Combination of Substrate with Enzyme When There Are Two Points of Contact. Part (d) shows formation of enzymatically inactive complex with two substrate molecules.

The maximal property is crucial in distinguishing the model contemplated in Fig. 11.1 from the case discussed earlier (especially for reversible reactions in Chapter 3) where the E-S complex is converted into an E-P complex. If we identified C_1 with the former and C with the latter, the first three reactions of 11.A would be the same, even though here we want to interpret the stages as stages of binding whereas in the other model we interpret them as binding followed by chemical conversion.)[2]

If this is how enzymes do in fact work, the immediate question is why we do not always observe the maximum. There are various possible answers. One is that reactions like 4 actually never occur, in which case it must be admitted that the kinetic model alone contributes nothing to the evidence for the structural properties of binding sites. Another is that reaction 4 sometimes fails due to steric hindrance or repulsion. Still another is that most kinetic studies of enzymes fail to push substrate concentrations high enough to observe the maxima. The last two hypotheses are surely plausible. There must be some substrates so formed that the first linkage hinders access to the second binding site. Moreover, everyone must have seen dozens of experimental reports in which the actual points recorded on graphs of v against S stop as soon as the rate shows signs of beginning to level off.

Now if the third hypothesis—that the failure to observe the maxima comes from the failure to look for them—is correct, the problem is simple: Some experiments must be repeated. However, if the second hypothesis is correct—that in the majority of cases the logically and chemically plausible binding of a second molecule to the second site does not occur because of steric hindrance or repulsion—the problem is more serious. Offhand I can see two possible approaches (other than to say that the model is untestable). One is to introduce synthetic substrate analogues—molecules in which the probable second group is smaller (or larger), or in which charged or polar elements which might be responsible for repulsion are replaced or covered. Such experiments should be performed not only in systems where no maxima are observed in the natural substrates, but also in systems where maxima are observed but where suitable structural alterations in the substrate might abolish them.

The alternative—the use of a variety of inhibitors rather than of substrates—has been tested. A particularly striking example is the work of London *et al.* on inhibitors of prostatic acid phosphatase (4, 5, 6). The papers should be read carefully; but the essence of the result, as far as our present discussion goes, can be stated simply: For a large variety of strong, moderate, and weak inhibitors, the strength of inhibition and the capacity to protect the enzyme against thermal denaturation correlate sharply with stereochemical features that would permit or forbid association with a set of binding sites. Moreover, the same stereochemical features (and measurements) are found in the substrates.

[2] As Hearon points out (2), both are unimolecular reactions of an enzyme-substrate complex, and they will affect the kinetic relations in exactly the same way. (See also (3).)

If a double unipolar complex can be formed, it is *a priori* possible that it can be formed randomly rather than according to a compulsory order, and that the bipolar complex C can also be formed randomly. The rate expression resulting from this generalization is a ratio of a quadratic in S to a cubic in S; and the location of the maximum is messy. However, the assumption that the two reactions leading to the double unipolar complex are separately in equilibrium (the weakest additional assumption we can introduce) leads to a rate expression formally similar to 11.1.

The theme is susceptible to further embroidery. For example, we may study the effects of **quasi-symmetry**. This is best illustrated by Fig. 11.2,

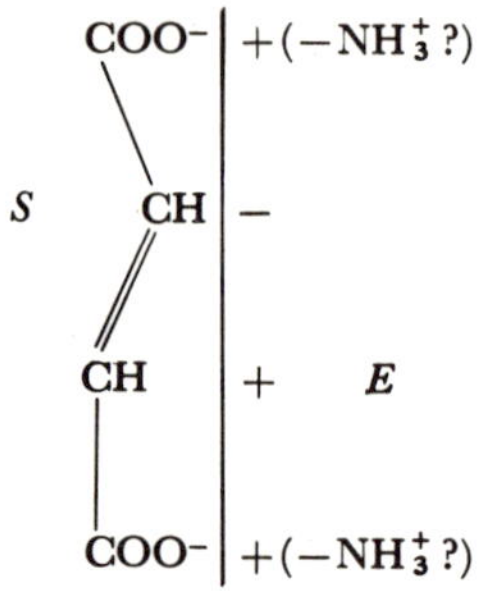

Fig. 11.2. Sketch of Structure of Active Center of Fumarase, as Suggested by Massey.

which illustrates Massey's idea of the structure of fumarase (7). This model admits of the possibility that fumarate or malate (binding through the carboxyl groups) may form "wrong" unipolar and bipolar as well as double unipolar complexes (see Fig. 11.3)—i.e., the carboxyls might be bound in such a way that, when the bipolar complex is completed, the essential catalytic groups (the two middle ones in the figure) will be juxtaposed to the wrong portions of the substrate. This is as true of fumarate, with its *trans* configuration, as it is of the more obviously asymmetrical malate; and Fisher *et al.* have furnished direct evidence for stereospecificity of the reaction (8).

If all the possible "right" and "wrong" steps are written out, and the quasisymmetry is introduced by identifying the constants of all homologous steps that involve only the binding groups of substrates (e.g., the carboxyls), the result is again a rate expression of the form of 11.1.

A further elaboration is the extension of the schemes to bimolecular reactions. Nothing very startling comes out of this, however; it is simply more complicated algebraically. There is a maximum with respect to each substrate; but in the ordered bimolecular case the maximum for the first substrate is a constant, while for the second substrate it is a function of the concentration of the first substrate. The reason for this is obvious: The order being compulsory, the second substrate can produce an effect only after the first phase (binding substrate 1) is completed, and so will function in a way that depends on the concentration of the first substrate.

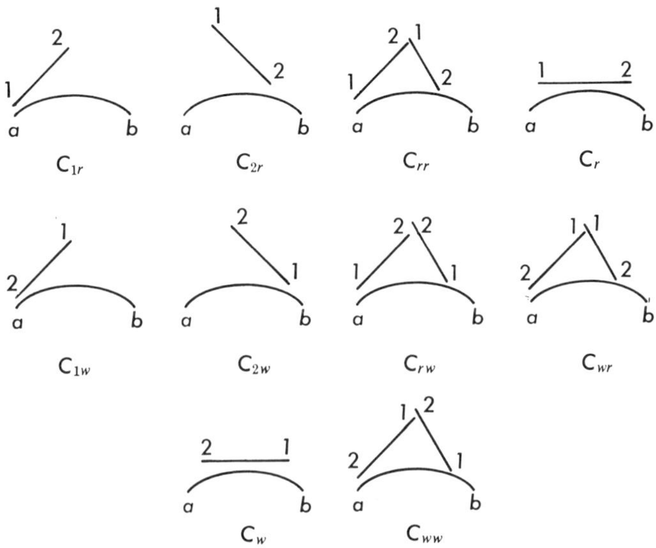

Fig. 11.3. Various Types of Complex Formation between Enzyme and Substrate in a Quasi-symmetrical System.

The generalization from binding sites of two groups to higher multiplicities is obvious. Ogston appealed to threefold binding sites (9) to account for certain apparent anomalies in the way acetate enters the tricarboxylic acid cycle, and in effect demonstrated that a symmetrical molecule, citric acid, can behave like an asymmetrical molecule when combined with an enzyme. Fig. 11.4 illustrates the model. In view of the tetrahedral structure of the C* carbon atom, a plane through the tertiary hydroxyl and carboxyl groups is a plane of symmetry for the molecule. But suppose that carboxyls 1 and 2 bind to groups *a* and *b*. If group *c* is specific for the hydroxyl 3, and if the third bond is required for enzyme activity, the molecule is effectively asymmetrical, since there is only one right-left orientation that can lead to enzyme activity.

This **three-point landing** model, like the two-point model, admits at least theoretically of ternary complexes and incorrect as well as correct binding orientations. (Quaternary complexes seem rather unlikely on steric

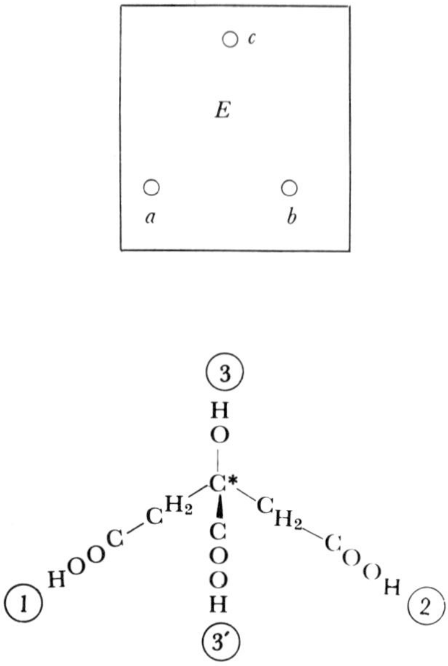

Fig. 11.4. Diagram Illustrating Principles of Three-point Contact between Enzyme and Substrate. The substrate shown is citric acid.

grounds.) The formal kinetic properties of the model should therefore resemble those of 11.1.

We initiated this section by seeking kinetic methods for confirming—and perhaps carrying farther—an already highly plausible case for multiplicity of binding groups on enzymes. The results can only be called non-contradictory but inconclusive. When the occurrence of a maximum in the v-S curve, of the kind predicted by the theory, is observable, well and good; but it is clear that the failure of such observations is so easily explained that no one is likely to abandon the model on account of such a failure.

Even granted all the positive observations we would like, the possibility that purely kinetic analysis might teach us more than we knew at the beginning seems a forlorn hope. For example, the improbability of finding three substrate molecules bound to one enzyme site means that we may not be able to distinguish kinetically between the two-point and the three-point model. Moreover, we cannot really distinguish the two-point model as we visualized it in Fig. 11.1, where the second substrate molecule inhibits enzyme activity competitively (blocks a binding site needed by the first substrate molecule) from an "uncompetitive" model in which the second substrate molecule does not compete for the second binding group, but affects the enzyme conformation or charge distribution in such a way that the second binding step cannot be completed, or if completed does not lead to turnover.

What a survey of work in this and related areas does indicate, however, is that kinetic methods may be of the greatest value when combined with specific and quantitative structural information about a variety of possible substrates and effectors. In particular, such work as that of London *et al.* (4, 5, 6), already mentioned, suggests that an analysis (armed with such independent structural information) of enzyme **stability** may be profitable. We therefore turn to some brief considerations about protein structure and stability.

PROBLEMS: 1. (a) Develop the rate expression for the two-point random case.

(b) Demonstrate the validity of the statement in the text, that the equilibrium approximation applied to the formation of C_{12} leads to an equation of the form of 11.1.

(c) Is there any other not too unreasonable approximation that would result in a similar form?

2. Develop the rate expression for the quasi-symmetrical case.

3. Develop the rate expression for the ordered bimolecular case.

4. Demonstrate the validity of the statement in the text about the equivalence of the competitive two-point case and the "uncompetitive" autoinhibition case.

STRUCTURAL AND FUNCTIONAL STABILITY OF ENZYMES

Let us begin with a quick-and-dirty summary of the currently accepted picture of protein structure, restricted to so-called "globular" proteins. (For more detail, and especially for a quantitative treatment, see refs. 10 and 11.)

1. All proteins consist of one or more unbranched polypeptide chains, in which the amino acid residues are linked (except for proline) by peptide bonds involving the α amino groups (**primary** structure)

2. Portions (in principle, all) of a polypeptide chain have a **secondary** structure—specifically, the α helix proposed by Pauling and Corey (12). The nonhelical portions of a chain are usually randomly coiled, although special circumstances might cause them to be fully extended. The helical structure involves the formation of hydrogen bonds between the carbonyl and imino groups of non-adjacent peptide links in the primary structure.

3. If we visualize the secondary structure of a protein chain crudely as being like auto coil springs linked endwise by wires, we may imagine this structure further folded into a compact **tertiary** structure. This structure is

determined by the occurrence of covalent bonds and noncovalent bonds. The former are —SS— bonds between cysteine residues. The latter include hydrogen bonds and/or ionic bonds (in Scheraga's terminology, "interactions between polar groups" (11)) and hydrophobic bonds.

4. Two or more nonidentical peptide chains with their own secondary and tertiary structure may be linked by covalent and noncovalent bonds as in 3.

5. Tertiary structures (as described in 3 and 4) may be further polymerized by noncovalent bonds, forming a **quaternary** structure (e.g., haemoglobin). It is such polymers that are contemplated as the source of multivalent enzymes in the Monod effector theory described in Chapter 8.

6. Hydrophobic bonds may involve hydrocarbon portions of amino acid residues, such as the side chains of leucine, isoleucine, and valine, aromatic and heterocyclic groups such as those of tyrosine, phenylalanine, proline, and tryptophane, or the methylene carbons (α carbons) of the polypeptide backbone.

7. Hydrophobic groups tend to minimize contact with water while polar groups frequently tend to maximize such contact. The resulting tendency is towards the formation of compact structures with hydrophobic cores and polar surfaces in aqueous solution.

8. Conformation, with its secondary, tertiary, and even quaternary components, is substantially determined by the primary structure (13).

The disruption of normal conformation without the rupture of covalent main-chain bonds is **denaturation.** It follows from the description of conformational factors that denaturation will not be an all-or-none process, but may occur in stages. Evidence for the occurrence of more than one stage is available (14) (see also comment by Tanford (15)).

It also follows that denaturation may not only be induced by elevated temperature (increasing the energy of the thermal vibrations of participants in noncovalent bonds and therefore the probability of bond breakage), but also by organic solvents and by substances that affect the activity of water, such as urea and guanidine.

Consequences for the relation of structure to enzyme activity and to the stability of enzyme activity also are evident. For instance, it follows from paragraph 8 that conformation is probably unique. It is also clear that the groups involved in substrate binding and reaction catalysis need not be immediate neighbors in the primary structure. Thus two significant side chains might be adjacent on a helix, hence four or five units apart in the main chain; they may be neighbors on two helices that are joined by tertiary linkages; or they might even be on two monomers of a quaternary structure. Thus it is possible (though it does not seem to be the case) that the search for characteristic main-chain subsequences might be either futile or irrelevant. It is certainly possible that the relevance of such sequences is indirect—that perhaps just one residue participates in the function of the enzyme, while its neighbors are involved in fixing the conformation and hence in apposing that residue to another active residue that is quite remote in the main chain.

It is of major interest to ask how much denaturation is required to affect or to abolish the activity of an enzyme, and what the kinetic properties of such activity loss may be. In particular, it is of importance to know whether the activity of a specific enzyme molecule changes progressively during denaturation, so that the turnover number, for example, varies with time, or whether the activity is unchanged up to a point and then abruptly drops to zero.

Unfortunately, except for a plethora of data on the time curves of activity loss, experimental answers to these questions are not plentiful. It is known that an average of two or more of the four —SS— bridges of pancreatic ribonuclease must be split (by reduction) (16) to affect activity, and that, conversely, the restoration of activity by controlled oxidation lags behind —SS— bridge formation and optical rotation (17). On the other hand, loss of activity occurs at urea concentrations at which spectral changes are not yet detectable (18), although much higher urea levels are required for complete abolition of activity.

Rather good methods for studying protein conformation are now available (19); but they have been applied to the correlation of conformation and enzyme activity in very few cases.

The equilibrium theory of the relation between such forms of a polypeptide as the helix and the random coil is excellent, but far too elaborate to bear repetition here; the reader is referred to the original papers (20, 21, 22). A kinetic theory of such phenomena is virtually nonexistent. We may try, however, to see what a rather crude approach tells us.

Suppose that each enzyme molecule contains a single bond or structure whose disruption is sufficient to abolish the activity. If the attack on the bond, whether thermal or chemical, is random, with a probability of rupture per unit time of k, the probability that the bond is intact (hence the fraction of activity surviving) is e^{-kt} after t units of time of exposure to attack. A graph of the logarithm of the fractional surviving activity at time t against t is a straight line, with slope $-k$.

If each molecule of enzyme contains n identical bonds or structures of this sort, all of which must be cleaved to inactivate, the fraction of activity surviving is calculated as follows: The probability that a given unit **is** destroyed is $1 - e^{-kt}$; the probability that all n of them are destroyed is $(1 - e^{-kt})^n$ if they are independent; and so the probability that not all n are destroyed, hence the fraction still active, is $1 - (1 - e^{-kt})^n$. This expression is formally identical with the expression for fraction of a population surviving a given dose of radiation (23); and its semilogarithmic plot has a marked shoulder (lag period), whose magnitude depends on n, before it asymptotically approaches the line $\ln n - kt$. An obvious generalization of this expression can be written for the case in which there are m classes of sensitive bonds, each with its own value of n and k.

It should therefore be easy to detect a requirement for the breakage of a number of bonds in excess of one for inactivation of an enzyme, and to estimate at least roughly the number of bonds and their sensitivity. The astounding fact is that virtually every investigation of the kinetics of denaturation in which enzyme activity was used as the measure has given a single-hit type of curve (linear semilog plot), sometimes with modifications that do not affect the conclusion. This gives one, as the French say, furiously to think.

The obvious conclusion is that every enzyme possesses one crucial bond whose destruction is necessary and sufficient for the abolition of activity. If one thinks this proposition over, it is clear that this bond or structure cannot be destroyed very late in the process of denaturation: The imagination staggers at the thought of the active center of the enzyme going cheerfully about its business while the whole protein molecule is crashing about its ears. But compare this argument with the evidence already cited (16) that most of the activity of ribonuclease (ca. 80%) survives the reduction of two out of four —SS— bridges, and the even stronger statement by Kalnitsky that one-third of the activity survives splitting of 3.9 out of 4 bridges (24). Or compare that evidence with the other data cited (18), according to which ribonuclease in less than $4M$ urea loses activity before physical indices of denaturation are apparent.

It is obvious that a key point here is the problem of the order of denaturation stages. Our simple probabilistic model treats the sensitive bonds or structures as independent in their reaction to attack by any agent. We must ask whether this is at least roughly correct, or whether there is something like a compulsory order of denaturation events.

Consider the following scheme:

(11.B) $$E \xrightarrow{1} E_1 \xrightarrow{2} X$$

where E_1 is normal in its enzymatic activity while X has lost all activity. The reactions have been taken as irreversible for the sake of mathematical simplicity, and also because (if the ribonuclease reconstitution experiments may serve as a guide) the reverse reaction in a purified system is quite slow.

The differential equations are solved in an elementary fashion, and the surviving fraction of the initial enzyme activity E_0 is:

(11.2) $$f = (E + E_1)/E_0 = (k_2 e^{-k_1 t} - k_1 e^{-k_2 t})/(k_2 - k_1)$$

For small values of t we may apply the series approximation to the exponentials; and we must keep the quadratic terms, inasmuch as the linear terms cancel. Hence

(11.3) $$f \sim 1 - \tfrac{1}{2}k_1k_2t^2$$

Taking logarithms, and applying the logarithmic series approximation:

(11.4) $$\ln f \sim -\tfrac{1}{2}k_1k_2t^2$$

This is an inverted parabola, with zero slope at the origin. Thus, if a step that does not destroy enzyme activity is a compulsory precursor of the step that does, there will be an initial shoulder in the semilog plot. In the case of very large k_1 or k_2 or both, of course, the shoulder might be of such short duration as to pass undetected; for then the slope $-k_1k_2t$, although zero for $t = 0$ exactly, might be appreciable for t very small but nonvanishing.

No such argument is required if we adopt an alternative scheme:

(11.C) $$E \xrightarrow{1} E_1 \qquad E \xrightarrow{3} X \qquad E_1 \xrightarrow{2} X$$

Running through the same procedure as before, we have for small t:

(11.5) $$\ln f \sim -k_3t$$

This result is independent of the sequence of reactions 1 and 2, as may be confirmed by putting, for instance, $k_2 = 0$ and repeating the derivation.

The graph does not have a constant slope, however. It is easy to show that the asymptote is:

(11.6) $$\ln[(k_2 - k_3)/(k_2 - k)] - kt \qquad \text{if } k_2 > k = k_1 = k_3$$

$$\ln[k_1/(k - k_2)] - k_2t \qquad \text{if } k_2 < k$$

In the first instance the slope of the asymptote is greater than the initial slope; in the second, it may be greater or less. Correspondingly, the intercept of the asymptote is positive in the first case, while in the second case it may be of either sign.

Since inactivation rate constants are generally not very large as rate constants go (less than 1/min and frequently very much less) (3, 14), the argument for patching up model 11.B is not very strong. Excluding this, then, the simple exponential loss of activity apparently implies that denatured but still active precursors are not compulsory steps in the loss of activity. Such precursors are not excluded, however; and comparison of 11.5 and 11.6 shows that the (rate-limiting) reactions leading through such precursors will eventually show up in the slope of the asymptote.

The theory suggests, for the interpretation of the results with reductive inactivation of ribonuclease (which, it must be remembered, takes place in urea solution), that the crucial structure can in fact be attacked at a small but finite rate (k_3) before major damage to the conformation by way of —SS— bridge cleavage has occurred, with a far more rapid attack (rate k_2) possible after such cleavage has gone far enough.

Interpretation of the initial slope demands caution, however. What we know of the nature of protein conformation and of the complexity of enzymatic "active centers" suggests that the crucial structure, even under one set of conditions, might be violated in more than one way (by rape or by sodomy, so to speak). But if we were to add to scheme 11.C a fourth reaction, $E \xrightarrow{4} X$, with rate constant $k_4 \neq k_3$, the initial slope would simply become $k_3 + k_4$, and this sum, replacing k_3 everywhere in the equations, would not be analyzable into its terms by rate measurements alone.

It is of interest to ask, before we take leave of this topic, whether the numerical values of inactivation parameters reveal anything about the nature of the sensitive structures. Unfortunately, the answer seems to be categorically in the negative. London's figures (5) are probably typical: The enthalpies and entropies of activation are extremely large, and are, I suppose, what one would expect for major disruption of large, complex, and highly ordered structures.

There seem to be two difficulties here. One of them is that the most sensitive methods used for studying overall conformation may well be irrelevant to the structure of the active center. Thus, if one reads the Nelson and Hummel paper (18) carefully, it becomes obvious why loss of activity at low urea concentrations occurred in the absence of observable physical indices. The method of difference spectra depends on changes in the absorption of aromatic residues, many of which, as mentioned earlier, are buried in the hydrophobic core. The addition of urea results in an immediate spectral change attributable to the effect of the altered dielectric constant of the medium on the absorption of the exposed aromatic residues, followed by slow changes which reflect the unfolding of the protein. For urea concentrations less than $4M$, the slow progressive changes do not occur. Thus it is obvious that the interesting small and early changes in conformation might not be reflected in the difference spectrum at all, and in any case would be correlated with activity changes only if the affected aromatic residues form a part of the active center. (It should be noted, incidentally, that control experiments proved that the activity changes were not artefacts due to the influence of urea on substrate binding.)

On the other hand, direct measurements of activity loss are usually conducted in normal media **after** inactivation in the absence of substrate, so that small, readily reversible changes in enzyme activity are almost certain to be missed or at least blurred.

Whether the technical difficulties in the way of clean experiments on activity loss in the incubation mixture can be overcome I do not know. In any case, it might be useful to see what the kinetics of such a system would be like. Let us consider the scheme:

(11.D) $$E + S \underset{-1}{\overset{1}{\rightleftharpoons}} C \qquad C \xrightarrow{2} E + P \qquad E \xrightarrow{3} E_d \qquad C \xrightarrow{4} E_d + S$$

We have simplified the system by treating the inactivation reactions as irreversible; and we assume that inactivation of C causes the loss of the bound S molecule.

We shall also assume that the transient reactions leading to the pseudo-stationary state are much more rapid than the inactivation reactions, so that we can write $C = ES/K$, $K = (k_{-1} + k_2)/k_1$ as usual.

What we should not assume is that S can be treated as a parameter in the differential equation for inactivation, since we are rather explicitly considering that the inactivation process is somewhat slow (which we aim to insure by controlling the concentration of denaturant). We must therefore deal with the simultaneous pair of differential equations:

(11.7) $$dE_d/dt = k_3 E + k_4 C = k_3 E + k_4 ES/K$$

$$dS/dt = -k_2 C = -k_2 ES/K$$

We eliminate dt between the equations, and solve for E_d as a function of S:

(11.8) $$E_d = (k_3 K/k_2) \ln(S_0/S) + (k_4/k_2)(S_0 - S)$$

where S_0 is the initial value of S (amount added to the incubation mixture).

Using the conservation equation for enzyme:

(11.9) $$E_t = E + C + E_d = E(1 + S/K) + E_d$$

we eliminate E_d from 11.8, and get the rate:

(11.10) $$v = k_2 C = k_2 ES/K = [VS/(K + S)] \times [1 - (k_3 K/V) \ln(S_0/S) - (k_4/V)(S_0 - S)]$$

The first factor in square brackets is what the rate would be if there were no inactivation.

To get the parameters in this term by using a control system lacking the denaturant chemical is likely to involve us in arguments about the effect of factors like dielectric constant on V and K. It is simpler to note that the extra terms in the second bracket reduce to zero as S approaches S_0. Hence, if the experimental curve of v against S is extrapolated back to S_0, the limit of the rate will be $v_0 = VS_0/(K + S_0)$. If various values of S_0 are employed in parallel incubation mixtures, the usual reciprocal plot, $1/v_0$ against $1/S_0$, will give the desired V and K values.

Armed with this information, we can proceed to analyze the second factor in 11.10 (which may be facilitated by plotting $v/[VS/(K+S)]$, rather than v, against S). The techniques are familiar (see Chapter 2). For $S \sim S_0$, the expansion of $-\ln(S_0/S) = \ln(S/S_0) = \ln[1 - (S_0 - S)/S_0]$ gives the linear approximation: $1 - (k_3 K/VS_0)(S_0 - S) - (k_4/V)(S_0 - S)$. For very small values of S, the last term of the bracket approaches the constant value S_0 while the logarithm approaches infinity; the logarithm therefore may be taken as the approximate asymptotic form of the function. Plotting the early data against $S_0 - S$ gives the slope $(k_3 K/S_0 + k_4)/V$, while plotting later data against $\ln(S_0/S)$ should also give a straight line with slope $-k_3 K/V$. K and V being known, the values of k_3 and k_4 are obtainable from the two slopes.

If all goes well—if the effects are not too small to detect, and the artefacts do not flourish too rankly—it is still extremely improbable that the values of the parameters will point unambiguously to certain structures without the help of other information. What one might hope for, however, is to find values that differ considerably from those estimated in the conventional types of experiment. If, for example, the ΔH^* and ΔS^* values turned out to be much lower than the customary huge ones, this would seem to me quite interesting and informative. If the active regions of enzymes lie on their "surfaces," as it would seem they must, there should be at least some enzymes in which these regions can be disrupted without the incidental generation of a great deal of irrelevant disorder. (Obviously, the failure to find any such instances even with adequate experimental methods, or their consistent presence in some and consistent absence in other enzymes, would still be theoretically significant.)

To make the argument as concrete as possible, let me illustrate by comparing some equilibrium parameters (which are easier to calculate than the activation parameters). Imagine, as part of an enzyme, an α helix of about 20 residues, joined to the rest of the protein covalently by a few residues of random coil, and anchored in a fixed position relative to the structure by side-chain hydrogen bonds.

Suppose first that the active center consists of some groups on the distal end of the helix and some on nearby sites on the rest of the protein. The active center can be split if the side-chain hydrogen bonds are cleaved, permitting the helix to rotate away from the rest of the protein. The entropy of this reaction consists primarily of the entropy of the rotational degrees of freedom. If the helix can be treated as a two-dimensional rotator, $\Delta S \sim R \ln(IkT/\hbar^2)$, where $\hbar = h/2\pi$, and k and h are the Boltzmann constant and the Planck constant respectively, while I is the rotational moment of inertia of the helix. This gives an estimate of ca. 20 entropy units, half that amount if we can treat the helix as a one-dimensional rotator, plus another 3–5 units per side-chain bond split (11). The values are upper bounds, since the rotation of the helix is not wholly free.

Suppose on the other hand that the groups of the active center reside on the surface of the helix (i.e., are made up of suitable side-chains of the residues in the helix), and that the loss of activity is due to the transition to a random coil. It would be difficult to convert only a part of the helix to random coil, since it takes about 3 turns to make a stable helix, and the entire helix contains only about 5 turns. With an estimated entropy change for the helix-coil transition of about 6 entropy units per residue, the total entropy change in this case would be 120 entropy units. But even if we imagine that we could unravel only 2–3 turns and leave the rest of the helix intact, we would have 50–70 entropy units involved.

In short, in so far as we may extrapolate from such calculations to the corresponding activation values, it seems clear that there should be active structures which can be destroyed at a relatively low entropy cost, and perhaps others which demand the generation of much more disorder. The data available so far, if they tell us anything at all about the structure of active centers, may reflect only the latter kind. Similar considerations apply to enthalpy values.

The use of urea, guanidine, and various nonpolar compounds such as dioxane and methanol as chemical denaturants is so widespread as to warrant a few words about mechanisms. One of the most elaborate studies of the kinetics of urea and guanidine effects is that of Simpson and Kauzmann already cited (14). The immediately striking feature is that the rate of denaturation (measured by change in optical rotation) of ovalbumin depends very strongly on the urea concentration (apparent order of about 15). This led the authors to the theory that urea reacts, stoichiometrically and in successive stages, with the protein, each added urea molecule raising the rate of denaturation by a constant factor. The proposed interpretation of this formal theory (which can be made to fit the data pretty well) is that urea attacks the peptide links of the α helix and, in opening up one turn, combines stoichiometrically with 16 carbonyl and imino groups. A similar idea was proposed by Peller (22), who however attributes the high apparent order of reaction to the cooperative nature of the helix-coil transition.

This view presents some difficulties, as Tanford has observed (15). One would expect the bond between a carbonyl or imino group and a urea molecule to be similar to that between either of them and a water molecule; and it is therefore not at all obvious why urea should contribute about 10 kilocalories per mole per urea molecule to the bound activation energy of denaturation, which is what one calculates from the proposed model.

The possibility that urea acts by raising the dielectric constant of the medium, and hence by decreasing the strength of electrostatic bonds in the protein, was ruled out by the fact that the rate of denaturation is unaffected by pH between 7 and 9, and by a rough calculation which suggests (to me, not altogether convincingly) that dilute salt solutions should be as effective denaturants as concentrated urea solutions under the dielectric constant mechanism—which in general they are not.

A second difficulty with the theory is that urea is a perfectly good denaturant for a minimally helical (25) molecule like that of ribonuclease. This suggests a possible alternative; that urea and guanidine affect the hydrophobic bonds which are so important in protein conformation. To understand how this can occur, it is necessary to recall that nonpolar solutes increase the degree of "ice-like" order in liquid water (26, 27). This increase of order implies a decrease of entropy, hence an increase in the free energy change of solution, and accounts for the relatively low solubility of nonpolar compounds in water, and for the tendency of nonpolar groups of proteins to associate preferentially with one another rather than with the aqueous medium.

Addition to the medium of a component that decreases the order of the solvent should contribute an entropy term favorable to the exposure of the nonpolar groups of a protein. That urea and guanidine may fulfil this specification is suggested—but only suggested!—by their extremely high solubility. But the kind of evidence one would like—e.g., increased water solubility of hydrocarbons in the presence of urea—has not come to my attention. However, some quite recent and rather elegant work by Hammes *et al.* (28, 29) does indicate an influence of both urea and guanidine on the structure of water.

If the hypothesis is correct, and if the entropy increase per mole (of urea or guanidine) is S_u, then the total entropy increase due to C moles is approximately CS_u, and the entropy factor of the denaturation rate constant is proportional to $e^{CS_u/R}$. Thus the high apparent reaction order observed for these compounds may in reality be exponential. It is impossible to recalculate the published values (14) with any accuracy, since the figures are reduced in size and devoid of a coordinate grid; but a rough attempt suggests that a linear plot of log k against C would fit the points about as well as the published graph of log k against log C.[3]

The effects of nonpolar and weakly polar solvents may be similar. It is certain that such solvents must influence the tendency of the hydrophobic bonds in a protein to break—not by increasing the entropy of the solvent but by furnishing a solvent whose entropy need not be increased, and secondarily by furnishing a solvent whose energy of interaction with the hydro-

[3] This is not the whole story, of course. The rate of ovalbumin denaturation in urea varies with temperature, and has a minimum at 20°C (14). Enthalpic factors in the rate must therefore be significant; and the entropy must also vary with temperature, in view of the effect of temperature on the structure of water. Since the data concern a secondary stage of denaturation, it is reasonable to suppose that some hydrogen bonds are broken as well as hydrophobic bonds. A complete theory (which I am not yet able to present) must take all these factors into account.

phobic groups compares favorably with that of the groups with one another. It is interesting that the exponential effect of solvent concentration, which we expect if the argument is valid, is indeed observed for dioxane (London, personal communication). The slope of the log k curve is not significantly affected by temperature, suggesting that the major influence of the solvent is on the activation entropy rather than the enthalpy—a conclusion which was to be expected if indeed hydrophobic bonds are being exchanged for other hydrophobic bonds, both sets of bonds being in any case rather weak.[4]

If hydrophobic bonds in proteins were necessarily confined to a hydrophobic core, whose disruption would necessitate wholesale disorder, the foregoing discussion would have little or no interest for the problem of the structure of the active center. However, Scheraga (11) has demonstrated the feasibility and indeed the likelihood that hydrophobic bonds may involve the nonpolar carbon atoms of polar side-chains (e.g., lysine), and that these may bind to the α carbons of the main chain as well as to purely hydrophobic side-chains. It follows that some hydrophobic bonds may lie at or close to the protein surface. If this is so—I revert to my previous thesis—the controlled application of agents like urea and guanidine on the one hand or of nonpolar and weakly polar organic solvents on the other hand might conceivably yield information about early, limited, and enzymatically significant phases of denaturation.

Problem: Prove that, if more than 2 steps are required to reach an enzymatically inactive form, $\ln f$ will initially vary as a correspondingly higher power of t than the second power.

PROPOSED MECHANISMS OF ENZYME ACTION

In order to see the various proposals about the mechanism of enzyme catalysis in some kind of perspective, it will be useful—if not exactly thrilling—to review the elementary facts about chemical bonds, their stability, and the nature of chemical reactions.

[4] I must note that this is not London's interpretation of his data, which concentrates on the effect of dioxane on the dielectric constant of the solution. London's argument depends on an ingenious calculation: From available data, he calculates the decrease of dielectric constant at various temperatures (relative to 41.4°C), and the dioxane concentration equivalent to this decrease; from this and his own data, he "corrects" the values of k at higher temperatures on the zero-dioxane graph. Thus "factoring out" the effect of dielectric constant, he observes a marked decrease in the slope of the log $k - 1/T$ curve, and so argues that the dielectric constant is a major contributor to thermal denaturation. While I would not at all argue that the dielectric constant is irrelevant, I find this calculation not altogether convincing, for this reason: It seems to me that one could duplicate the calculation with any solution parameter which decreases with temperature and with dioxane concentration (e.g., surface tension), eliminating the parameter between the two relations to get a "dioxane equivalent of temperature." If the argument were valid, moreover, it would have to be applied also to the curves in the presence of dioxane; hence the identity of the slopes remains to be explained.

1. The atoms joined by a covalent bond are held together by forces which are analyzed in detail in quantum theory (30, 31), and about whose special properties I shall for the moment say nothing, except that they depend on electron pairs shared by the atoms.

2. At all temperatures, the bound atoms execute motions relative to the center of mass of the molecule to which they belong. Thus the atoms may vibrate along the conventional bond direction as if they were joined by a spring, or the pair of atoms may rotate relative to other bonds.

3. The stability of a covalent bond depends essentially on a balance between the forces holding them together and the thermal motions of molecule and solvent molecules, since the latter, by collision, may increase the vibrational amplitude and energy.

4. In addition to the immediate bonding forces attributable to shared electron pairs, forces due to other parts of the molecule (if any) act on the atoms joined by a particular bond. Some of these forces may be favorable to the stability of the bond itself, some unfavorable.

5. A chemical reaction may involve (a) formation of a bond between two previously independent atoms; (b) cleavage of a bond; or (c) most frequently, cleavage of some existing bonds and formation of other new bonds. In all types of reaction, the electrical and mechanical factors play significant roles. Thermal energy of vibration assists in overcoming the energy barrier in bond cleavage; so do attractions from without and repulsions from within the molecule. In the formation of a new bond, the energy barrier due to the electrostatic repulsions of the electron shells is partly overcome by kinetic energy of the reactants when and if they collide, and partly by attractive forces between the reacting atoms, attractive forces due to other parts of the reacting molecules, and intramolecular forces tending to weaken the repulsions.[5]

My reason for beginning with these review items from Chemistry 100 is that virtually all proposed mechanisms for enzymatic catalysis concentrate on some of the reaction factors mentioned while ignoring others. All the theories are partial, and it is well to keep this in mind when one is apparently confronted by choices among them. It is possible to find strong experimental support for all the models; and it is also possible, in every instance, to find experiments which are not explained by a particular model, or explained only by some very adroit quibbling. At the risk of my well-earned reputation for holding strong (and loud) opinions, I shall not declare any one model the winner. In the present situation, on the contrary, it is my considered view that a little eclecticism would be very helpful.

[5] I have expressed these points in classical rather than quantum-mechanical terms for the sake of simplicity; translation would not affect their relevance or significance.

I classify the models as static, quasistatic, and dynamic; and this classification refers to the way in which they ignore or take account of **nuclear** displacements or movements in the catalytic process, as opposed to electronic displacements and time-independent electrical forces. The categories are conveniences, not absolutes; and the models are not generally pure—assignments are based on the emphasis they place on one or another aspect of the reaction process.

Static (Chemical) Models

The basic assumption of the static models is that enzyme-catalyzed reactions are governed by the electrical forces known to govern ordinary chemical reactions, especially those that form the subject-matter of organic chemistry.[6] The theory of such reactions rests on a certain amount of semi-empirical fudging in addition to a perfectly sound quantum-mechanical foundation; but it is well supported by experimental evidence. It is also by now so universally taught as to permit me simply to refer to suitable books for review, and to take at least general familiarity on the part of the reader for granted (32, 33, 34).

The role assigned to the enzyme, in addition to the binding of the substrates, is that of furnishing the helping groups, the generalized acids and bases, so familiar in organic reactions. Such groups may act as attacking groups, forming intermediates more reactive than the original compound, as in the base hydrolysis of carboxylic esters; they may attract or combine with leaving groups, as with the hydroxyl-containing solvents used so frequently in the S_N reactions of sulphonates or halides; they may combine with attacking groups and increase their reactivity, as with the $AlCl_3$ enhancement of the electrophilicity of acyl or alkyl halides in the Friedel-Crafts reaction.

The classical example of this sort of analysis is the work of Wilson and his associates on the mechanism of action of acetylcholinesterase (35).[7] Another beautiful series of analyses comes from Koshland (36), who approached the problem through such examples as the stereochemistry of enzymatic hydrolyses by various glycosidases, the role of group specificity in such reactions, the exchange reactions of maltose and sucrose phosphorylase, and so forth.[8]

Koshland has pointed out (37), however, that such considerations do not always convincingly account for the **quantitative** efficiency of enzymes

[6] This view is by now so familiar that it is perhaps difficult to realize that it was once a bold and speculative theory.

[7] These papers are also a paradigm of the approach to mechanism by way of kinetics, including the analysis of pH curves and the application of substrates and substrate analogues of known structure.

[8] In the first edition of this volume, I expounded Koshland's reasoning at considerable length. It is now so familiar a part of biochemistry that I content myself with this allusion.

as catalysts, if we simply estimate rates from the observed rates with organic catalysts of low molecular weight. An obvious omission is the difference between reaction probabilities involving random favorable orientations of substrates and catalysts in solution and those where favorable orientations have been guaranteed by fixed relations in the enzyme surface.

Koshland has developed a simple but, it seems to me, roughly adequate theory (38) of this effect. What makes the theory specially interesting is that the calculated effectiveness of enzymes, if we ignore catalytic groups and simply treat the enzyme surface as a sort of breadboard for the assembly of reaction components, actually comes out lower than that in free solution. This is not due to an inherently lower efficiency of the enzyme, but to the low enzyme concentrations that can reasonably be assumed on the basis of experience. With the same enzyme concentrations, however, and a few catalytic groups which demand favorable orientations, the ratio of the catalyzed reactions in solution and on the enzyme surface are altogether different; and, with not too implausible assignments of numerical values, enormous enhancements of rates, resembling those actually observed, are predicted.

In essence, the situation is this: If a reaction requires the occurrence of a number of events (favorable spatial relationships), the rate of reaction is roughly proportional to the **product** of the probabilities of these events, and is therefore quite sensitive to any influence (organization on the enzyme surface) which increases all the probabilities at once.

It is true, as Jencks (39) has pointed out, that reactions in solution involving five molecules are so slow as never to have been observed, so that the calculated favorable **ratio** (enzymatic to nonenzymatic rate) may have little significance in absolute terms. It seems to me, however, that this affects Koshland's particular calculation but not the underlying theoretical principle. Let me put the matter somewhat differently, in view of the fact that Koshland's method does not deal directly with the intrinsic rate constants. Suppose we consider only one catalyst (which I presume Jencks would not regard as unrealistic). At Koshland's figures of $10^{-3}M$ for the concentration and 10 for the orientation factor, the catalyzed reaction rate in solution would be $\sim 10^{-6}$ times smaller than the spontaneous rate if the catalyst had no influence on the intrinsic rate constant. By the definition of catalysis, I suppose it would not be outrageous to imagine that the observed catalyzed rate in solution is ten times the uncatalyzed rate. This means that the intrinsic rate must have been raised by a factor of 10^7. If we now put the catalyst in a suitable position on the enzyme, and so raise the probability of a proper orientation of substrates and catalyst to about unity, we can get the full benefit of this rate effect on each enzyme molecule. (And the point of Koshland's outrageous stipulation of three catalysts is of course that it pays to do on the enzyme what it would admittedly be worse than futile to try in solution—to use more than one catalytic group and get essentially the full benefit of their concerted actions.)

Quasistatic Models

By "quasistatic" I mean to imply that the possibility of atomic as well as electronic movement is taken into account, but that it is the altered positions of the atoms rather than their intervening states of motion that are deemed significant. Such ideas are not to be excluded even from the static chemical models. Thus it is well known that the arrangement of the atoms in the transition state or activated state of certain reaction is often quite different from the usual arrangement; for example, in substitution reactions on tetrahedral carbon atoms the activated state involves a shift of the leaving group and a conversion of the three remaining groups to a planar instead of a tetrahedral arrangement. What makes such a model quasistatic is that the new arrangement is material, not the motion.

The first of the two major quasistatic models is perhaps the oldest known theory of catalysis in a new incarnation. The activation of hydrogen in catalyzed hydrogenations was explained on the ground (40) that the catalyst surface was covered with binding groups which almost but not quite fit the bond distance in the hydrogen molecule, so that the bound molecules were placed under stress. This theory made its way into the literature of enzymology (41, 42), and then (except for a discussion in the first edition of this book) disappeared for a number of years.

Currently one of its most eloquent and reasoned spokesmen is Jencks, whose admirable discussions the reader is urged to see for himself (39, 43). A major argument for considering such a theory is the existence of numerous reactions for which plausible static chemical mechanisms are difficult to find without much question-begging. The essence of the model is a flat contradiction of the old idea that the enzyme surface fits its destined substrate perfectly—the template concept. What is proposed in its place is that the substrate does not quite fit the enzyme, but can be bound in such a way as to strain it. The result is a system whose free energy is greater than it would have been if the substrate could have been bound with perfect fit; the activation energy for reaction is decreased by a corresponding amount.[9]

This model, which was perhaps no more than a wild guess when it was introduced to explain catalytic hydrogenation, has now acquired a highly

[9] But I must point out in passing that the interpretation of Hofstee's data, cited by Jencks, is subject to the cautionary remarks of Chapter 2. Varying turnover numbers for a series of substrates, with little variation of K values, does not conclusively prove that there is no variation of affinity constants.

respectable pedigree from organic chemistry, which, for example, has long recognized the instability of rings with too few members, has ascribed the instability quite explicitly to strain, and has rationalized the notion of strain in the usual quantum-mechanical terms. The ancestry of the strain idea shows up quite clearly in Bruice's "steric compression" (44), in which the reactivity of model compounds is used in support of an esteratic enzyme model in which an enzyme carboxyl group participates in the formation of an anhydride-like intermediate.

If indeed resemblance between bound substrate and the transition state for the reaction is the key to the strain effect, one would expect substrate analogues naturally resembling the transition state to be bound very tightly, hence to be excellent inhibitors, since the binding energy need not go into energy of strain. This prediction has been verified in a number of instances (43).

When Wilson (35) compared the actions of acetylcholinesterase on its natural substrate and on the analogue acetyldimethylaminoethanol, and took into account the dimensions of the two compounds, he was led to remark, concerning the effect of the extra methyl group of the substrate, that "for this methyl group to come into close proximity with the enzyme would appear to require a reshaping of the enzyme"; and his study of the effectiveness of ternary and quaternary ammonium compounds as inhibitors led him to the same conclusion. He also saw quite clearly the relevance of possible induced conformational change to the physiological rôle of acetylcholine.

More recently (and more persistently) Koshland (37, 45) has advocated this notion—that the presence of the substrate may lead to a change in the conformation of the enzyme. The kind of evidence that led Koshland to this view is the failure of muscle phosphorylase (unlike sucrose phosphorylase) to catalyze phosphate exchange with glucose-1-phosphate in the absence of the second substrate, the failure of amylomaltase to hydrolyze the close substrate analogue and effective competitive inhibitor α-methylglucoside, and the failure of hexokinase to act as an ATPase. The apparent common factor in these cases is that the negative results are correlated, not with steric hindrance or alien bonds, but with the absence of a second substrate (or in the amylomaltase case with the absence of an apparently irrelevant part of a substrate). What Koshland wants to conclude is that the missing factors are responsible for an **induced fit**—a change in the conformation of the enzyme leading to a suitable alignment of catalytic groups, and hence to reaction.

This induced-fit model, the second major quasistatic model, encounters more difficulty than the strain theory, because it demands the kind of experimental evidence that is not easy to obtain and still less easy to interpret unambiguously, and because it cannot appeal to a large body of simpler models with well-explored theoretical consequences and experimental confirmation. The attempt to interpret changes in slope of Arrhenius plots of rate data (logarithm of rate or rate constant against $1/T$) as evidence for conformational transitions is beset by the possibility of other and equally plausible interpretations, as the discussion of Koshland's 1959 paper shows. Jencks (43) cites some evidence for substrate-induced changes in enzymes; some of these seem fairly direct, others only hopeful possibilities.

Jencks has also proposed a novel version of the conformational change theory (43), which relates it to the strain model and offers a possible advantage. If the substrate will bind only to a deformed state of the enzyme, the energy of strain in the enzyme may be converted, after binding, to energy of strain in the substrate. Thus, in addition to Koshland's proposed correct alignment of catalytic groups, there is another way in which conformational change may enhance reaction rates.

Dynamic Models

The logical synthesis and extension of the two quasistatic models is dynamic. The realistic picture of enzyme and substrate alike, as I pointed out earlier, is of systems of atoms in **continual** thermal motion. These motions, moreover, are not random, but characteristic and reproducible because of the way they interact with the intramolecular bonding forces; they are mirrored, for example, in the infrared spectra of the molecules. It is reasonable to ask whether the motions themselves may contribute to the enzymatic catalysis.

As early as 1947, Fano (46) proposed a very simple but by no means uninstructive model. His substrate was a diatomic molecule, his enzyme two large cubes hinged together at one pair of edges and so capable of rotating freely (but not through a complete revolution) around the hinge. The two binding sites on the enzyme were located on the two cubes (much as in the rotating helix inactivation model discussed earlier in this chapter). Fano proposed that the two-point binding would then occur sequentially: First one binding site would be occupied by half of the diatomic substrate; then, when the two cubes happen to rotate in such a way that they are in apposition, the second part of the substrate binds to the second site—and locks the cubes together! Fano then calculates the entropy decrease due to the lost rotational freedom, on the assumption that the system behaves like a one-dimensional rotator. Finally, he observes, "Brownian fluctuations of the pressure of water molecules on the large faces of the blocks tend to tear the bond..."

What Fano was trying to do was to answer the question I put earlier: Why are enzymes large molecules? The partial answer he was proposing in the work just described was simply that a very large rotator could lose (on substrate binding) and regain (on substrate splitting) a very appreciable rotational entropy, which would therefore act to drive the turnover step.

With more realistic protein models based on more information, Fano's elementary approach does not become ridiculous; rather, further possibilities open up. A major current protagonist of dynamic models is Morris London; and I shall now briefly summarize his reasoning.

London's studies of the denaturation of prostatic acid phosphatase (5) revealed a marked pH dependence, with maximum stability near the pH region of maximum activity. The data resemble those of Lewis (47) on ovalbumin. They suggested to London, as similar data have to others (18), that electrostatic repulsions play a role in denaturation.[10]

London's already-cited studies (4, 6) on stabilization of the enzyme by substrates and by suitable inhibitors led him to the picture of a binding site with four components, divided into two pairs lying on either side of a "critical seam" (presumably of hydrogen bonds). The boundary between two adjacent turns of a helix would of course exemplify such a seam; but obviously there are still other possibilities (recall the rotating helix of the earlier discussion of inactivation, where the critical seam would be the crevice between the helix and the rest of the enzyme).

Where London went a step further was in proposing that the protein is in a constant state of complex oscillation, some of the force constants being contributed by hydrogen bonds, electrostatic repulsion, and so on, while the rest have their usual sources in the covalent binding forces. In particular, he would have it that the structure of the active site (visualized as two or more turns of an α helix, with side chains contributing the binding and catalytic groups) participates in this oscillatory behavior. In this way, London believes, the thermal oscillations contribute to the energy of activation.

It is obvious, I think, that this proposal embodies strain and conformational change in one model. It is also obvious that, whether the specific model turns out to be sound and usable or not, its general assumptions are more realistic than those of the quasistatic and static models, and must assuredly be taken into account in the eventual satisfactory theory.

[10] These inferences were greatly strengthened by data on effects of ions other than H^+. London began with a simplified but representative version of charge distributions at various pH values (Figs. 11.5–7); the hatched area is postulated to be the charge range for activity, the blackened areas are the regions of denaturation. Comparison of the figures makes it obvious how pH affects activity and denaturation. But, if ions other than H^+ can be bound, it must be possible to shift the charge distribution at a given pH till it resembles that at another pH, and hence either to increase or to decrease activities and denaturation rates, depending on the pH at which the experiment is performed (example in Fig. 11.8, at pH 4). The ion effect must be uncompetitive, in contrast to the effects discussed earlier; and London demonstrated this from the kinetic data.

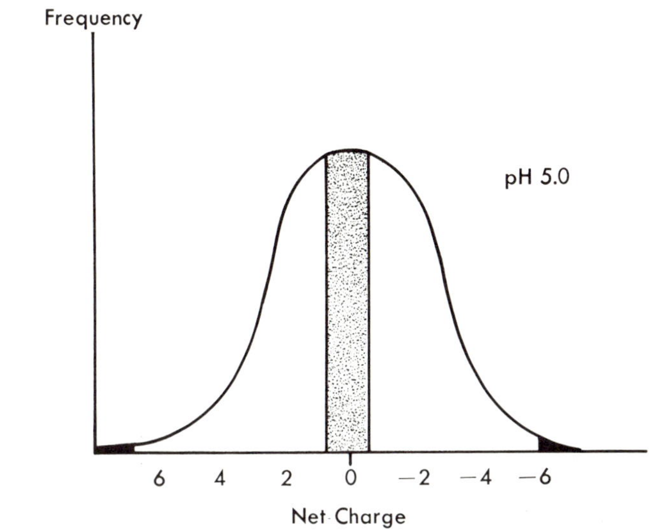

Fig. 11.5. Charge Distribution of Population of Enzyme Molecules at pH of Maximum Activity. Molecules in shaded area are enzymatically active, molecules in blackened areas unstable.

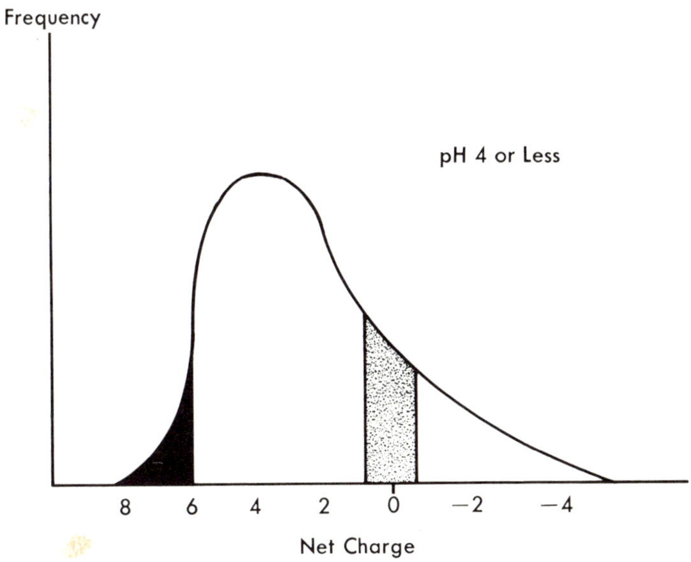

Fig. 11.6. Similar Curve Appreciably Below pH of Maximum Activity.

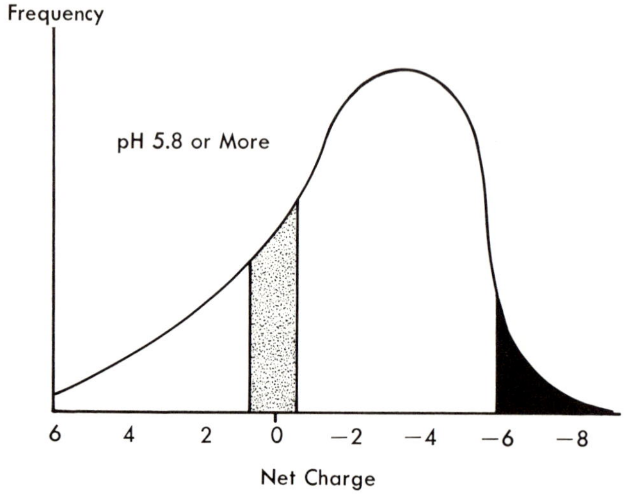

Fig. 11.7. Similar Curve Appreciably Below pH of Maximum Activity.

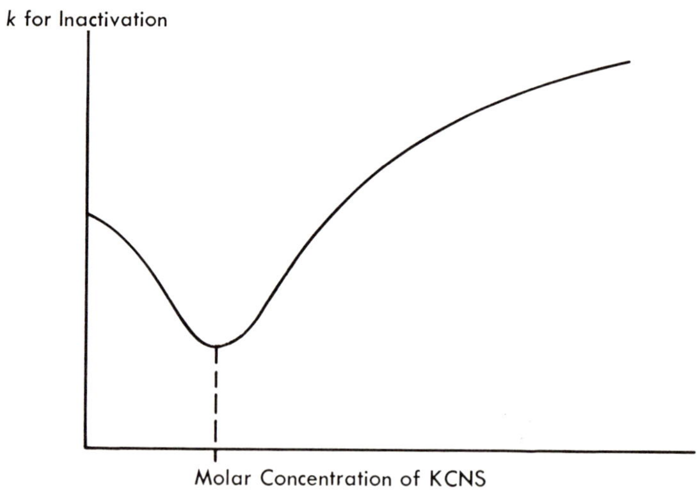

Fig. 11.8. Effect of CNS^- on Rate of Thermal Inactivation of Prostatic Acid Phosphates when pH Is Below Optimum, with the Charge Distribution Illustrated in Fig. 11.6.

Jencks, whose taste for logic admirably exceeds even his passion for his own favorite theory, arrived at the conclusion that an "oscillating enzyme" would be the ideal model (43); and he considered London's proposal. He backed away from it for reasons that I do not consider adequate. To put the issues as clearly as possible, let me quote Jencks: "The difficulty with such a mechanism lies in the specification of the forces which could set such an oscillation in motion and a mechanism to prevent the damping out of the oscillation by random collisions of small molecules and by energy transfer within the protein itself."

This objection might be meaningful if it were supposed that the oscillation had to be an unusual one, which would not normally be likely to occur in the molecule. In that case, one would certainly be inclined to ask: What starts it going, and what keeps it going? But if I understand London's admittedly rather brief reference to thermal vibrations correctly, he is contemplating the normal modes of the molecule.

Now the normal modes, as I pointed out near the beginning of this section, are due to the interaction of the thermal agitation of the atoms composing the structure with the various forces that hold the structure together. The protein molecule may be viewed as an assembly of coupled vibrators and rotators. Due to the coupling, as well as to the variety of atom pairs and hence of forces, the vibrational and rotational spectrum is complex; its modes are not simply those of an assembly of independent vibrators and rotators. Energy circulates through such a structure in a way too complex for us to describe at present.[11]

One of the characteristics of such a large and complex assembly is that it is to a degree self-stabilizing. To state the matter rather crudely: If one were to damp one element, or in general if one were momentarily to interfere with one element so as to change its frequency, the circulation of energy in the structure due to coupling would ensure re-excitation; and the re-excitation would necessarily be in one of the normal modes, since these are determined by the force constants and moments of inertia.

If such an assembly is immersed in an assembly of thermally agitated smaller particles, as is the case with a protein in solution, there will be a continual exchange of energy between the coupled structure and the moving particles. On the average, if the temperature of the whole system is constant,

[11] A well-known example, excessively simple but instructive, is a system of two pendula on a common slightly elastic support. If one is started swinging while the other is left at rest, the second pendulum will eventually begin to swing, while the first will decrease in amplitude, till the second is in full swing when the first is at rest. This two-way exchange of energy between the two units will continue (modified only by damping due to friction in the suspensions).

there will be no redistribution of energy; but there will of course be local effects as the energy of collisions of small particles is converted into vibrational or rotational energy in the coupled structure. Some of the energy will be distributed through the structure; but eventually it will manifest itself either as energy of collisions returned to the particles by a portion of the structure, or as energy of vibrational or rotational states above the ground state, which will eventually be emitted as infrared radiation while the affected parts of the structure return to the ground state, or at least to a lower state.

Electronic transitions, which require energies of the order of 10,000 calories per mole, are extremely unlikely to occur in such a system at moderate temperatures.

Precisely how or to what degree the vibrational and rotational energy of such a coupled system will contribute to activation of the substrate must be considered an open question. A contribution to strain is obviously one of the possibilities.

Perhaps the most intriguing if also the most speculative possibility is the occurrence of a resonance mechanism. The bound substrate is a small assembly of coupled oscillators. The atoms constituting the bond that is supposed to react may be considered as an oscillator which is subject to a periodic external force—the periodic field of the oscillating enzyme region to which it is bound. A property of oscillating systems under periodic fields of force is that the amplitude of oscillation varies inversely as the difference between the frequency of the external field and the natural frequency of the oscillator. If the frequencies are almost the same, the amplitude will approach infinity. To put it another way, if a substrate oscillator and an enzyme oscillator are matched in frequency—are, as it were, tuned to one another—energy transfer between them will be maximal. The tuned enzyme oscillator will serve to funnel thermal energy from the whole system of enzyme plus medium into the substrate oscillator. I am putting the matter in an oversimplified way, of course, in the interests of clarity and brevity alike. But something like the model just described seems physically reasonable.[12]

Obviously, we are in no position at this time to state, in even one instance, whether part of the active center of a given enzyme is in fact approximately tuned to a natural frequency of the relevant substrate bond. We are still at at the stage of being quite pleased with ourselves if we can decide with some assurance, in a few cases, what species of amino acids are involved in the active center, and if we can perhaps come close to a guess about the arrangement of those amino acids. Moreover, the extreme messiness of infrared

[12] A quantum-mechanical version of this model is possible; and energy transfer will occur most readily when differences in energy levels of the quantized oscillators are most closely matched.

spectroscopy in aqueous media constitutes a serious obstacle to direct testing of the hypothesis.

For the moment, however, the important point is that Jenck's theoretical objection can be met, and that there is an oscillatory mechanism that at least deserves serious consideration and testing. That this mechanism alone can account for enzymatic catalysis I do not for an instant believe. Until an adequate quantitative theory is worked out, however, I see no way to set an upper bound to the oscillatory contribution. A single vibrational quantum amounts to about 600 cal/mole; but if the mean lifetimes of excited states and ground states in the enzyme and substrate oscillators are comparable, the transfer of more than one quantum—to which I have referred in classical terms as "resonance"—seems possible.

MISCELLANEOUS PROBLEMS

1. Why are Enzymes Proteins?

The variety of catalysts and catalytic model systems of low molecular weight (48) raises the problem of why biological catalysts invariably are proteins. At one time a pat reply was that this is necessary to keep them from leaking out of the cell. Our current knowledge of transport processes, together with such phenomena as transformation by DNA, suggests that size is perhaps one of the least important of the factors that keep molecules inside or outside of cells.

One obvious answer is suggested by the discussion of the dynamic model in the last section: Only a large and complex molecule could even conceivably serve as an oscillatory catalyst, since it is the size and complexity that buffer, as it were, against random disturbance of the essential properties. The Fano (46) rotational entropy model may also be relevant, since the magnitude of the effect depends strongly on the size of the molecule.

Another answer is suggested by specificity considerations: No other type of macromolecule is capable of the variety of local configurations that can be imagined to exist in proteins. An adequate discussion of just how great the variety might be is beyond our scope here; but the reader is invited to get a start on the problem by comparing the number of ways in which, say, 100 places of a sequence can be filled by selecting elements from 20 classes with the number available using 4 classes.

An ingenious theory by Kirkwood (49) suggests still another possibility. The essence of Kirkwood's theory is that the mobility of protons is so great that protein charge distributions are continually fluctuating. As a result, even electrically symmetrical molecules will have momentary dipoles, capable of interacting with the complex of active site and substrate.[13] The effect, as Kirkwood shows, is to increase the turnover constant of the enzyme by increasing the polarity of the bound substrate. The effect is essentially nonspecific; and the mechanism is not intended to explain the specificity of enzymes, but rather their catalytic powers—or at least part of those powers.

Kirkwood did not view his theory as an answer to our chief question, which he did not consider at all. (He was fascinated by the fact that his rate enhancement factor had a pH maximum; but this circumstance does not seem to give a good explanation of enzymatic pH curves.) However, it is obvious that Kirkwood's derivation involves an averaging process whose validity would be somewhat questionable if the number of units involved were not sufficiently large. It therefore seems reasonable to assert that the Kirkwood mechanism illustrates the point of having enzymes that are proteins.[14]

2. Sequences Implicated in Activity

There is an accumulating mass of evidence which supports the idea of characteristic amino acid subsequences involved in enzyme activity and located in the active center. The data have been reviewed by Koshland, to whose discussion I refer the reader (37).

The most significant evidence about the composition of the active region is that obtained from labelling of one or more amino acids (e.g., by means of a part of the substrate). Evidence from identification of pK values (Chapter 10) tells us that a given amino acid must be dissociated (or must not be dissociated, as the case may be) for activity to occur; but this alone does not establish its proximity to the substrate or its role in the catalysis.

On this basis, serine has been established as an amino acid that must be in contact with and somehow react with the substrate in many enzyme systems. By means of the usual techniques, sequences of amino acids including the essential serine have been isolated and analyzed; and it is a fact that these sequences bear a remarkable resemblance to one another in a series of proteases, esterases, and mutases. In each case, serine is flanked on the left (i.e., on the amino end) by a dicarboxylic acid and on the right by a small neutral amino acid (glycine or alanine).

[13] The reasoning, as applied to charge-dipole interactions, is similar to the reasoning used in deriving the well-known London forces which account for van der Waals interactions of uncharged and nonpolar molecules (50).

[14] The question "Why are enzymes proteins?" has an odd ring; and I hope that no one supposed me to be interpreting it teleologically. For those with an appetite for rigor, let me rephrase it thus: "What selective advantage is conferred upon organisms whose enzymatic catalyses are mediated by proteins?" So stated, of course, the question has been answered only in part; it seems intuitively clear that efficient enzymes are good for their possessors, but the thesis is arguable.

These nearest neighbors are part of the active region by definition (proximity); but their function is unknown. As I have pointed out earlier, they might well be involved in determining the proximity of another fold of the protein to the serine-containing fold. In this connection, I draw attention to Lanni's study on the compositionally unusual neighbors of the bridge-forming cysteines of ribonuclease (51); I have commented elsewhere (52) on the possibility that they direct correct —SS— bonding.

Evidence from photo-oxidation and pH studies indicates the implication of one histidine residue in each of the above cases. In trypsin at least this histidine does not occur within six residues of the serine on either side; similar information is not available for the other enzymes. From this it is clear that the histidine is quite probably on another fold of the protein, brought into proximity to the serine region by the folding.

The situation, as Koshland has remarked (37), makes a certain kind of rough sense. For the reactions catalyzed do have common features in esterases and phosphate-transfer enzymes: an attacking hydroxyl group, a leaving alcoholic group, a central atom doubly bonded to oxygen. This consideration and the range of the experimental evidence make it at least plausible to suppose that the configuration of histidine near a serine with characteristic nearest neighbors is relevant to the mechanism of action of certain classes of enzymes.

The specificity differences among these enzymes is interesting in itself, since it means that amino acid residues located close to the specific portions of the substrates, rather than to the crucial region of bond breakage and formation, must be involved. Koshland points out an immediate inference from this, for which there is indeed a little experimental support—that it should be possible to modify the specificity of such enzymes without risking the activity.

3. The Identity of E

In the introduction I mentioned rather briefly the question whether the enzyme molecule released after catalysis is in fact identical to the enzyme molecule before substrate is bound. Let me try to formulate this a little more sharply.

Every assembly of molecules at a given temperature is characterized by a distribution of molecular energies (as well as of local bond energies within molecules)—some molecules are "hotter" than others, at least at some times. The distributions are stabilized, as a rule, by collisional exchanges of energy with vessel walls, solvent, etc., so that, while a given molecule may not be at the same energy level all the time, the fraction of molecules at any level remains the same.

When a reaction is catalyzed, and products appear, energy is released in the form of heat. In reactions with an overall $\Delta H > 0$, heat is absorbed, not released, if we consider average properties of the assembly as a whole. Locally, however, this must be true: The step from *E-S* to *E-P* complex necessarily involves a release of free energy (namely, the free energy of activation for the reverse process), some if not all of which may be enthalpic. The same argument holds for the step from *E-P* to $E + P$.

What happens to the heat released? Some of it may appear in the products as kinetic energy of translation, or as bond energy, some of which may later be transmitted to the solvent medium by collision. Some of the energy is transferred to the enzyme locally, and eventually this will be distributed through the enzyme, and perhaps in part transmitted to the medium. Moreover, if the overall reaction absorbs heat, processes of heat absorption (from the medium via the enzyme) will have taken place; and there is no reason to doubt that everything will add up—for the assembly. If net heat is released by the reaction, the thermostatic water bath will eventually absorb it, the temperature of the enzyme-substrate-medium assembly will remain constant, and so one would expect the energy distribution of the enzyme system to remain constant. Thus, though the enzyme molecule that emerges from a catalytic event is not energetically identical with the one that entered the event, we might expect that the assembly as a whole will be unchanged.

One little question remains, however. One can imagine a situation in which an enzyme absorbs enough energy locally to initiate denaturation before it can be redistributed. Since denaturation absorbs a large amount of heat, this might seem very unlikely; but it must be remembered that the usual ΔH measurements for denaturation correspond to great disruption of conformation, while the enthalpy per hydrogen bond (to take one possible instance) is only about 1500 cal/mole. It is thus conceivable that activity might change the energy distribution and activity distribution, temporarily if not permanently.

The answer to this speculative question hinges in part on the time it takes for locally absorbed energy to be redistributed as compared with the time required for a given degree of denaturation. We know from an estimate by Eigen and Hammes (53) that the helix-coil transition can occur in 10^{-7} sec; and a lower limit for the time is probably given by the time required to break a single hydrogen bond, which is of the order of or perhaps a little less than 10^{-11} sec. What we do not know, and what I see no way to calculate accurately at the moment, is the mean lifetime of the excited state in a protein oscillator. *Faute de mieux*, it seems reasonable to assume that the estimate given by Polanyi and Wigner (54) for the recurrence time of an energy state in an isolated oscillator:

$$\tau = Ae^{E/RT},$$

where E is the energy and $A \sim 10^{-13}$–10^{-14} sec, could be taken as an upper bound.[15] If so, with an E of several times the energy of the vibrational ground state—that is, roughly several times RT—, one would get an estimate of at most 10^{-11}–10^{-12} sec.

From this one would guess that the probability of local denaturation by absorption of energy released in a catalyzed reaction is very small. It would be interesting, however, to have a more solidly founded estimate.

CONCLUDING REMARKS

There can be little doubt that no single one of the proposed mechanisms for enzymatic catalysis is operative everywhere and at all times. The activity of catalytic groups, favorable orientation of the substrate with respect to these groups, the stressing of the substrate into a shape closer to that of the transition state, and mobilization of thermal energy through specific channels in the enzyme all seem plausible and useful mechanisms.

A major deficiency in our understanding of enzymatic mechanisms lies in the complete absence of any unified theory, no matter how primitive, and in particular of any quantitative theory. We look at a particular situation, and say that the actions of certain kinds of catalytic groups would obviously be highly relevant to the enhancement of the reaction rate. Or we look at other reactions, and say that the usual types of electronic pushes and pulls seem very unlikely to account for catalysis with these particular substrates, and look around for other alternatives, such as strain. But we do not have even the crudest attempt at a theoretical treatment which might lead us to assesss in this or that particular case, **how much** each of the possible mechanisms, contributes to the enhancement of reaction rate. In the absence of such a theory, progress beyond our present condition, in which the various proponents of single mechanisms conduct monologues, is likely to be slower than it should be.

I am well aware of the enormous difficulties that beset the path to a rigorous and exact theory. However, those who followed my reasoning in Chapter 1 will realize that one generally arrives at a rigorous and exact theory by starting, as best one may, with a nonrigorous and inexact theory, finding out what is wrong with it, and building in the needed improvements as fast as one's imagination and one's mathematical powers will permit. The history of science does not consist exclusively of revolutions.

I am also well aware that there are many gaps in our experimental knowledge. But I think I have made evident my belief that the function of theory is to lead experiment, not to follow it. Crucial experiments will be

[15] The mean lifetime can hardly exceed the recurrence time of a state; and one would expect coupling to diminish the lifetime by facilitating energy transfer.

performed when someone produces a theory—no matter how defective—that points to a crucial experiment. Such gains as we have made in this area up to the present have rested heavily on such contributory theories as are available—the theory of absolute reaction rates, the semiquantitative theories of valence, bond stability, and reactivity in theoretical organic chemistry, the Pauling model of protein structure and the resulting theories of the helix-coil transition, the various theories relating macromolecular structure to such indices as optical rotation and ultraviolet and infrared spectra, modern theories of water structure and hydrophobic binding, and so forth.

As one who has witnessed the introduction into molecular biology of analytical paper chromatography and effective column chromatography, of nucleic acid hybridization methods, of density labelling and density gradient centrifugation, of adequate methods for the X-ray analysis of protein structure, and all the other wonders of our contemporary laboratories, I am acutely aware of the importance of being able to take ideas from the desk to the bench with effective techniques. But I am also convinced that we are, at the moment, in remarkably good shape technically—and I strongly suspect that the next significant advances are likely to come when someone at a desk says, "We need to know such-and-such, and I don't see how we are going to find it out." Accordingly, while the potential gains from detailed comparative study of specific enzymes are far from being fully realized, I feel justified in emphasizing the importance, for further progress in the understanding of enzymatic catalysis, of improvement in the quality of our mathematical theories.

REFERENCES

1. Tsuboi, K. K., and Hudson, P. B., Arch. Biochem. Biophys. *55*, 191 (1955).
2. Hearon, J. Z., Physiol. Revs. *32*, 499 (1952).
3. Fisher, H. F., J. Theoret. Biol. *3*, 484 (1962).
4. London, M., McHugh, R., and Hudson, P. B., Arch. Biochem. Biophys. *73*, 72 (1958).
5. London, M., Wigler, P., and Hudson, P. B., Arch. Biochem. Biophys. *52*, 236 (1954).
6. London, M., McHugh, R., and Hudson, P. B., Arch. Biochem. Biophys. *55*, 121 (1955).
7. Massey, V., Biochem. J. *55*, 172 (1953).
8. Fisher, H. F., Frieden, C., McKee, J. S. M., and Alberty. R. A., J. Am. Chem. Soc. *77*, 4436 (1955)
9. Ogston, A. G., Nature *164*, 180 (1949).
10. Cecil, R., in "The Proteins" (Neurath, H., ed.), vol. I, 380, Academic Press, New York, 1963.

11. Scheraga, H. A., in "The Proteins" (Neurath, H. ed.), vol. I, 478, Academic Press, New York, 1963.
12. Pauling, L., and Corey, R. S., Proc. Nat. Acad. Sci. *37*, 235 (1951).
13. Anfinsen, C. B., in "Basic Problems in Neoplastic Disease" (Gellhorn, A., and Hirschberg, E., eds.), 112, Columbia Univ. Press, New York, 1962.
14. Simpson, R. B., and Kauzmann, W., J. Am. Chem. Soc. *75*, 5139 (1953).
15. Tanford, C., "Physical Chemistry of Macromolecules", Wiley, New York, 1961.
16. Sela, M., White, F. H., Jr., and Anfinsen, C. B., Science *125*, 691 (1957).
17. Anfinsen, C. B., Haber, E., Sela, M., and White, F. H., Jr., Proc. Nat. Acad. Sci. *47*, 1309 (1961).
18. Nelson, C. A., and Hummel, J. P., J. Biol. Chem. *237*, 1567 (1962).
19. Schellman, J. A., and Schellman, C., in "The Proteins" (Neurath, H., ed.). vol. II, 1, Academic Press, New York, 1964.
20. Zimm, B. H., and Bragg, J. K., J. Chem. Phys. *31*, 526 (1959).
21. Gibbs, J. H., and DiMarzio, E. A., J. Chem. Phys. *30*, 271 (1959).
22. Peller, L., J. Phys. Chem. *63*, 1194, 1199 (1959).
23. Atwood, K. C., and Norman, A., Proc. Nat. Acad. Sci. *35*, 696 (1949).
24. Kalnitsky, G., J. Cell. Comp. Physiol. (Supp. 1), 211 (1959).
25. Doty, P., in "Biophysical Science" (Oncley, J. L., ed.), 107, Wiley, New York, 1959.
26. Némethy, G., and Scheraga, H. A., J. Che. Phys. *36*, 3382, 3401 (1962).
27. Némethy, G., and Scheraga, H. A., J. Phys. Chem. *66*, 1773 (1962).
28. Hammes, G. G., and Swann, J. C., Biochem. *6*, 1591 (1967).
29. Hammes, G. G., and Schimmel, P. R., J. Am. Chem. Soc. *89*, 442 (1967).
30. Kauzmann, W., "Quantum Chemistry", Academic Press, New York, 1957.
31. Daudel, R., Lefebvre, R., and Moser, C., "Quantum Chemistry: Methods and Applications", Interscience, New York, 1959.
32. Gould, E. S., "Mechanism and Structure in Organic Chemistry", Holt, Rinehart, & Winston, New York, 1959.
33. Kosower, E. M., "Molecular Biochemistry", McGraw-Hill, New York, 1962.
34. Ingraham, L. L., "Biochemical Mechanisms", Wiley, New York, 1962.
35. Wilson, I. B., and Bergmann, F., J. Biol. Chem. *185*, 479 (1950); *186*, 683 (1950); Bergmann, Wilson, and Nachmansohn, D., J. Biol. Chem. *186*, 693 (1950); Wilson, Bergmann, and Nachmansohn, J. Biol. Chem. *186*, 781 (1950); Wilson, J. Biol. Chem. *197*, 215 (1952).
36. Koshland, D. E., Jr., in "The Mechanism of Enzyme Action" (McElroy, W. D., and Glass, B., eds.), 608, Johns Hopkins Press, Baltimore, 1954.
37. Koshland, D. E., Jr., in "Advances in Enzymology" (Nord, F. F., ed.), 45, Interscience, New York, 1960.
38. Koshland, D. E., Jr., J. Theoret. Biol. *2*, 75 (1962).
39. Jencks, W. P., Ann. Rev. Biochem. *32*, 639 (1963).
40. Balandin, A. A., Z. Physik. Chem. (B) *2*, 289 (1929).
41. Quastel, J. H., Biochem. J. *20*, 166 (1926).
42. Haldane, J. B. S., "Enzymes", Longmans, Green, London, 1930.
43. Jencks, W. P., in "Current Aspects of Biochemical Energetics" (Kaplan, N. O., and Kennedy, E., eds.), 273.

44. Bruice, T. C., and Pandit, U. K., Proc. Nat. Acad. Sci. *46*, 402 (1960).
45. Koshland, D. E., Jr., in "Symposium on Enzyme Reaction Mechanisms" (Hollaender, A., ed.), 245, Wistar Inst., Philadelphia, 1959.
46. Fano, U. J., Chem. Phys. *15*, 845 (1947).
47. Lewis, P. S., Biochem. J., *20*, 978 (1926).
48. Langenbeck, W., in "Advances in Enzymology" (Nord, F. F., ed.), vol. XIV, 163. Interscience, New York, 1953.
49. Kirkwood, J. G., in "The Physical Chemistry of Enzymes", 78, Faraday Soc. Discussion # 20 (1956).
50. London, F., in "Structure and Moleuclar Forces in (a) Pure Liquids and (b) Solutions", Faraday Soc. Discussion (1936).
51. Lanni, F., Proc. Nat. Acad. Sci. *46*, 1563 (1960).
52. Reiner, J. M., "The Organism as an Adaptive Control System", Prentice-Hall Inc., Englewood Cliffs, New Jersey, 1967.
53. Eigen, M., and Hammes, G. G., in "Advances in Enzymology" (Nord, F. F., ed.) vol. XXV, 1. Interscience-Wiley, New York, 1963.
54. Polanyi, M., and Wigner, E., Z. Physik. Chem. (A) *139*, 439 (1928).

Appendix

SOLUTIONS TO SOME PROBLEMS

I have included a number of problems in this book; for, just as mathematical power increases with knowledge, so mathematical facility increases with application. Some of the problems are almost trivial classroom exercises; some are almost small research problems. All of them are as typical as possible of the sort of activity that you will get into it if you become seriously involved with kinetic analysis.

In this appendix, I present solutions, or methods of solution, for some—not all—of the problems that appear in the text. By this device I hope to encourage and occasionally to enlighten the serious reader. Though "looking in the back of the book" is viewed with contempt in certain quarters, I believe it is not unreasonable for someone who has worked hard at a problem to get at least a hint as to whether his efforts have been crowned with success. Also, since I do not subscribe to the pedagogical theory which apparently assumes that everyone should be forced to recapitulate the entire intellectual history of the human race, I think that there may legitimately arise occasions when you have done your best and are compelled to give up; and at that point you may benefit almost as much by seeing how the thing is done as by having done it yourself.

Having due regard to human frailty (i.e., being lazy and therefore understanding laziness), I have not given solutions to *all* the problems—an ancient but not altogether dishonorable textbook device. It should ordinarily be true that, if you can answer the problems solutions for which are furnished, you can solve the others. If you are baffled, lay the problem aside and try it again another day. The old wheeze about your brain working while you sleep is not without justification—provided you have furnished it something to work on while awake. As a last resort, if hopelessly frustrated, write the author a nasty letter or a polite request for enlightenment—or possibly both.

Chapter 1, p. 27

1. Solution to every other problem.

$$y = (ax^2 + bx)/(x + c) = (ax^2 + bx)/c(1 + x/c)$$
$$= (ax^2 + bx)(1 - x/c + \cdots)/c.$$

$y = 0$ for $x = 0$, $y \cong bx/c$ near origin.

$$\begin{aligned} y &= (ax^2+bx+d)/(x+c) = (ax^2+bx+d)/c(1+x/c) \\ &= (ax^2+bx+d)(1-x/c+\cdots)/c \\ &= (ax^2+bx+d-ax^3/c-bx^2/c-dx/c+\cdots)/c. \end{aligned}$$

$y \cong d/c + (b - d/c)x/c$ near origin.

$$\begin{aligned} y &= [\ln(a+bx)]/(x+c) = [\ln(a)+\ln(1+bx/a)]/c(1+x/c) \\ &\cong [\ln(a)+bx/a](1-x/c+\cdots)/c. \end{aligned}$$

$y \cong \ln(a)/c + [b/a - \ln(a)/c]x/c$ near origin.

$$\begin{aligned} y &= (a+bx)^{1/2}/(x+c)^{1/2} \\ &= a^{1/2}(1+\tfrac{1}{2}bx/a+\cdots)(1-\tfrac{1}{2}x/c+\cdots)/c^{1/2}. \end{aligned}$$

$y \cong (a/c)^{1/2}[1 + \frac{1}{2}(b/a - 1/c)x]$ near origin.

$$y = (ax^2+bx)(x^2+cx) = ax^4+acx^3+bx^3+bcx^2.$$

$y \cong bcx^2$ near origin.

2. With a constant in the numerator, and $x + c$ in the denominator, the function is linear near the origin. If the denominator is $x^2 + c$, the binomial expansion of that will give a constant and no other powers less than the second. Hence the lowest power near the origin will be the first if such a power appears in the numerator, otherwise the second.

Problems Chapter 1, p. 28 (asymptotic behavior)

PROBLEM 2:

$$\begin{aligned} (ax^3+bx^2)/(x+c) &= (ax^3+bx^2)/x(1+c/x) \\ &= (ax^2+bx)(1-c/x+\cdots) \cong ax^2+bx-acx-bc. \end{aligned}$$

The asymptote is a parabola.

PROBLEM 4:

$$\begin{aligned} (ax^2+b)/(x^2+c) &= (ax^2+b)/x^2(1+c/x^2) \\ &= (a+b/x^2)(1-c/x^2) \cong a+b/x^2-ac/x^2-bc/x^4 \\ &\cong a+(b-ac)/x^2. \end{aligned}$$

PROBLEM 6:

$$\begin{aligned} (a+bx)^{1/2}/(x+c) &= x^{1/2}(b+a/x)^{1/2}/x(1+c/x) \\ &\cong (b+\tfrac{1}{2}a/x)(1-c/x)/x^{1/2} \\ &= (b+\tfrac{1}{2}a/x-bc/x-\tfrac{1}{2}ac/x^2)/x^{1/2} \\ &\cong b/x^{1/2}. \end{aligned}$$

And so on. Notice that the number of terms you keep depends somewhat on the amount of information you want. In the fourth and sixth problems, for instance, the limit as x approaches infinity is a in one case and 0 in the other. But this does not tell you in what manner these limits are approached. It is of some value on occasion to know that the function of the fourth problem approaches a along a curve that looks something like a sharpened rectangular hyperbola, and that the function of the sixth problem approaches 0 along a curve that looks like a rather weakened hyperbola. Remember, it is also true that $a(1 - e^{-bx})$ approaches a in the limit, and that e^{-bx} approaches zero; but their behavior while approaching the limit is different from that of the first two functions. The difference may not seem important, since these problems are artificial; but in a real biological problem the fate of a theory might hang on such differences.

Chapter 1, p. 32

1. Answers to second and fourth examples.

$$y = (ax^2 + bx + c)/(x - d).$$

Expansion near $x = 0$ gives

$$y = -[c + (b + c/d)x]/d.$$

For large x, factor x out of denominator and divide the factor into numerator:

$$y = (ax + b + c/x)/(1 - d/x).$$

As x approaches infinity, y approaches the asymptote:

$$y = ax + b:$$

also as x approaches minus infinity.

The zeroes of the function are those of the numerator:

$$-b \pm (b^2 - 4ac)^{1/2},$$

both of them negative for positive values of the constants.

The extrema are at:

$$x_e = \{ad \pm [a^2d^2 + a(c + bd)]^{1/2}\}a,$$

one being positive and one negative.

The second derivative equals $(ad^2 + c + bd)/(x - d)^3$, which has the same sign as $x - d$—negative for $x < d$, positive for $x > d$.

The function has a pole at $x = d$. It is obviously n egative for $0 < x < d$, positive for $x > d$.

The function therefore consists of two branches, separated by the line $x = d$. The first branch, to the left of $x = d$, is concave down, with a maximum to the left of the y axis. If the zeroes are real $(b^2 - 4ac \geq 0)$, the function rises above the x axis between them, and the maximum is between them. The upper branch is concave up, has a minimum above $x = d$. The lower end of the lower branch and the upper end of the upper branch approach the linear asymptote $y = ax + b$, the former from below, the latter from above.

The fourth problem differs from the second in the substitution of $-b$ for b. Immediate consequences of this are:

a. Both zeroes are positive.

b. The slope at $x = 0$ need not be negative; this depends on the sign of $c - bd$.

c. The extrema may be located on either side of the y axis as before, or both on the positive side, or they may be imaginary, depending on the sign of the discriminant $ad^2 + c - bd$.

d. The sign of the second derivative depends not only on that of $x - d$ but also on that of $ad^2 + c - bd$, being positive when these have the same sign and negative when they have opposite signs.

Considering the case where the lower extremum (maximum) lies to the left of the y axis as before ($ad^2 + c - bd > 0$, $c - bd > 0$), it is evident that the zeroes of the function cannot result from the rising of this maximum above the x axis, since both zeroes are now positive. The zeroes, if real, must therefore involve dipping of the upper extremum (minimum) below the x axis.

If the lower extremum lies between $x = 0$ and $x = d$, either extremum might conceivably produce the zeroes. To find out what actually happens, apply the condition for the lower branch to cross the axis:

$y > 0$ for $x = x_{e-}$ (x_{e-} refers to minus sign before radical).

Squaring to remove the radical sign, and simplifying, we find:

$$b^2 > 4ac,$$

which is the condition for the zeroes to be real. Thus the zeroes will be associated with the lower branch of the function whenever they exist, as long as the extrema also exist.

Finally, consider the case where the extrema are imaginary ($ad^2 + c - bd < 0$, which entails $c - bd < 0$). Testing the sign of the second derivative as usual, we find that it is positive for $x < d$ and negative for $x > d$. This suggests that the lower branch of the curve rises, curving upwards, from negative to positive values of the ordinate (and abscissa), until it approaches the line $x = d$ asymptotically above the x axis. The second branch is asymptotic to $x = d$ below the x axis, and rises, curving downward. The tails of both branches are asymptotic to $y = ax - b$, the lower branch above the asymptote, the upper branch below the asymptote. This would separate the two zeroes of the function, one being at the crossing of the x axis by the lower branch, the other at the crossing by the upper branch.

It is easy to verify the positions of the branches relative to $x = d$. Set $x = d - z$, where z is a very small quantity (implying that x is just below d). Substituting in y, we get:

$$y = (ad^2 - 2adz + az^2 - bd + bz + c)/(-z).$$

Neglecting powers of z in the numerator, $y \cong (ad^2 + c - bd)/(-z) > 0$. Setting $x = d + z$, we find similarly that $y < 0$ just above $x = d$. From these findings, and the fact that the branches are asymptotic to the negative and the positive limbs of $y = ax - b$, and that the branches are continuous (no poles other than $x = d$), it follows that our proposed description of the curve is correct.

A special case would arise if d were identical with one of the roots of the numerator. This would mean:

$$\pm(b^2 - 4ac)^{1/2} = 2ad - b,$$

from which (comparing signs on the two sides) it follows that this could occur only if $d > b/2a$ in the case of the larger root, or if $d < b/2a$ in the case of the smaller root. Squaring to remove the radical, we find the relation among the constants which must hold for either root to equal d:

$$2ad^2 + c - bd = 0.$$

In either case, the function would take the form:

$$y = a(x - r),$$

where r is one of the roots of $ax^2 - bx + c = 0$.

2. Consider the second example of problem 1. If we substituted $-d$ for d (as we did $-b$ for b in the fourth example), the pole would come at $x = -d$. The upper extremum might come at negative x values (if $c < bd$). Imaginary extrema are possible. The general effect would be that of moving the configuration of the second or fourth example to the left. The result would differ from that of the fourth example in that the zeroes could not be positive.

If in the same example we were to substitute $-a$ for a, the zeroes would now come on opposite sides of the y axis, while both extrema would be positive. The general effect would be that of crowding the lower branch closer to the line $x = d$. However, the direction of the asymptote is reversed ($y = -ax + b$), and the directions of curvature of the two branches are reversed; the lower branch has a minimum and the upper a maximum if the extrema are real. The upper branch lies wholly below the x axis, as may be demonstrated by inspecting the sign of the upper y_e. For

$$dy/dx = [(x - d)(-2ax + b) - (-ax^2 + bx + c)]/(x - d)^2;$$

hence at $dy/dx = 0$ we have $y_e = (-ax_e^2 + bx_e + c)/(x_e - d) = -2ax_e + b$. But the upper $x_e > d$, hence $y_e < -2ad + b$. Since the extrema are real only if $ad^2 > c + bd$, it follows that $b < ad - c/d$, hence $-2ad + b < -ad - c/d < 0$.

In a similar way, other functions may be explored as the parameters are permitted to change sign. In some cases the results will not differ radically in pattern, in others the shapes of the curves may be greatly altered.

Chapter 1, p. 35

1. $y = F(x)$ is symmetric about the line $x = x_0$ if for every point on the graph a given distance above $x = x_0$ there is a corresponding point (same y value) the same distance below x_0:

$$(z)[F(x_0 + z) = F(x_0 - z)].$$

Let the larger abscissa be $x = x_0 + z$. Then $z = x - x_0$, so $x_0 - z = x_0 - (x - x_0) = 2x_0 - x$. Substituting in the symmetry condition:

$$(x)[F(x) = F(2x_0 - x)].$$

Chapter 2, p. 55

(a) $dC_1/dt = k_1E^2 - (k_{-1} + k_2S)C_1 + k_{-2}C_2;$
$dC_2/dt = k_2SC_1 - (k_{-2} + k_3)C_2;$
$E_t = E + C_1 + C_2.$

(c) $dC_1/dt = k_1SE - (k_{-1} + k_2)C_1 + k_{-2}C_2;$
$dC_2/dt = k_2C_1 - (k_{-2} + k_3)C_2;$
$E_t = E + C_1 + C_2.$

Chapter 8, p. 158

PROBLEM 2

If C_x is formed but cannot turn over, K_4 remains finite but $k_6 = 0$. From 8.3 it follows that $u_x = K_1$, independent of X. The second term of the numerator of 8.2 drops out; the surviving third term is linear rather than quadratic in X, as is the last term of the denominator. The rate v varies from $k_2SE_t/(K_1 + S)$ at $X = 0$ to $k_7S/[K_5 + S) + (K_3K_5 + k_7S/k_3)S/K_1K_4]$ as X approaches infinity. This differs from the complete system (8.2) in the second term of the denominator of the asymptotic form. The second term being quadratic, the asymptotic rate will have a maximum as a function of S, and will decrease to zero as S approaches infinity. Hence, even if the condition for activation by X is fulfilled in the complete system for all S, it will not be fulfilled in this special case for sufficiently large S values.

The vanishing of k_6 implies the vanishing of c and γ in 8.4 and 8.5. Hence overshoot and undershoot are no longer possible.

If C_x were also not formed, we would have $1/K_4 = 0$; and the asymptotic form of the $v - X$ curve would reduce to that of the complete case. However, overshoot and undershoot would still be excluded according to 8.4 and 8.5. If both numerator and denominator of $v(X)$ are linear, in short, the sign of dv/dX cannot change.

PROBLEM 4:

The effect is readily seen by applying the strategies discussed in Chapter 1: Solve for C_x before C, and use the pseudostationary equation for C_x to simplify that for C; proceed similarly with C and E_x. Then it is readily evident that, as in the case developed in the text, $C_x = CX/K_4$, $C = SE/u_x$, and $C'_x = E_x S/K_5$. But, whereas the case of the text gives $E_x = EX/u_s$, the new system gives $E_x = EX/K_3$—that is, the u_s of the text reduces to its constant term K_3, and is no longer a function of S. The major effect of this change is on the argument associated with 8.6. If $u_s = K_3$, S/u_s no longer approaches a finite limit as S approaches infinity. The last term of 8.6 is therefore no longer the only quadratic term; and inhibition by an activator at intermediate activator concentrations requires not merely that $k_6 - k_2$ be negative and S large, but also that the various parameters be such that, for large S, the negative third term of 8.6 shall outweigh the positive first term:

$$(k_7 - k_2)K_1/K_3 K_5 < (k_2 - k_6)/K_4$$

A similar argument holds for the possibility of activation by an eventual inhibitor.

The inequalities governing the occurrence of overshoot or undershoot, however, are not affected by the altered stoichiometry.

Chapter 8, p. 164

PROBLEM 2:

If we get the second derivative of 8.12 with respect to x, and set it equal to zero, for the case $n = 2$, $u = 0$ (2-place system, no activator present), we get ($z = 1 + x$):

$$a^2 + 3az - 3az^2 - z^3 = 0$$

with $a = L(1 + w)^2$. The root of this equation, considered as a function of x, gives the inflection point of 8.12. The polynomial is evidently positive for $x = 0$, provided $a > 1$, which is reasonable to assume. As x increases, the cubic term eventually becomes significant, and there will be at least one real value of x for which the polynomial vanishes. As w increases from 0, however, a increases; and in view of the fact that the positive constant term is a^2, and the coefficient of the positive linear and quadratic terms is proportional to a, the value of x required to make the polynomial vanish will increase—i.e., the inflection point will move to the right.

PROBLEM 4:

To the notation of Chapter 5 we must add C_{jk}, meaning the concentration of enzyme molecules combined with j molecules of substrate and k molecules of inhibitor. Obviously, if inhibitor is microscopically competitive as specified, k cannot exceed $n-j$. Evidently the usual recursive form of the pseudo-stationary equations for the C_{jk} will yield:

$$C_{jk} = C_j I^k / q_k$$

where q_k bears the same relation to the products of inhibitor constants that p_j bears to products of binding constants.

The conservation equation will now be:

$$E_t = \sum_{j=0}^{n} \sum_{k=0}^{n-j} C_{jk}.$$

If we assume all binding and inhibitor constants microscopically identical, and carry out the usual transformations involving the statistical factors of rate and other constants, we have:

$$p_j = M^j \binom{n}{j}; \qquad q_k = N^k \binom{n-j}{k}$$

where N is the intrinsic inhibitor constant (analogous to M). The summation over k must be carried out first, since its upper limit and coefficients are functions of j. We get at this step:

$$E_t = \sum_{j}^{n} C_j (1+w)^{n-j}$$

where $w = I/N$. Then, treating C_j the same way (using the simplified form of p_j just given), we get:

$$E_t = C_0 (1 + w + x)^n.$$

The numerator of v is not different from that of Chapter 5, in view of the assumption that a single bound molecule of inhibitor inactivates the enzyme. Hence the expression for the rate in the presence of inhibitor becomes:

$$v = nE_t m_a x (1+x)^{n-1} / (1+w+x)^n$$

For $w \neq 0$, no factors cancel between numerator and denominator, so that this expression, even with all intrinsic constants the same, does not reduce to the comparable expression for $n = 1$.

PROBLEM 5:

If groups not bound by inhibitor turn over normally with rate constant m_a, the numerator of v is:

$$\sum \sum m_a C_{jk}$$

where C_{jk} is as in problem 4. Applying the reasoning of Chapter 5 to evaluate this sum, we get:

$$v = nE_t m_a x(1 + w + x)^{n-1}/(1 + w + x)^n = nE_t m_a x/(1 + w + x)$$

The expression is that for a classical competitive inhibitor acting on an enzyme at concentration nE_t.

PROBLEM 6:

Deal with the R_j series as in problems 4 and 5, defining R_{jk} by analogy with C_{jk}, and otherwise proceeding as in the usual treatment of the allosteric system. Notice how this system differs from the simplified version of Monod's inhibitor-activator system as given in the text, where inhibitor combines only with members of the T series.

Chapter 8, p. 209

Protection by inhibitor in allosteric system:

The mechanics of the problem present nothing new. But give some thought to possible variants of the assumptions. Does substrate as well as inhibitor protect any bound group against denaturation, or any polymer containing at least one bound group? What assumptions about turnover of inhibitor-bound groups or polymers are required (and plausible)?

Chapter 9, p. 233

PROBLEM 1:

Solve $dC/dt = 0$ and $dS/dt = 0$ simultaneously; show that the only solution is 0, 0. Proceed with the routine as shown in the text, dropping product terms as in 9.11. Get λ and show that both values are negative, and that they are real (discriminant > 0).

PROBLEM 2:

If we hopefully* treat overshoot in terms of the approximation 9.11, we

* Attend, dear colleagues, to the usage! "Hopefully" is used to denote a state of mind or emotion, not to designate a desideratum. I know not whom to hold responsible for the chronic mistranslation of **hoffentlich**; but specimens of it have proliferated upon the unhappy carcase of the scientific literature like lice, and I am thoroughly sick of it. It may be wounding to patriotic pride to acknowledge that German frequently is richer in idiomatic resources than English. Nevertheless, the disadvantage is not dispelled by turning good German into bad English.

While we are on such a topic, let me plead with the authors and copy editors of textbooks to take cognizance of the fact that "motivate" describes an effect that we **hoffentlich** have on students, not something that we do to a subject-matter.

Lexicographers, confronted with the inexhaustible stream of linguistic garbage that issues from the "media", are sunk in hopeless ignominy and lost to shame. Need we follow them into the abyss before we are even pushed?

ask for the condition that an extremum of u_1 may exist—that is, we set $du_1/dt = 0$, solve for t, and inquire what conditions the parameters must satisfy in order that t shall be real and positive. The solution is given in terms of a logarithm of a negative quantity times A_{12}/A_{11}. It follows at once that A_{12}/A_{11} must be negative in order that the argument of the logarithm be positive, which in turn is required in order that the logarithm be real. The reader is urged to pursue the matter further. For t to be positive, we demand (since $\lambda_1 > \lambda_2$) that the argument of the logarithm exceed 1 in order that t be positive. I suggest a return to the textbooks to find out why $\lambda_1 > \lambda_2$, and how the A_{ij} are expressed in terms of the kinetic parameters and of the initial conditions. Now express the conditions for the extremal t to be real and positive in terms of these basic quantities, and see how much you can say about them.

If our optimism is more limited, and we ask what can be said about the exact form 9.8 (for $n = 2$), we must be content with broader conclusions. Suppose $u_1(0) = X_1(0) - \bar{X}_1 < 0$, so that the X_1 curve will rise from its lower previous steady state to a higher new steady state, or u_1 will rise from its negative initial value to the eventual 0. A maximum in the time curve will occur if u_1 rises above the zero level and then drops to it. This demands that there exist a finite time T at which $u_1(T) = 0$ while the corresponding value of $du_1/dt > 0$. This seems evident from the figure (9.3(a)); but it can be supplied with firmer ground. Every calculus text contains Rolle's Theorem, which states in effect that, if a continuous function vanishes at the end-points of an interval, it must have an extremum somewhere inside the interval. Thus if $u_1(t)$ vanishes at infinity, and also at some finite time T, then $du_1/dt = 0$ for some time T' such that $T < T' < \infty$.

Thus if the reader encounters a specific instance of 9.8 for $n = 2$, he can set $u_1 = 0$ in the equations, and see whether the signs of the parameters permit this simultaneously with $u_2 \neq 0$ and $du_1/dt > 0$. This will evidently depend on the sign of $u_2(T)$ as well, hence on whether the old steady state of X_2 is above or below the new steady state.

Can anything be said about the general equation 9.8 for $n = 2$, that might bear on the problem? It is evident that, if $u_i = 0$ is a stable critical state of the system, the values of the b_{ij} must be such that the λ_m are negative. The reader will see, if he has carried out the calculation, that this implies, for instance, that $b_{11} + b_{22} < 0$ and $b_{11}b_{22} - b_{12}b_{21} > 0$, which in turn implies that, if b_{11} and b_{22} have opposite signs, so must b_{12} and b_{21}. Now at $t = T$, the equations 9.8 reduce to: $du_1/dt = b_{12}u_2$, $du_2/dt = b_{22}u_2$ by virtue of the condition $u_1(T) = 0$. Hence, for $du_1/dt > 0$, we must have b_{12} and u_2 of the same sign. If $u_2(0) > 0$ and u_2 is dropping to a lower steady state, then $du_2/dt < 0$ and so b_{22} must have a sign opposite to that of b_{12}. This is so unless u_2 has already reached an extremum (undershoot) previously, in which case there is a $T'' < T$ such that $u_2(T'') = 0$ and $du_2(T'')/dt < 0$, a condition which would permit us to examine the constants b_{11} and b_{21}. Moreover, the initial slopes of the curves must have signs opposite to the initial signs of the u_i, which, using 9.7 in the form $u_i(0) = X_i(0) - \bar{X}_i$, permits us to place restrictions on the a_{ijk} and the b_{ij} in terms of the initial conditions.

The reader is urged to carry on with this game as far as he can—to see what kinds of situations are admitted or excluded. For example, can u_1 and u_2 both exhibit overshoot in the same game (not necessarily at the same time T!)? Can one of them overshoot and the other undershoot?

Chapter 9, p. 234

PROBLEM 1:

Writing $v_1 = V_1S/(K_1 + S)$, and using the transformation $y = (v_1 - V_2)(P_1 + K_2) + V_2K_2$, we get:

$$dy(1 - V_2K_2/y) = (v_1 - V_2)^2\, dt$$

the solution of which is:

$$y - y_0 - V_2K_2 \ln(y/y_0) = (v_1 - V_2)^2t$$

By virtue of the properties of the logarithm, as discussed in Chapter 2 in connection with equation 2.17, it is evident that t approaches infinity as y approaches 0. But $y = 0$, by definition, is the same as:

$$P_1 = K_2\, v_1/(V_2 - v_1)$$

If $v_1 - V_2 > 0$, there is no positive value of P_1 for which $y = 0$. For large values of P_1 (and hence of y), the linear term greatly exceeds the logarithm, and asymptotically

$$P_1 = (v_1 - V_2)t$$

The same points are evident from the differential equation itself. For if $v_1 > V_2$, *a fortiori* $v_1 > V_2P_1/(K_2 + P_1)$. Hence $dP_1/dt > 0$; and P_1 is always increasing. For large P_1, $P_1/(K_2 + P_1)$ approaches 1, and the differential equation is asymptotically $dP_1/dt = v_1 - V_2$.

Chapter 9, p. 236

PROBLEM 2:

Substitute into 9.19, using 9.17 and 9.18, the new variable $u = B_0 - \bar{B}_0$, where $\bar{B}_0$ is the solution of the steady-state condition for 9.19. Eliminate from the substituted 9.19 all the terms that vanish in the steady state, and drop the term in u^2. What remains is

Chapter 10, p. 276

PROBLEM 1:

Taking the derivative of $1/V'$ (or $\log(1/V')$) and setting it equal to zero (clearing of fractions in h):

$$\tau_1(h^2 + K_2 h) = \sigma_2 K_2(K_1 + h)$$

The parameters τ_1 and σ_2 cancel whenever they are equal, regardless of their numerical value.

PROBLEM 2:

If all complexes are active (i.e., all of C is active), $v = k_2 C$. Use 10.9, 10.13′, and 10.16, and rearrange into the conventional form $V'S/(K'' + S)$, whereupon it will be found that V' is independent of h.

PROBLEM 4:

Correction in press: The equation under the given activity specification is quadratic, not cubic (see Problem 1). But try something like: $\tau_1 = 1$, $\tau_2 = 1$, $\sigma_3 = 1$.

PROBLEM 6:

In the general derivation in the text, we assumed that it might be reasonable to specify that a certain number of ionizable groups in a given dissociation set might be ionized and a certain other number of such groups unionized. This is perfectly reasonable, even though all the groups have the same pK: it amounts to specifying a range on the titration curve of the set. If there are r groups in a set, the specification "τ, σ" for that set implies that activity will occur if the number of ionized groups lies between τ and $r - \sigma$.

A specification like "τ_2, σ_1" is more doubtful, although it cannot be ruled out conclusively. Remember that we have numbered the sets in increasing order of pK; since pK $= -\log K$, this means that the numbering goes in decreasing order of K. Thus we would expect most of the members of set j to dissociate before any of set $j + 1$, unless the pK values are extremely close.

Practically speaking, proton fluctuations will affect the estimation of the parameters from pH-activity data. If we accept Laidler's idea of a combination of positive and negative groups for activity as roughly correct, we must of course also stipulate a suitable distance between the groups (and perhaps a suitable spatial relation to still other groups). If the negative group is drawn from a set which, at a given pH, is not highly ionized, only a very few of the fluctuating configurations available will have the right spatial relations at any given moment, and the probability of activity will be very low—even though only one ionized group would suffice for activity if it were the right one. Thus we should expect pH-activity data to overestimate τ values, at least as compared with our intuitive expectation of the number of charged groups present in the enzymatically active configuration.

Looking in the same way at the set from which the positive charge is to be selected, we can see again that σ must be higher than the actual number of positive charges in the active configuration. In short, the pH-activity curve will be sharper and narrower than one would expect in the absence of fluctuations, where the probability of getting the right group with the right charge out of r groups would be roughly τ/r resp. σ/r.

Chapter 11, p. 303

PROBLEM:

Add, for example, another intermediate E_2 and a reaction step 3. Solve the system of equations in the conventional way, apply the exponential series approximation to $(E+E_1+E_2)/E_0$, and simplify. (Note that $E(0) = E_0$, $E_1(0) = 0$, $E_2(0) = 0$.) Now both the first-degree and the second-degree terms cancel, leaving $f \sim 1 - k_1k_2k_3t^3/6$. Repeat for the next stage (one more intermediate and reaction step).

One would suppose that there should be a more elegant and general method than this brutal plodding. So far, I have failed to discover one. If any reader hits on one, I hope he will enlighten me.

Index